Computer-Aided Design of Optoelectronic Integrated Circuits and Systems

James J. Morikuni

Motorola
Schaumburg, Illinois

Sung-Mo Kang

Department of Electrical and Computer Engineering
University of Illinois at Urbana-Champaign

To join a Prentice Hall PTR internet mailing list
point to http://www.prenhall.com/register

Prentice Hall PTR
Upper Saddle River, New Jersey 07458
http://www.prenhall.com

Library of Congress Cataloging-in-Publication Data

Morikuni, James J.
    Computer-aided design of Optoelectronic integrated circuits and
systems / James J. Morikuni, Sung-Mo Kang.
        p.   cm.
    Includes bibliographical references and index.
    ISBN 0-13-264433-9
    1. Optoelectronic devices--Design and construction--Data
processing. 2. Integrated circuits--Design and construction--Data processing.
3. Computer-aided design. I. Kang, Sung-Mo, 1945-
II. Title.
TA1750.M67 1997
621.3815'2--dc20

Editorial/production supervision: *Nicholas Radhuber*
Manufacturing manager: *Alexis Heydt*
Acquisitions editor: *Russ Hall*
Cover design: *Bruce Kenselaar*
Cover design director: *Jerry Votta*

© 1997 by Prentice Hall PTR
Prentice-Hall, Inc.
A Simon & Schuster Company
Upper Saddle River, New Jersey 07458

The publisher offers discounts on this book when ordered in bulk quantities.
For more information, contact:

        Corporate Sales Department
        PTR Prentice Hall
        One Lake Street
        Upper Saddle River, NJ 07458

        Phone: 800-382-3419, Fax: 201-236-7141
        E-mail: dan_rush@prenhall.com

Printed in the United States of America
10 9 8 7 6 5 4 3 2 1

ISBN 0-13-264433-9

Prentice-Hall International (UK) Limited, London
Prentice-Hall of Australia Pty. Limited, Sydney
Prentice-Hall Canada Inc., Toronto
Prentice-Hall Hispanoamericana, S.A., Mexico
Prentice-Hall of India Private Limited, New Delhi
Prentice-Hall of Japan, Inc., Tokyo
Simon & Schuster Asia Pte. Ltd., Singapore
Editora Prentice-Hall do Brasil, Ltda., Rio de Janeiro

TABLE OF CONTENTS

PREFACE

*T*he proliferation of optical and fiber-optic communications has created a need for efficient and accurate CAD tools for the design of optoelectronic integrated circuits and systems. In the electronic world, highly advanced CAD tools exist for the design, analysis, and simulation of nearly every aspect of integration, ranging from process to device to circuit to system. When CAD tools are properly utilized, it is often possible to produce successful designs after only one design iteration. Given the considerable time and cost associated with unnecessary design revisions, CAD tools have proven themselves invaluable to electronic designers. A similar framework for optoelectronics, however, remains to be developed.

Computer-Aided Design of Optoelectronic Integrated Circuits and Systems addresses this issue and illustrates the use of modeling and simulation in optoelectronic design and analysis. The material covered in this book includes several popular methods for optoelectronic simulation, analytical and numerical optoelectronic circuit-level models, examples of the use of such models, tools for the simulation of complete optical links and systems, techniques for mixed-mode simulation of optoelectronic devices and circuits, and methods for the mixed-level simulation of optoelectronic integrated circuits and systems. Since most of the methods used in optoelectronic simulation are based on those used in electronic simulation, this book also includes a chapter that reviews the most widely used electronic simulation techniques.

The goal of this book is to provide a much-needed starting point for the construction of an optoelectronic CAD infrastructure. While the field of optoelectronics has progressed significantly over the years, it has still not realized its full potential. When compared to the history of electronics, it is clear that the next logical step in the evolution of optoelectronics is the development of a stable modeling and simulation framework. In the early days of the electronic world, IC layout was done manually with Rubylith and electronic circuits were designed on breadboards. Compare this with the situation today, in which entire systems can be designed completely by computer; in fact, the complexity of modern electronic circuits and systems has exploded to the point that they cannot be designed *without* computers. Similarly, while the design of optoelectronic integrated circuits and systems can be merely expedited by CAD tools today, it will soon become dependent on them.

This book is targeted toward a very broad audience. To ensure maximum possible coverage, enough basic information is included for this volume to be useful to engineers with various backgrounds. Examples of those who would benefit from this book are the silicon circuit designer who has no optoelectronic design expertise, the optoelectronic device designer who has little experience designing integrated circuits, the traditional electronic CAD tool developer who has no knowledge of optoelectronics, or the optoelectronic designer who has no CAD tools with which to build circuits and systems. This volume is also instructive to those who have little or no background in the field and wish to obtain a fundamental understanding of the design issues involved in optoelectronic integration.

Computer-Aided Design of Optoelectronic Integrated Circuits and Systems is ideal for use by the practicing engineer or as an advanced textbook or reference for graduate-level courses on optoelectronic design, analysis, modeling, or simulation.

ACKNOWLEDGMENTS

*T*he compilation of this book involved the work and contributions of a great number of people. We would like to take this opportunity to thank all of those who aided in the production of the final manuscript.

First, we would like to acknowledge the funding of the National Science Foundation (NSF) and the Advance Research Projects Agency (ARPA). Their establishment of the NSF Engineering Research Center (ERC) program, the NSF Center for Compound Semiconductor Microelectronics (CCSM), and the ARPA Center for Optoelectronics Science and Technology (COST) at the University of Illinois at Urbana-Champaign provided the type of multidisciplinary environment that is crucial to the type of research presented in this book. The work detailed here involved the contributions of several different research groups, ranging from device to circuit to system, an interaction that would have been difficult to accomplish without these programs. We would like to thank Prof. S. G. Bishop, Director of the CCSM and COST, for his leadership in this effort, as well as the Department of Electrical and Computer Engineering at the University of Illinois at Urbana-Champaign for its continuous support.

We would like to express our thanks to the many researchers who developed the models and tools upon which this book is based. Among these (in order of appearance) are Prof. A. T. Yang, now at the University of Washington, for his development of the iSMILE circuit simulator; Dr. D. S. Gao, now at Sun Microsystems, for his work on the multiple quantum-well laser diode model; J. P. Bianchi, now at GE, for the work that he did on the MSM model; Dr. K. Cioffi, now at Rockwell International, and Prof. T. N. Trick, of the University of Illinois at Urbana-Champaign, for their development of the HEMT model; A. Xiang, of the

University of Illinois at Urbana-Champaign, for his work on the embedded MSM model; S. Javro for the development of the single solution laser diode model; B. Onat and Prof. M. Selim Ünlü, of Boston University, for their mixed-mode device-circuit simulator; J. Pang, now at Motorola, for his work on the table-based laser diode and HBT models; Dr. D. H. Cho, now at Hyundai, for his work on the table-based FET model; and Dr. B. K. Whitlock, now at Cypress Semiconductor, and Prof. E. Conforti, of the University of Campinas, Brazil, for their work on the iFROST system simulator. Most of these researchers were, at one point, affiliated with the University of Illinois at Urbana-Champaign; their contributions are gratefully acknowledged.

We also owe a great deal of appreciation to Prof. I. Adesida, Prof. J. J. Coleman, Prof. M. Feng, and Dr. J. W. Lockwood, all of the University of Illinois at Urbana-Champaign. The work of their research groups provided the background for many of the simulation examples presented in this book.

In addition, the research presented in Chapter 6 was supported by the National Science Foundation under Grant No. ECS 9309607. B. Onat and M. S. Ünlü of Boston University acknowledge this funding.

The continuous support, encouragement, and technical guidance of our colleagues were critical during the preparation of this book. The insightful comments of Dr. A. Dharchoudhury, now at Motorola; Prof. Y. Leblebici, of the Istanbul Technical University, Turkey; and P. V. Mena of the University of Illinois at Urbana-Champaign were invaluable in the completion of this book. We would also like to thank L. Lin for his contributions to the preparation of the manuscript.

Finally, we would like to express our gratitude to our families. Their understanding and support throughout the course of this undertaking made this project possible; their patience during particularly busy periods was instrumental to the timely completion of this work.

James J. Morikuni
Schaumburg, Illinois

Sung-Mo (Steve) Kang
Urbana, Illinois

1

INTRODUCTION

As the limitations of semiconductor and fabrication technologies become apparent, optics is rapidly becoming an attractive alternative to conventional electronics; this is especially true for interconnections. Optics has many advantages over electronics including speed as well as immunity to electromagnetic interference (EMI), crosstalk, and parasitic electronic effects, such as capacitance and inductance. Since current technology is overwhelmingly electronic, a complete conversion to optics is neither desirable nor feasible. Rather, a cost-effective and efficient interface between optics and electronics must be created. Systems which employ both optical and electrical devices have come to be referred to as *optoelectronic* systems. More specifically, integrated circuits which incorporate both electrical and optical devices are known as OEICs or OptoElectronic Integrated Circuits.

1.1 Optoelectronic Concepts

The primary application of optoelectronics is presently in the well-established area of long-distance fiber-optic telecommunications (Figure 1.1). Fiber-optic systems convert electronic signals to optical signals for transmission, then convert them back to electronic signals at the receiving end. In this fashion, the superior transmission qualities of optics are exploited while compatibility with existing electronic devices is retained. The concept of optical communication can also be applied on a smaller scale; indeed, optical interconnections/interconnects on the computer-to-computer network level have achieved some degree of success (Figure 1.2) [1.1]. While present-day Ethernet-based computer networks operate at about

1

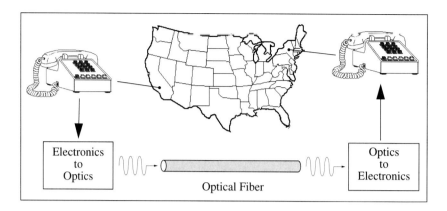

Figure 1.1: Long-distance fiber-optic telecommunications.

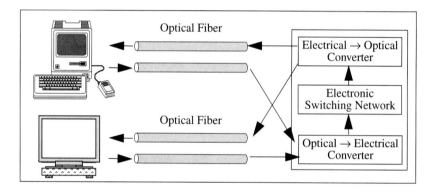

Figure 1.2: Computer-to-computer network-level optoelectronics.

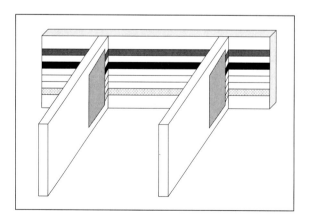

Figure 1.3: Board-to-board optical interconnect/bus/backplane.

10 Mb/s, optoelectronic fiber-based networks function at speeds over 1 Gb/s. Working one level below optical networks, several groups have investigated optical buses or backplanes, parallel paths along which many light signals could travel (Figure 1.3) [1.2], [1.3]. Optical interconnect can even be achieved on the interchip or intrachip level [1.4]. Application of optoelectronics to interchip communications directly addresses conventional electrical interconnection problems such as interference between mixed-mode signals, power dissipation in the I/O pad drivers, and the "I/O pin explosion," in which the required amount of I/O increases faster than the available area for I/O pads. Optical clock distribution is also an area of interest [1.5]; since the difference in propagation time between optical signals is nearly negligible, clock skew can be virtually eliminated.

Whether the application is long-distance fiber-optic transmission, computer-to-computer optical networking, board-to-board, chip-to-chip, or even gate-to-gate optical interconnects, every optoelectronic system consists of three basic components: the transmitter (the electro-optic converter), the transmission medium, and the receiver (the opto-electronic converter). While the optical fiber has become the medium of choice for transmission, both lasers and LEDs can be used as optical sources, depending on the application. Similarly, many devices exist for optical detection, including (but not limited to) the MSM, PIN, and the APD photodetectors.

1.2 Computer-Aided Design

When the concept of optoelectronics was first conceived, the main research emphasis was on the development of process technologies for individual devices. Much effort has been expended over the past several decades perfecting the technologies behind semiconductor lasers and detectors, as well as optical fibers and waveguides. In addition, there has been significant research into the development of III-V compound semiconductor electronic devices, not only because of their compatibility with optical materials, but also because of their inherent speed.

Once it became clear that optoelectronic devices could be successfully fabricated, researchers began investigating the possibility of optoelectronic systems. As an example, consider the field of long-distance fiber-optic telecommunications, which grew rapidly around optoelectronic technology. Other examples are SONET (Synchronous Optical NETwork), FDDI (Fiber Distributed Data Interface), B-ISDN (Broadband-Integrated Services Digital Network), and Fibre Channel. The high-speed optoelectronic devices available today, along with the superior transmission properties of optical fibers, give system designers the tools to build multi-Gb/s networks.

The state of optoelectronic circuit and system design today parallels that of the electronic world over a decade ago. As in the early days of electronic design, optoelectronic circuit and subsystem design, as well as systems integration, tends to take place in a trial-and-error manner; designs are conceived, then built in a "breadboarding" fashion. Designs that do not work are modified, upon which another prototype is built. Usually, several iterations must occur before an acceptable result is achieved. In *electronic* circuit and system design, CAD (Computer-Aided Design) and EDA (Electronic Design Automation) tools are used to reduce the design cycle and the number of required iterations; however, similar tools do not exist for *opto*electronic devices and systems. As a compromise, modeling and simulation of the *electrical* portion of an optoelectronic design is often performed and the *optical* portion is left unaccounted for. As an example, consider the design of a laser-based optoelectronic transmitter. Circuit simulation tools are often used to design and simulate the transistors comprising the electrical circuit which drives the laser, but no such steps are taken to model the laser itself. At best, the *electrical* properties of the laser diode, such as its equivalent input impedance, are included in the design, the electro-optic conversion being completely neglected. This approach often necessitates several design iterations; for example, the laser driver and the laser diode often work independently but do not operate properly when connected together. Since no tools exist to predict the *integrated* behavior of the laser and its driver ahead of time, the designer's only recourse is to go through another design iteration. A similar case arises in the design of an optoelectronic receiver, in which the transistors comprising the electronic preamplifier can be simulated, but the optical properties of the photodetector cannot. While this method is clearly inefficient, the lack of availability of suitable CAD tools leaves little alternative.

The situation only becomes worse when the integration of optoelectronic *systems* must be considered. Although circuit simulation can be used to some extent in optoelectronic *sub*system design (e.g., transmitter, receiver), its role in system design is very minimal. Unlike the electronic world, which has a wealth of system-level design, analysis, and simulation tools, very little, if any, equivalents exist for the systems integration of optoelectronic designs. In these cases, the "breadboarding" nature of the design becomes even more intense. With no simulation tools, it is difficult for the system designer to predict the performance of the entire link, to formulate specifications for the devices that comprise the link, and to identify potential bottlenecks in the link. For lack of a better approach, the systems engineer often obtains whatever discrete optoelectronic components are available at the time and "assembles" his system on prototype PC boards. Only when the system fails to meet the desired performance does the systems engineer have the insight to drive device designers to make improvements.

It would be incorrect, however, to state that absolutely no tools exist for optoelectronic modeling and simulation. As mentioned previously, the main focus of optoelectronic research over the past several decades has been on the development of optoelectronic *devices*. Accordingly, the CAD tools that have been developed have been clearly geared toward device simulation. In fact, many general-purpose device simulators exist to model detailed material and physical and geometrical aspects of single semiconductor devices. There are even device simulators that have been developed specifically to model semiconductor laser diodes [1.6]. Device simulators traditionally model semiconductor devices in two or even three dimensions; the basic approach taken is to discretize all relevant semiconductor equations in space, then to solve the equations at each point in space by brute-force finite difference calculations. Since the simulations are based on fundamental semiconductor equations, they are usually quite accurate, lending considerable insight into device design. However, the computationally intense nature of solving the entire set of semiconductor equations at each discretized point in space can result in simulation times on the order of hours. When the design of single devices is being conducted, such times are often viewed as an acceptable trade-off for computational accuracy. However, when circuits and systems consisting of such devices must be modeled, these long simulation times are clearly unaccceptable. Consider the analogous situation in the electronic world; while device simulation is often used to design and optimize experimental transistor structures, it is wholly inadequate for modeling VLSI circuits. A device-level simulation of a VLSI circuit would most likely require years to complete, even on a supercomputer.

1.3 Book Overview

Since a significant framework already exists for optoelectronic device-level modeling and simulation, the goal of this book is to illustrate methods by which optoelectronic integrated *circuits* and *systems* can be modeled, designed, analyzed, and simulated through the use of CAD tools. As emphasized in the previous section, few such tools exist for optoelectronic simulation; thus, a significant amount of effort is expended in this book describing simulators that are capable of accommodating optical and optoelectronic devices. Highlighted in Chapter 3 is iSMILE, a general-purpose circuit simulator that allows for the incorporation of new devices, and highlighted in Chapter 8 is the iFROST system simulator, which is used to model complete optoelectronic links.

It should be noted that commercial general-purpose modeling and simulation tools frequently appear on the market. While the tools described in this book were state of the art at their inception, it is quite possible that commercial tools may someday prove to be superior. It is to be emphasized, however, that the goal of this book is not to promote a given simulator; rather, it is to investigate the simulation *methodologies* required for optoelectronic design. Although better tools will always be in development, the techniques used in their development remain relatively constant. As an added advantage, the tools discussed in this book were developed at University of Illinois at Urbana-Champaign; thus, the authors had full access to the simulator source code. This provided much more insight in both the methods and implementation of optoelectronic simulators than would be afforded by the reverse engineering of a compiled commercial tool.

Once the optoelectronic simulation infrastructure is developed, the task turns to the development of appropriate models for optoelectronic devices. Electronic circuits and devices are characterized by voltages and currents; the modeling of optoelectronic devices requires not only node voltages and branch currents, but also some method of representing light (optical power). The physics that govern the conversion of electricity to light and light to electricity must also be included. This book examines a multitude of optoelectronic device models, as well as models for electronic devices that are often used in conjunction with optoelectronic devices.

Finally, this book also discusses several novel approaches to optoelectronic device simulation. While there are few optoelectronic CAD tools in existence, those that are available usually rely on extensions to *standard* electronic simulation methods. For example, device-level optoelectronic simulation can be performed by simply adding additional equations describing the optical properties of the device to the chosen set of semiconductor equations. Once this is accomplished, spatial discretization and finite-difference computation occur in a manner similar to that of conventional electrical device simulation. There are, however, several novel electrical simulation methods which are often overlooked when attempting to establish an optoelectronic simulation framework. In this book, several such approaches are also discussed, including the mixed-mode simulation of optoelectronic devices and circuits, the table-based numerical simulation of laser diodes, and the mapping of conventional optoelectronic device equations into forms that can be easily simulated.

As mentioned throughout this introduction, a tremendous infrastructure currently exists for electronic simulation. Accordingly, Chapter 2 presents a review of the various methods traditionally used in electronic CAD tools. Topics covered range from device simulation, to circuit simulation, to system simulation. This chapter provides good background for the rest of the book and provides ample opportunities for the reader to anticipate various aspects of electronic CAD which can be leveraged for optoelectronic simulation.

In Chapter 3, the discussion moves toward optoelectronic circuit-level simulation. The chapter begins with an introduction to model-independent simulation through a detailed description of the iSMILE circuit simulator. This tool allows for the implementation of new device models without requiring modification of the existing simulator source code. Since the modeling and simulation of optoelectronic integrated circuits require the development of new optoelectronic device models, a model-independent simulator, such as iSMILE, is crucial. The remainder of the chapter contains a description of several models that have been implemented in iSMILE, including models for the quantum-well semiconductor laser diode, the MSM photodetector, and the HEMT and MESFET compound semiconductor transistors.

Chapter 4 contains several simulation examples that utilize the simulator and models presented in Chapter 3. The designs contain both electronic (transistors) and optoelectronic (emitters and detectors) devices, making them good examples of optoelectronic integrated circuits and subsystems. The two main areas of focus in this chapter are the design and simulation of optoelectronic transmitters and receivers. As mentioned in Section 1.1, every optoelectronic system consists of a transmitter and receiver; thus, Chapter 4 provides good insight on how simulation can be used to design these essential building blocks of any optoelectronic system. The design examples of Chapter 4 also include descriptions about the various trade-offs involved in optoelectronic subsystem design. Indeed, because this book also contains a good deal of information on optoelectronic design and analysis, the book was titled "Computer-Aided Design of . . ." rather than "Modeling and Simulation of . . ."

Chapter 5 contains a description of alternate methods for implementing equivalent-circuit models into existing simulators. The first approach illustrated is the embedding of an MSM model directly into the simulator source code. While Chapter 3 emphasizes that this approach is clearly undesirable, it nevertheless remains one of the most popular approaches, and is thus, worthy of mention. This example also depicts a situation in which the direct-embedding method, although involved, is the optimum approach to new-model implementation. Also contained in Chapter 5 is a novel approach to modeling laser diodes. The technique presented in this chapter circumvents many of the problems associated with the equivalent-circuit modeling of laser diodes by transforming the fundamental laser device equations into a form more amenable to circuit simulation.

Chapter 6 was contributed by B. Onat and M. S. Ünlü of Boston University's Department of Electrical, Computer and Systems Engineering. While the authors of this book chose to write a continuous, rather than edited, text, they felt that the work of Onat and Ünlü was unique enough to warrant inclusion. The approach detailed in this chapter is the mixed-mode simulation of optoelectronic devices and circuits. It was mentioned previously that device simulation can often consume several hours, rendering it useless for circuit- and system-level simulation. However, in many designs, there are many electronic devices (transistors), but only one optical device (laser, LED, or detector). In these situations, the long simulation time resulting from device-level analysis of the single optoelectronic device is often an acceptable trade-off for accuracy. In the mixed-mode environment, complex devices, such as emitters and detectors, are modeled at the device level, while better understood devices, such as transistors, are modeled at the circuit level. The unique feature of the work presented in this chapter is that the device and circuit simulators are completely encompassed in one simulation framework. Conventional approaches to mixed-mode optoelectronic simulation involve performing the device simulation first, then porting the simulation results to a circuit simulator.

In Chapter 7, another unique approach to optoelectronic simulation is presented. The work illustrated here is a table-based numerical approach to the circuit-level modeling of laser diodes. This technique relies not on device physics, but rather on measured laser data as the basis for its equivalent-circuit model. Through numerical interpolation algorithms, the table-based method is able to model the laser diode with no inclusion of device physics. This approach has three main advantages. First, the physics behind semiconductor lasers can change significantly with minute changes in the laser geometry, growth, processing, or structure, especially when dimensions on the submicron level are approached. This unfortunate fact of nature creates the need for a modified laser model for each type of laser. Although the field of semiconductor lasers is well established, it is continually evolving. Laser structures which appear promising one day are often realized as useless the next. The significant amount of time that is required to implement a new laser model can, thus, be totally wasted when device researchers decide to abandon one laser structure in favor of another. Table-based simulation provides a method for the rapid prototyping of lasers; only when a given type of laser diode is deemed useful is it necessary to invest the time required for a full physics-based laser model. Second, it is often the case that the physics governing a new type of laser are not fully understood. This is especially true when completely new structures or geometries are tested. Since a numerical model does not require any device physics, table-based models can accurately represent device behavior without a knowledge of the mechanisms involved. Third, numerical modeling relies on sets of device data. Often, device laboratories experience delays due to equipment failure, shortage of materials, or scheduling problems. When used in conjunction with a device simulator such as MINILASE [1.6], a table-based laser model can be used to prototype the circuit-level behavior of advanced laser diode structures *before they are even fabricated*. In this manner, the impact of new lasers on integrated circuits, subsystems, and systems can be anticipated before costly research efforts are undertaken.

Finally, Chapter 8 covers the topic of optoelectronic system simulation. In this chapter, the iFROST simulator is depicted as a method for modeling complete optical links. While there are few tools available for optoelectronic circuit simulation, there are even fewer available for optoelectronic system simulation. Through the use of system simulation, not only can the system-level performance of discrete devices be predicted but the reverse process can be performed as well: given a set of system specifications, a system simulator can also be used to determine the required levels of performance of the individual devices comprising the link. In this fashion, both "top-down" and "bottom-up" design methodologies can be used. System simulation does not concern itself with the detailed physics of the individual devices; rather, it focuses on the terminal, or "black box," behavior of these elements. The terms "macromodeling," "behavioral modeling," and "functional simulation," are often used as synonyms for system simulation. By considering only system-level attributes of discrete optoelectronic devices, complex system-level analyses can be performed, including total system speed, bandwidth, and signal integrity. Also included in this chapter is a mixed-level circuit-system simulation interface. There are often situations in which a macromodel does not contain enough detail; macromodels traditionally rely on manufacturer data sheets or simplistic measurements for the required device input parameters. When these are inadequate, it is often desirable to first simulate the device at the circuit level; once detailed circuit-level simulation results are obtained, they are interpreted and ported to the system simulator. This situation is analogous to the device-circuit mixed-mode environment mentioned previously.

1.4 References

[1.1] J. R. Sauer, "An optoelectronic multi-Gb/s packet switching network," University of Colorado OCS Technical Report 89-06, 1989.

[1.2] H. Tajima, Y. Okada, and K. Tamura, "A high speed optical common bus for a multi-processor system," *Transactions of the IEICE of Japan*, vol. E66, no. 1, pp. 47-48, 1983.

[1.3] A. W. Lohmann, "Optical bus network," *Optik*, vol. 74, pp. 30-35, 1986.

[1.4] K. W. Jelley, G. T. Valliath, and J. W. Stafford, "High-speed chip-to-chip optical interconnect," *IEEE Photonics Technology Letters*, vol. 5, no. 10, pp. 1157-1159, 1992.

[1.5] B. D. Clymer and J. W. Goodman, "Optical clock distribution to silicon chips," *Optical Engineering*, vol. 25, no. 10, pp. 1103-1108, 1986.

[1.6] G. H. Song, K. Hess, T. Kerkhoven, and U. Ravaioli, "Two dimensional simulation of quantum well lasers," *European Transactions on Telecommunications and Related Technologies*, vol. I, no. 4, pp. 375-381, 1990.

2

CONVENTIONAL SIMULATION METHODS

The simulation of semiconductor devices can be performed at various stages of integration, ranging from modeling of the materials that constitute the wafers and epitaxial layers, to simulation of device processing technology, to simulation of device structures, circuits composed of such devices, and systems composed of such circuits. This chapter will focus on simulation methods used in the latter three; that is, it will concentrate on techniques for the design of devices, circuits, and systems. The discussions presented in this chapter contain a brief review of conventional simulation methods for *electronic* design, which will then serve as background for the rest of the book, which focuses on *optoelectronic* design. The purpose of this chapter is not to provide a comprehensive description of these methods and their related tools, but rather to highlight those aspects that are directly pertinent to subsequent chapters. Indeed, entire books have been written about the subjects covered in each section of this chapter.

2.1 Circuit Simulation

Silicon integrated circuit technology has matured to the point that an engineer can design complex ICs, consisting of thousands or millions of transistors, starting with only a set of characteristic parameters for a typical device. In the case of an MOS transistor, these parameters would include the threshold voltage, channel-length modulation factor, and various capacitances, among other information. These parameters are typically measured for a given device technology by process engineers in the fabrication facility, then distributed to circuit designers who insert them into a standard MOS model in a standard circuit simulator (such as SPICE [2.1]). The parameters are usually measured for a specific device geometry (i.e., gate width and length in the case of an MOS transistor). Since the standard MOS models are sophisticated enough to scale transistor performance with device size, the circuit designer is left with only two tasks: determining the size of each MOS transistor and how the transistors should be interconnected. Simulation time and complexity increase with increasing transistor count; thus, when simulating large ICs, circuits are typically partitioned into several blocks, each of which is simulated separately. The behavior of the IC as a whole can be evaluated by the use of system-level tools, which will be discussed in Section 2.3.

The fact that a single set of device parameters is adequate for VLSI design is a testament to the uniformity of modern fabrication technologies. With wafers eight inches in diameter and larger currently in use today, if uniformity of device parameters both across the wafer and between wafers cannot be guaranteed, circuit design cannot proceed. Since *some* fluctuation of parameters is inevitable, the statistical variations of device parameters from their mean values are often measured by process engineers. This information can then be used by the circuit designer to ensure that his designs operate not only under the nominal device parameters, but also over the (statistically defined) *spread* in each parameter. Various design tools and methodologies exist [2.1] – [2.3] to ensure that a given circuit design, with a given set of nominal device parameters, will operate properly when statistical variations of these parameters are taken into account. Typically, the user specifies a given figure of circuit performance (such as gain, bandwidth, etc.), and the tool simulates the circuit over the *range* of each device parameter. If the circuit cannot meet the specifications for a particular value within the range, the simulator automatically resizes the transistors until the specifications are met. The end result is a circuit design (i.e., a list of transistor sizes and interconnections) that produces a given level of performance under all statistically probable ranges of each device parameter. Such practices are referred to as statistical design and are beyond the scope of this chapter.

The most widespread circuit simulator is SPICE (Simulation Program with Integrated Circuit Emphasis), developed at the University of California at Berkeley [2.4]. While entire books have been written about SPICE [2.5], [2.6], only a brief review of its operation will be presented here; interested readers can consult [2.4] – [2.6] for an in-depth tutorial. Since the inception of SPICE in the 1970s, several commercial SPICE-like simulators have been marketed which have better convergence, more functionality, and more models than Berkeley SPICE; however, the basic operation of all of these simulators is the same.

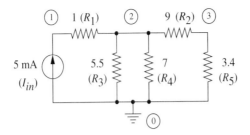

Figure 2.1: Sample circuit topology.

As a prelude to the iSMILE circuit simulator, which will be presented in Chapter 3, a short description of SPICE basics will be helpful. The input to SPICE is a file containing a description of the circuit to be simulated (which is called the netlist), a list of the (previously mentioned) device parameters corresponding to each model, and a command which dictates the type of analysis to be performed. While SPICE and its derivatives can perform many types of analyses, only the three basic ones will be considered in this section: DC, AC, and transient analysis. In order to construct the netlist, all of the nodes in the circuit to be simulated must be identified. In Figure 2.1, a simple resistive network is depicted with four nodes (in SPICE, node 0 is taken to be ground). Since resistors are simple enough to describe with only one parameter, there is no need for the separate list of model parameters mentioned previously; rather, the value of the resistance is simply appended to the resistor description, as shown in Figure 2.2. The general syntax of the netlist dictates that for each line, the type of device is declared first, followed by the node connectivity, followed by the device value. For example, the first line of the netlist describes a resistor, named R1, which is connected between nodes 1 and 2, with a value of 1 Ω. The last item in the netlist is a current source with current flowing from node 0 to node 1, with a value of 5 mA. The first letter of the name of each device corresponds to the type of device that it describes; R denotes resistor, C denotes capacitor, L represents inductor, and so forth. A complete list of all of the elements built into SPICE and its derivatives can be found in [2.4] – [2.6].

```
R1  1 2 1
R2  2 3 9
R3  2 0 5.5
R4  2 0 7
R5  3 0 3.4
Iin 0 1 5e-3
```

Figure 2.2: Netlist portion of input file.

```
.MODEL   MOD1 NPN   (IS=1e-15 BF=100 VAF-200 ISE=1e-12
+   BR=0.1 VAR=200 ISC=1e-12 RE=1 RC=10 RB=100
+   VJE=0.6 VJC=0.5 CJE=2e-12 CJC=2e-12 CJS=1e-12)
```

Figure 2.3: `.MODEL` portion of input file.

The second major portion of the netlist is a listing of the parameters for the device models. This is generally referred to as a `.MODEL` statement or a `.MODEL` "card," a terminology which dates back to the days when each instruction in a program was encoded on a punched card. In fact, the input file is sometimes referred to as the input "deck" for this reason. The resistor and current source models required only one input parameter. More complex models, on the other hand, require several parameters. Consider the Gummel-Poon model for the BJT, which has 41 input parameters in SPICE. Figure 2.3 depicts a sample `.MODEL` card for such a model. It should be noted that not all 41 of the parameters are listed; those that are not explicitly specified take on default values. The parameters listed in the `.MODEL` card of Figure 2.3 are, in order, the transport saturation current, ideal maximum forward beta, forward Early voltage, B-E leakage saturation current, ideal maximum reverse beta, reverse Early voltage, B-C leakage saturation current, emitter resistance, collector resistance, base resistance, B-E built-in potential, B-C built-in potential, B-E zero-bias depletion capacitance, B-C zero-bias depletion capacitance, and zero-bias collector-substrate capacitance [2.1].

Finally, the type of analysis must be specified. While there are many different analyses in SPICE, ranging from noise to Fourier to pole-zero analysis, the three that will be of interest in this book will be the DC, AC, and transient analyses which perform sweeps of DC currents/voltages, AC small-signal frequency, and time, respectively. Each input deck requires one analysis card; these cards take the forms depicted in Figure 2.4 [2.1]. In general, the starting point, stopping point and increment of the sweep are specified for each analysis. For DC analysis, the name of the source to be swept must also be specified (in Figure 2.1 it would be the current source). The AC analysis card also requires the type of frequency variation; the options are decade, octave, and linear variation. In each case, the number of points

```
.DC source_name start stop increment
.AC freq_variation number_of_points start_freq stop_freq
.TRAN time_step stop_time start_time
```

Figure 2.4: SPICE analysis cards.

per interval must also be specified. The .TRAN card assumes that the start time is zero unless explicitly specified.

There are many other "." cards including .IC for initial conditions, .OPTIONS for simulation options, .TF for transfer function analysis, etc. [2.4]. However, as emphasized throughout this chapter, only those functions critical to an understanding of the rest of this book will be accentuated.

2.1.1 DC analysis

The most fundamental analysis is DC, for both AC and transient analysis use DC analysis first to determine the operating point of the circuit. Using Figure 2.1 as an example, there are three node voltages (plus ground). The system depicted could be written using Kirchhoff's current law:

$$\frac{V_1 - V_2}{R_1} = I_{in}$$

$$\frac{V_2}{R_3 \| R_4} + \frac{V_2 - V_1}{R_1} + \frac{V_2 - V_3}{R_2} = 0 \qquad (2.1)$$

$$\frac{V_3}{R_5} + \frac{V_3 - V_2}{R_2} = 0$$

which can be written in matrix form as

$$
\begin{matrix}
V_1 & V_2 & V_3
\end{matrix}
$$

$$
\begin{bmatrix}
\dfrac{1}{R_1} & \dfrac{-1}{R_1} & 0 \\[2mm]
\dfrac{-1}{R_1} & \left(\dfrac{1}{R_1} + \dfrac{1}{R_3 \| R_4} + \dfrac{1}{R_2}\right) & \dfrac{-1}{R_2} \\[2mm]
0 & \dfrac{-1}{R_2} & \dfrac{1}{R_2} + \dfrac{1}{R_5}
\end{bmatrix}
\cdot
\begin{bmatrix}
V_1 \\[2mm] V_2 \\[2mm] V_3
\end{bmatrix}
=
\begin{bmatrix}
I_{in} \\[2mm] 0 \\[2mm] 0
\end{bmatrix}
\qquad (2.2)
$$

or equivalently as

$$\overline{Y}\overline{V} = \overline{I} \qquad (2.3)$$

where the matrix $\overline{Y}$ represents the resistive components, the vector $\overline{V}$ represents the unknown node voltages, and $\overline{I}$ is a vector of current sources. Because of the form of (2.3), $\overline{Y}$ is also known as the nodal admittance matrix [2.7]. By observing the matrix entries that result from the resistors R_1 and R_2, a general rule can be deduced: every occurrence of a resistor between nodes V_+ and V_- results in a matrix entry given by (2.4). That is, for every occurrence of a resistor R between nodes V_+ and V_-, the four matrix entries in (2.4) are simply added to the appropriate positions in the system $\overline{Y}$ matrix. This simple representation does not appear for resistors R_3, R_4, and R_5 in this example because node 0 (V_0) was not included in the matrix since it was known to be ground or 0 V. For resistors between node V_+ and ground, the matrix entry $1/R$ is inserted at the intersection of the vertical and horizontal V_+ labels, as seen in the case of R_5 in (2.2).

$$
\begin{array}{cc} V_+ & V_- \end{array}
$$
$$
\begin{bmatrix} +\dfrac{1}{R} & -\dfrac{1}{R} \\ -\dfrac{1}{R} & +\dfrac{1}{R} \end{bmatrix} \cdot \begin{bmatrix} V_+ \\ V_- \end{bmatrix} \tag{2.4}
$$

The matrix entries in (2.4) are known as *stamps*. Rather than write Kirchhoff's current law for the system and build the system matrix from the resulting equations, the entire system $\overline{Y}$ matrix can be constructed directly from the netlist by simply inserting the appropriate stamps at the appropriate positions in the $\overline{Y}$ matrix. In the case of R_2, $V_+ = V_2$ and $V_- = V_3$; thus, $1/R_2$ was placed at the matrix locations (V_2,V_2) and (V_3,V_3) while $-1/R_2$ was placed at the locations (V_2,V_3) and (V_3,V_2), where $(x,y) \rightarrow (row,column)$. In general, the stamp for a current source of I_0 flowing from node V_+ to V_- is

$$
\begin{array}{cc} V_+ & V_- \end{array}
$$
$$
\begin{bmatrix} \end{bmatrix} \cdot \begin{bmatrix} V_+ \\ V_- \end{bmatrix} = \begin{bmatrix} -I_0 \\ +I_0 \end{bmatrix} \tag{2.5}
$$

In Figure 2.1, however, $V_+ = 0$ (ground), so only $+I_{in}$ appears in the matrix. Similar DC stamps exist for every circuit element that exists in SPICE.

Once the nodal admittance matrix has been constructed, SPICE determines the vector of unknown voltages $\overline{V}$ by solving (2.2) through standard matrix techniques, such as Gaussian elimination. The details of such methods are left to standard linear algebra textbooks such as [2.10], while the *numerical* solution of these matrices can be found in [2.7] – [2.9].

The preceding DC analysis was based on the assumption that all of the circuit elements were linear. When nonlinear elements are used, the situation is complicated significantly. The general approach to solving for the DC voltages in a nonlinear circuit is to first set up the system equations in a manner similar to the linear case. The next step is to make an initial guess at $\overline{V} \equiv \overline{V}_0$, then to linearize the nonlinear elements about the initial voltages $\overline{V}$. The nodal admittance matrix is then stamped with the linearized elements and solved via Gaussian elimination as in the linear case. The resulting $\overline{V}$ is compared with the initial guess $\overline{V}_0$; if the two vectors are within some tolerance, then the system is deemed to be solved. If the vectors differ appreciably (i.e., if they are outside the predetermined tolerance), then a new initial guess $\overline{V}_0$ is determined as a function of the solved $\overline{V}$ and the gradients of the nonlinear elements evaluated at $\overline{V}$.

This process is referred to as Newton-Raphson iteration and can be described more formally by assuming that the nonlinear elements under question can be expressed as

$$I = I(V) \tag{2.6}$$

For the simple resistor depicted previously,

$$I(V) = \frac{V}{R} \tag{2.7}$$

However, resistors can be both linear and nonlinear, with (2.6) being the most general description. By taking the linear terms of a Taylor expansion, (2.6) can be represented as

$$I(V) = I(V_0) + \left[\frac{dI}{dV}\right]_{V_0} \cdot (V - V_0) \tag{2.8}$$

which is the "linearized" version of the nonlinear equation (2.6) evaluated at the initial guess. Since the derivative evaluated at V_0 is a constant, equation (2.8) can be rewritten as

$$I(V) = \left[\frac{dI}{dV}\right]_{V_0} \cdot V + \left\{ I(V_0) - \left[\frac{dI}{dV}\right]_{V_0} \cdot V_0 \right\} \tag{2.9}$$

This is clearly seen to represent a resistive circuit with a constant current source as depicted in Figure 2.5. The linearized circuit is then stamped into the nodal admittance matrix, and the system is solved through conventional matrix techniques. Once the solution is obtained, the resulting value of V is compared with the initial guess V_0. If the difference between the quantities is less than a predetermined tolerance, then convergence is obtained; if not, then the new value of V is used as the initial guess, the derivative is used to guide the next solution, and the process is repeated until convergence is achieved.

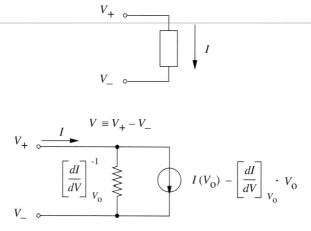

Figure 2.5: Original nonlinear (top) circuit and linearized (bottom) version.

While this example depicted a single device, the Newton-Raphson method depicted in (2.8) can be extended to systems of circuit elements as shown in (2.10), where the Jacobian matrix $\bar{J}$ is the generalized form of the derivative as given by (2.11).

$$\bar{I}(\bar{V}) = \bar{I}(\bar{V}_o) + [\bar{J}]_{\bar{V}_o} (\bar{V} - \bar{V}_o) \tag{2.10}$$

$$\bar{J} = \begin{bmatrix} \dfrac{\partial I_1}{\partial V_1} & \dfrac{\partial I_1}{\partial V_2} & \cdots & \dfrac{\partial I_1}{\partial V_n} \\[2ex] \dfrac{\partial I_2}{\partial V_1} & & & \dfrac{\partial I_2}{\partial V_n} \\[2ex] \vdots & & & \vdots \\[2ex] \dfrac{\partial I_n}{\partial V_1} & \dfrac{\partial I_n}{\partial V_2} & \cdots & \dfrac{\partial I_n}{\partial V_n} \end{bmatrix} \tag{2.11}$$

Thus, both linear and nonlinear DC systems can be solved through use of conventional linear algebra techniques and the Newton-Raphson method. Through the .DC analysis in SPICE-like circuit simulators, DC sources and vectors of sources can be swept over arbitrary ranges, allowing for the determination of all of the system's DC currents and voltages. It should be noted that not all nonlinear DC problems can be solved by Newton-Raphson iteration, and that not all DC systems can be expressed as simply as the resistive matrix depicted in this section. Many algorithms exist for both the formation and the solution of the DC matrix; only a representative subset of these techniques was presented in this section. These methods, however, provide adequate background for an understanding of the rest of the book. Interested readers should consult the references for a more in-depth treatment of DC analysis, in general.

2.1.2 AC analysis

As mentioned previously, in circuit simulation, DC analysis is fundamental because it is also the basis for both AC and transient analysis. The first step in AC analysis is to determine the DC operating, or bias, point. This is accomplished by performing a DC analysis on the circuit, and is referred to as a DC *operating point* analysis. In Figure 2.6, an MOS inverter is depicted, along with its DC voltage transfer characteristic (obtained by DC analysis as discussed in the previous section). This same circuit can also be used as a small-signal voltage amplifier provided that the DC operating point is set somewhere on the transitional portion of the characteristic; that is, at a point where the slope is nonzero. (Clearly, small input fluctuations outside the sloped region will result in zero, or near-zero, output fluctuations). Thus, in order for the inverter to be used as an amplifier, V_{in} must be set to V_{bias}; one way to accomplish this is by constructing a resistive network around the inverter power supply V_{dd} as shown in Figure 2.7 [2.11]. The main task in the bias point design of the circuit in this figure then becomes the proper choice of resistors R_1 and R_2 such that $V_{in} = V_{bias}$. Consequently, as the first step in AC analysis, DC operating point analysis is used to determine the DC voltages at every

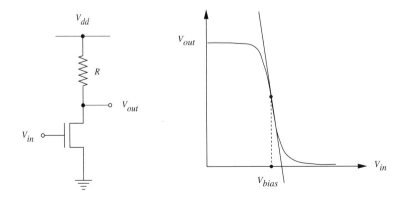

Figure 2.6: DC inverter circuit and its voltage transfer characteristic.

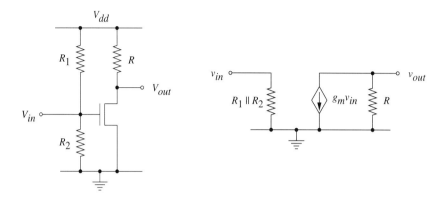

Figure 2.7: DC inverter with resistive network to set bias point (left) and small-signal AC linearized circuit (right).

node in the circuit, with the node voltage V_{in} being particularly important in this example.

Once the DC operating point $V_{in,o}$ has been established, the circuit is linearized about that point. That is, the nonlinear circuit elements are represented by the slopes of their DC characteristics evaluated at their DC bias points. This results in a linear circuit; in the case of Figure 2.6, this results in a common-source amplifier. The nonlinear DC I-V characteristic of the MOSFET is replaced by the popular small-signal model in Figure 2.7, where upper and lowercase voltages represent DC and AC values, respectively. In this representation, the nonlinear I-V relationship is replaced by a simple linear expression, namely

$$I_{out} = f(V_{in}) \quad \rightarrow \quad i_{out} = g_m v_{in} \qquad\qquad g_m \equiv \left. \frac{\partial I_{out}}{\partial V_{in}} \right|_{V_{in,o}} \qquad (2.12)$$

After the circuit has been linearized, the resulting system of equations is, as in the DC case, stamped into a matrix. Obviously, many of the element stamps for AC analysis are different than their DC counterparts. For example, the DC stamp for a capacitor is a matrix with all zeros since no current flows through a capacitor in DC. In AC analysis, however, the stamp for a capacitor is given by (2.13). This stamp is derived by formulating the circuit equations based on the relationship $I(\omega) = j\omega C \cdot V(\omega)$, which comes from the fact that time derivatives map into multiplication by $j\omega$ in the frequency domain. The stamp for a resistor, on the other hand, is the same for both AC and DC since ideal linear resistors have no frequency dependence.

$$
\begin{array}{cc}
V_+ & V_-
\end{array}
$$
$$
\begin{bmatrix}
j\omega C & -j\omega C \\
-j\omega C & j\omega C
\end{bmatrix}
\cdot
\begin{bmatrix}
V_+ \\
V_-
\end{bmatrix}
\tag{2.13}
$$

The matrix formulation for the circuit of Figure 2.8 is

$$
\begin{array}{ccc}
v_1 & v_2 & v_3
\end{array}
$$
$$
\begin{bmatrix}
\dfrac{1}{R} & \dfrac{-1}{R} & 0 \\[2ex]
\dfrac{-1}{R} & \left(\dfrac{1}{j\omega L_1}+\dfrac{1}{j\omega L_2}+j\omega C_1+\dfrac{1}{R}\right) & \dfrac{-1}{j\omega L_2} \\[2ex]
0 & \dfrac{-1}{j\omega L_2} & \dfrac{1}{j\omega L_2}+j\omega C_2
\end{bmatrix}
\cdot
\begin{bmatrix}
v_1 \\[2ex] v_2 \\[2ex] v_3
\end{bmatrix}
=
\begin{bmatrix}
i_{in} \\[2ex] 0 \\[2ex] 0
\end{bmatrix}
\tag{2.14}
$$

   The matrix is solved for each ω that falls within the frequency ranges specified by the .AC analysis command, as illustrated in Figure 2.4. The main assumption in AC analysis is that the input to the amplifier is a signal small enough that it does not shift the bias point of the amplifier, hence the name "small-signal analysis." It is important to realize that the DC operating point is calculated only once, at the beginning of the AC simulation. Hence, the effects of AC inputs large enough to perturb the circuit's bias point are not considered.

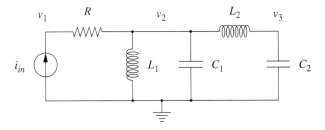

Figure 2.8: AC circuit containing resistors, inductors, and capacitors.

2.1.3 Transient analysis

Time-domain analysis is the most complex of the three analyses considered in this chapter. As with AC analysis, the first step in transient analysis is the determination of the DC operating point; from that point forward, the analysis is much more involved.

The general problem in transient analysis is to solve systems of differential equations of the form

$$\frac{d\bar{x}}{dt} \equiv \dot{\bar{x}} = \bar{f}(\bar{x}, t) \qquad \bar{x}(0) = \bar{x}_\mathrm{o} \tag{2.15}$$

The most obvious differential equations to be solved are those that include the current-voltage characteristics of capacitors and inductors:

$$i = C\frac{dv}{dt} \qquad v = L\frac{di}{dt} \tag{2.16}$$

As seen in Figure 2.4, the system must be solved over some time interval. The usual approach taken in transient simulation is to discretize the equations with respect to time over the specified interval and to solve the system of equations at each discrete time point. The spacing between each discrete time point is referred to as the *time step* and is commonly denoted as *h*. In the discretization of the system equations, the time step is often not uniform. That is, *h* is usually made small during periods under which the circuit undergoes rapid or drastic change, and *h* is made large during periods of relative inactivity (Figure 2.9). In this manner, the circuit can be simulated both accurately and quickly.

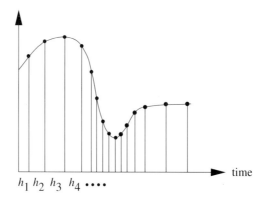

Figure 2.9: Variation of time step to accommodate rapidly changing regions.

Clearly, integration is necessary to solve the differential equations of (2.15) and (2.16). Since most sets of differential equations encountered in circuit simulation are not analytically solvable, numerical integration is used. There are many, many different methods for numerical integration; however, only a few of the most basic forms will be included here. As with previous sections, the purpose of the examples illustrated here is not to provide a comprehensive tutorial on transient analysis and numerical integration, but to provide enough background for the unfamiliar reader to understand the optoelectronic simulation methods presented in the rest of this book.

The most popular numerical integration schemes fit a class of algorithms known as *linear multistep methods*, so named because the solution of $\bar{x}$ at any given time point can be expressed as a function of the values of $\bar{x}$ and $\dot{\bar{x}}$ at both the present and previous time points. "One-step" methods depend only on the values of $\bar{x}$ and $\dot{\bar{x}}$ at the present time point and rely on no information from previous time points [2.12]. Linear multistep methods can be used to solve for the value of x at a given time point according to

$$x_n = \sum_{i=1}^{k} a_i x_{n-i} + h \sum_{i=0}^{k} b_i \dot{x}_{n-i} \tag{2.17}$$

with $a_0 = 0$ or equivalently,

$$\sum_{i=0}^{k} (a_i x_{n-i} + h b_i \dot{x}_{n-i}) = 0 \tag{2.18}$$

with $a_0 = -1$. In this formulation, the subscripts n and $n-i$ denote quantities evaluated at time points t_n and t_{n-i}, respectively. The index k denotes the number of steps in the linear multistep formula. For a solution of x which depends only on the present time point and the time point immediately preceding it, $k = 1$. For a solution which depends on the present and the previous *two* time points, $k = 2$. If the coefficient $b_0 = 0$, the equation is explicit and x_n can be easily determined in terms of its value(s) at previous time points. If $b_0 \neq 0$, the equation is implicit, with x_n being dependent on its values at both past and present time points. Since x_n appears on both sides of the equation when $b_0 \neq 0$, an iterative method, such as the Newton-Raphson technique must be employed to solve for x_n [2.12].

The main task in generating a linear multistep formula, then, becomes the proper choice of k and the a_i and b_i coefficients; each such choice affects various properties of the linear multistep formula. The methods used to generate these coefficients are not important for the purposes of this chapter; [2.7] – [2.12] contain a good description of this topic. There are several choices of the coefficients which are often used in circuit simulation:

If $k = 1$, $a_1 = 1$, $b_0 = 0$, and $b_1 = 1$, (2.17) becomes (2.19), which is known as the *forward Euler* formula.

$$x_n = x_{n-1} + h\dot{x}_{n-1} \tag{2.19}$$

If $k = 1$, $a_1 = 1$, $b_0 = 1$, and $b_1 = 0$, (2.17) becomes (2.20), which is known as the *backward Euler* formula.

$$x_n = x_{n-1} + h\dot{x}_n \tag{2.20}$$

If $k = 1$, $a_1 = 1$, $b_0 = 0.5$, and $b_1 = 0.5$, (2.17) becomes (2.21), which is known as the *trapezoidal rule*.

$$x_n = x_{n-1} + \frac{h}{2}(\dot{x}_n + \dot{x}_{n-1}) \tag{2.21}$$

There is also a class of algorithms, known as Gear's formulas, for which $k > 1$, resulting in a solution for x that depends on several previous time steps. Of special interest is the *second order Gear* formula in which $k = 2$, $a_1 = 4/3$, $a_2 = -1/3$, $b_0 = 2/3$, $b_1 = 0$, and $b_2 = 0$:

$$x_n = \frac{4}{3}x_{n-1} - \frac{1}{3}x_{n-2} + \frac{2}{3}h\dot{x}_n \tag{2.22}$$

The integration formulas presented here are, of course, only numerical approximations to the true solution. The *global truncation error* (GTE) is defined to be the difference between the actual and computed solutions at a given time point, while the *local truncation error* (LTE) is defined to be the difference between the actual and computed solutions at a given time point, assuming all previously computed points to be exact [2.12]. Reference [2.12] describes the LTE, then, as the amount of error committed in one time step. The amount of error incurred is one of the factors in the choice of a numerical integration algorithm; in [2.6] – [2.9] and [2.12], the derivation of the error incurred for each of the algorithms discussed is presented. Again, the details of these methods are not necessary for an understanding of the rest of this book.

Another factor in the choice of a numerical integration algorithm is stability. Even though the error at each time point (LTE) may be small, the total error (GTE) may actually become quite large if the LTEs gradually build up without bound [2.12]. The *stability* of a linear multistep formula can be studied by applying it to a test function [2.7]

$$\dot{x} = \lambda x \tag{2.23}$$

which results in (2.24), a linear difference equation. The stability of this difference equation can be analyzed by standard methods such as those found in [2.7] – [2.9] and [2.12].

$$\sum_{i=0}^{k} (a_i x_{n-i} + h \lambda b_i x_{n-i}) = 0 \tag{2.24}$$

While there are several methods for numerically evaluating derivatives such as $\dot{x}_n$, a popular approach is the divided difference technique. In [2.12], this finite difference interpolation method is described as

$$\frac{dx_n}{dt} \approx \frac{x_n - x_{n-1}}{t_n - t_{n-1}} \tag{2.25}$$

$$\frac{d^2 x_n}{dt^2} \approx 2 \cdot \frac{\dfrac{x_n - x_{n-1}}{t_n - t_{n-1}} - \dfrac{x_{n-1} - x_{n-2}}{t_{n-1} - t_{n-2}}}{t_n - t_{n-2}} \tag{2.26}$$

This algorithm can easily be extended to compute higher-order derivatives.

In this section, several methods for numerical integration were presented, and it was emphasized that the choice of integration scheme depends on various factors such as accuracy and stability. In general, each integration method has its own characteristics. For example, the forward Euler algorithm is easy to compute since it is explicit and depends only on values evaluated at previous time points; however, it tends to be fairly inaccurate [2.6] when compared to other methods. The trapezoidal rule is more accurate than the forward or backward Euler methods but is subject to spurious oscillations when the time step is chosen improperly. The Gear methods are more stable than the others but require more computational overhead [2.6]. Thus, in general, the choice of numerical integration algorithm also depends on the circuit to be simulated. The SPICE circuit simulator allows users the choice between the backward Euler, trapezoidal, and Gear algorithms. It is up to the user to decide which is best for a given application. In passing, it should be noted that the trapezoidal rule is chosen in SPICE as the default algorithm.

To understand how numerical integration schemes and transient analysis are applied to circuit simulation, consider the model for a capacitor, which can be described by

$$q = Cv \qquad i = C \frac{dv}{dt} \tag{2.27}$$

The first step in transient analysis is the determination of the DC operating point; that is, the circuit is solved in the DC domain with all capacitors open circuited and all inductors short-circuited. This serves as a description of the transient system at the time point $t = 0$. Next, as in DC and AC analysis, Kirchhoff's laws are used to express the equations describing the system. Consider the application of the trapezoidal rule (which is repeated here for convenience) to the capacitor of (2.27). In (2.29), the quantity x is replaced with the charge q, and the equation is rearranged so that the time derivative of q is on the left-hand side:

$$x_n = x_{n-1} + \frac{h}{2}(\dot{x}_n + \dot{x}_{n-1}) \qquad (2.28)$$

$$\dot{q}_n = \frac{2}{h}q_n - \frac{2}{h}q_{n-1} - \dot{q}_{n-1} \qquad (2.29)$$

Using the fact that the current is, by definition, the time derivative of the charge, and that $q = Cv$, (2.29) can be expressed as

$$i_n = 2\frac{C}{h}v_n - 2\frac{C}{h}v_{n-1} - i_{n-1} \qquad (2.30)$$

Since quantities at the previous time step $n - 1$ are known, (2.30) can be represented by the circuit of Figure 2.10. The equivalent-circuit representation of (2.30) is known as the companion model [2.12]. The companion models for each element are determined and stamped into the admittance matrix, in the same manner as in the DC and AC cases. Once this is accomplished, the matrix contains only linear elements and is solved by the same methods as used previously.

In general, capacitors can also be nonlinear:

$$q = f(v) \qquad\qquad i = \frac{dq}{dt} = \frac{df(v)}{dt} \qquad (2.31)$$

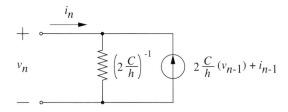

Figure 2.10: Transient companion model of a capacitor.

Following the previous steps, the trapezoidal rule can be applied to obtain

$$i_n = \frac{2}{h} f(v_n) - \frac{2}{h} f(v_{n-1}) - i_{n-1} \tag{2.32}$$

The linear resistor in Figure 2.10 is then replaced with a nonlinear resistor of value

$$\left[\frac{2}{h} f(v_n) \right]^{-1} \tag{2.33}$$

while the value of the current source becomes

$$\frac{2}{h} f(v_{n-1}) + i_{n-1} \tag{2.34}$$

The matrix is again formulated; however, since the matrix now contains nonlinear elements, Newton-Raphson iteration must be performed to determine the correct solution.

Transient simulation can be summarized as follows: first, the numerical integration scheme is chosen. Next, the DC bias point is determined and used as an initial guess ($t = 0$), then time is advanced to the first time point. The companion models are then determined for each of the circuit elements based on the numerical integration algorithm and the component equations, and the system matrix is created. The matrix is solved, either by linear or nonlinear methods, resulting in the solution of the circuit at the given time point. Time n is then incremented to the next time point $n + 1$ through the time step $t_{n+1} = t_n + h$, and the process is repeated until the end of the simulation interval is reached.

Because of the multiple steps that must be performed at each time point, transient analysis is time consuming. In addition, because of the inherent trade-offs involved in the choice of a numerical integration algorithm, the accuracy, stability, and convergence of a transient simulation may become issues. Also, the capability of a simulator to vary the time step so that radical changes are simulated at closer intervals plays a key role in the reduction of error in the simulation. All of these factors combine to make transient simulation the most involved of the three analyses discussed in this section.

2.2 Device Simulation

Taking a step back in the process, the individual transistors that comprise circuits are traditionally designed using device simulation. Although this book deals mainly with circuit and system simulation, a background of device simulation will be helpful to put the various levels of simulation into perspective. It will also aid in the understanding of Chapter 6, which deals with mixed-mode device-circuit simulation.

2.2.1 Device equations

At the device level, the fundamental equations of semiconductor physics are solved for a given set of materials, device geometries, doping profiles, and the like [2.13]. In [2.14], it is stated that the basic equations of semiconductor device operation can be placed into three groups: Maxwell's equations, current density equations, and continuity equations. The first group, Maxwell's equations, are given by (2.35) – (2.38).

$$\overline{\nabla} \times \overline{E} = -\frac{\partial \overline{B}}{\partial t} \tag{2.35}$$

$$\overline{\nabla} \times \overline{H} = \overline{J} + \frac{\partial \overline{D}}{\partial t} \tag{2.36}$$

$$\overline{\nabla} \cdot \overline{D} = \rho(x, y, z) \tag{2.37}$$

$$\overline{\nabla} \cdot \overline{B} = 0 \tag{2.38}$$

In these equations, $\overline{E}$, $\overline{H}$, $\overline{D}$, $\overline{B}$, $\overline{J}$, and ρ are the electric field, the magnetic field, the displacement vector, the induction vector, the conduction current density, and the total electric charge density, respectively. By using the relations $\overline{E} = -\overline{\nabla}V$ and $\overline{D} = \varepsilon\overline{E}$, (2.37) can also be expressed in the form of Poisson's equation

$$\nabla^2 V = -\frac{\rho}{\varepsilon} \tag{2.39}$$

The second set of semiconductor equations consists of the current density equations (2.40) – (2.42) [2.14].

$$\overline{J}_n = q\mu_n n\overline{E} + qD_n\overline{\nabla}n \tag{2.40}$$

$$\overline{J}_p = q\mu_p p\overline{E} - qD_p\overline{\nabla}p \tag{2.41}$$

$$\overline{J} = \overline{J}_n + \overline{J}_p \tag{2.42}$$

The quantities $\overline{J}_n$, $\overline{J}_p$, and $\overline{J}$ are the electron, hole, and conduction current densities, respectively, while n, p, μ_n, μ_p, D_n, D_p, and q are the electron and hole densities, the electron and hole mobilities, the electron and hole diffusion constants, and the electronic charge, respectively.

The third set of equations consists of the continuity equations

$$\frac{\partial n}{\partial t} = G_n - R_n + \frac{1}{q}\overline{\nabla}\cdot\overline{J}_n \qquad (2.43)$$

$$\frac{\partial p}{\partial t} = G_p - R_p - \frac{1}{q}\overline{\nabla}\cdot\overline{J}_p \qquad (2.44)$$

where G and R denote generation and recombination rates, respectively.

Similar equations exist for heterojunction structures commonly found in compound semiconductor devices (both electrical and optical). A separate description is necessary because of the energy band discontinuities introduced at heterojunctions. The discussion in this chapter, however, will be limited to the simulation of conventional electronic devices.

For particularly small-geometry devices, various quantum-mechanical formulas such as Schrodinger's equation are solved as well:

$$\left[-\frac{\hbar^2}{2m}\nabla^2 + V(\overline{r})\right]\cdot\psi(\overline{r}) = E\psi(\overline{r}) \qquad (2.45)$$

In (2.45), V is the potential energy as a function of position $\overline{r}$, m and E are the particle mass and energy, respectively, and ψ is the wave function. The most common application is to solve (2.45) self-consistently with Poisson's equation (2.39). For example, the charge in a quantum well can be determined by

$$\rho = q\sum_n N_n|\psi(z)|^2 \qquad (2.46)$$

where n is the energy subband level and [2.15]

$$N_n = \frac{m^*kT}{\pi\hbar^2}\ln\left[1 + \exp\left(\frac{E_f - E_n}{kT}\right)\right] \qquad (2.47)$$

The charge density computed from (2.46) and (2.47) is inserted into Poisson's equation (2.39) which is solved for the potential V. This new potential is then inserted back into Schrodinger's equation (2.45); the resulting wave function ψ is placed into (2.46), upon which a new charge density is computed. This process continues iteratively until the values of ψ and V no longer fluctuate appreciably. While many compound semiconductor devices are quantum mechanical in nature (laser, heterojunction transistors, etc.), quantum-mechanical effects must also be considered in conventional silicon transistors when the dimensions necessitate it. A good example of this is the deep-submicron MOS transistor, in which the gate can shrink to lengths on the order of tenths of microns.

At a level one step closer to the fundamental physics governing device behavior, the Boltzmann transport equation is often used to characterize the nonequilibrium distribution of electrons as a function of momentum, time, and space [2.13], [2.15], [2.16]. The Boltzmann distribution function is used to describe the probability of finding a particle at a given momentum and position $f(\bar{k}, \bar{r}, t)$ or over a finite range of momentum and space $f(\bar{k}, \bar{r}, t) d\bar{k} d\bar{r}$. In order to find the actual density of particles at a given point at a given time, the distribution function can be integrated according to [2.13]

$$n(\bar{r}, t) = \int f(\bar{k}, \bar{r}, t) d\bar{k} \tag{2.48}$$

The electrons in a semiconductor are acted upon by forces according to [2.16]

$$\hbar \frac{d\bar{k}}{dt} = \bar{F} \tag{2.49}$$

In this equation, the forces could be internal $\bar{F}_i$, such as collisions with phonons, or external $\bar{F}_e$, such as forces due to external fields. In an area about the particle position $\bar{r}$, the rate of change of the particle distribution is given by

$$\frac{df}{dt} = \frac{\partial f}{\partial t} + \frac{1}{\hbar} \bar{F} \cdot \bar{\nabla}_k f + \bar{v} \cdot \bar{\nabla}_r f \tag{2.50}$$

Equation (2.50) is known as the Boltzmann transport equation and is equal to zero since the total number of states in a crystal is constant [2.16]. The first term represents local changes in the distribution, while the second and third terms represent distribution changes due to forces and concentration gradients, respectively [2.16].

The effects of the internal and external forces are often separated for clarity. If the probability that a particle will be scattered, through an internal collision, from wave vector $\bar{k}$ to $\bar{k}'$ is denoted as $S(\bar{k}, \bar{k}')$, then the effect of the internal forces on the particle distribution can be described by (2.51) [2.13].

$$\frac{1}{\hbar} \bar{F}_i \cdot \bar{\nabla}_k f = \sum_{\bar{k}} S(\bar{k}, \bar{k}') f(\bar{k}') [1 - f(\bar{k})] - S(\bar{k}', \bar{k}) f(\bar{k}) [1 - f(\bar{k}')] \tag{2.51}$$

In this equation, the presence of the $1 - f$ terms is required by the Pauli exclusion principle, which forbids electrons to be in identical states. While these terms can often be omitted, in situations where the distribution becomes degenerate (i.e., heavy concentrations), they must be included. The scattering functions $S(\bar{k}', \bar{k})$ take on many different forms, depending on the type of scattering under consideration.

Because of its complexity, analytical solutions to the Boltzmann transport equation are not possible. Most analyses of the Boltzmann transport equation are performed by making simplifying assumptions on (2.50) and (2.51). For example, in (2.51), the scattering mechanism is often assumed to be elastic, randomizing, and/or independent of external forces. Another approach is to use the method of moments on the Boltzmann transport equation. In [2.13], [2.15] and [2.16], it is shown that the drift-diffusion, continuity, and energy/power balance equations can all be derived, using this method, as special cases of the Boltzmann transport equation.

2.2.2 Device simulation methods

Device simulation is typically performed in two or three dimensions. Thus, the spatial profiles of quantities such as the carrier concentrations and the electric field can be determined as functions not only of material and geometry, but of space, as well. Device simulation involves the solution of the previously mentioned equations simultaneously and self-consistently and is usually done by discretizing the differential equations and solving them through finite difference or finite element techniques. Since device simulation solves fundamental physical equations, the results obtained are usually quite accurate. However, since several dimensions of discretization are required (x, y, z, t), and since all equations must be solved at all grid points in space and time, device simulations typically consume large amounts of time and computing resources. Simulation times on the order of hours or days are not uncommon. Still, device simulation is the most accurate type of simulation and is indispensable for device design. Circuit simulation is essentially a trade-off between speed and accuracy. An entire circuit or subcircuit *could* be simulated at the device level; the results would be very accurate, but the simulation could take days, weeks, or even months to complete.

The first step in the finite difference technique is to spatially discretize the device to be studied, using a grid or mesh, which divides the device up into a collection of rectangles whose edges are usually parallel to the coordinate axes. The mesh spacing in any dimension can be either uniform, with an equal spacing between each grid point, or nonuniform. As with transient circuit simulation, it is often useful to use a nonuniform mesh over areas of the device where rapidly fluctuating values exist (Figure 2.11). While the finite difference method discretizes the device into rectangles, the finite element method uses nonuniform shapes such as triangles and quadrilaterals. These shapes provide a high degree of flexibility when analyzing devices or structures with irregular geometries [2.13]. While both the finite difference and the finite element methods have advantages in certain situations, the finite difference method tends to be a more popular technique for solving the semiconductor equations. Thus, the remainder of this section will concentrate on the finite difference method; this will be helpful for an understanding of the device-circuit simulation environment presented in Chapter 6. An application of the finite

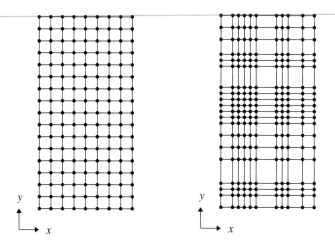

Figure 2.11: Spatial discretization: uniform (left) and nonuniform (right) grids.

element method is presented in Chapter 7; the interested reader is referred both to Chapter 7 and its reference list for more details on this simulation technique.

The second step in the finite difference simulation is the replacement of derivatives with their finite difference approximations. The basis for this substitution is the Taylor series representation of the function to be estimated:

$$f(x+\Delta x) = f(x) + \frac{df(x)}{dx}\Delta x + \frac{d^2 f(x)}{dx^2} \cdot \frac{(\Delta x)^2}{2!} + \dots + \frac{d^n f(x)}{dx^n} \cdot \frac{(\Delta x)^n}{n!} \quad (2.52)$$

If only first order terms are retained, then the finite difference representation of the derivative is given by

$$\frac{df(x)}{dx} \approx \frac{f(x+\Delta x) - f(x)}{\Delta x} \quad (2.53)$$

As with the various numerical integration techniques introduced in Section 2.1.3, there are many different ways to approximate the derivative, such as the forward difference method of (2.53) and the backward and central difference methods of (2.54) and (2.55), respectively [2.13].

$$\frac{df(x)}{dx} \approx \frac{f(x) - f(x-\Delta x)}{\Delta x} \quad (2.54)$$

$$\frac{df(x)}{dx} \approx \frac{f(x+\Delta x) - f(x-\Delta x)}{2\Delta x} \quad (2.55)$$

The third step in finite difference simulation is the solution of the discretized equations, subject to boundary conditions. The discretized equations are usually solved using the same techniques outlined in the transient simulation of circuits in Section 2.1.3 (i.e., backward Euler, forward Euler, etc.). Other more sophisticated numerical algorithms, such as the Crank-Nicolson scheme, are often used as well.

Thus, the three main steps involved in finite-difference device simulation are (1) mesh generation, (2) equation discretization, and (3) equation solution. Detailed examples of this process are presented in Chapter 6, which discusses device-level simulation within the context of a circuit simulator. Examples of stand-alone device simulators can be found in [2.17] – [2.19], in which the PISCES-II, MINIMOS, and FIELDAY tools, respectively, are described.

The device simulation methods just presented are generally applied to the "classical" semiconductor equations (2.35) – (2.47). Solution of the Boltzmann transport equation and its related expressions (2.48) – (2.51) is much more involved, often requiring techniques such Monte Carlo simulation. In the Monte Carlo method, a very detailed description of the device is given (such as the Boltzmann transport equation), then random statistical distributions and sequences are generated and used to simulate the behavior of individual particles within the device. The random distributions are used to mimic the effects of nature in stochastic events such as scattering and collisions. Predictably, with this level of detail, Monte Carlo simulations often require vast amounts of time and computing resources. A detailed treatment of Monte Carlo simulation is beyond the scope of both this chapter and this book (which focuses on circuit and system simulation). Interested readers are referred to [2.13] and [2.20].

2.3 System Simulation

Once circuits and devices have been successfully designed, their performance within the context of systems can also be evaluated. Again, due to the explosive proliferation of silicon technology, many simulation tools exist for such purposes. Because of the overwhelmingly large number of digital ICs in use today, most system simulation tools are capable of modeling only digital systems; examples are hardware description languages such as VHDL and Verilog [2.21] – [2.23]. System simulation is also referred to as behavioral or functional modeling since it is concerned more with the behavior or function of the system, rather than with the performance of the individual circuits and devices that the system is composed of. System simulation, then, treats circuits not as collections of transistors, but as "black boxes." For a given circuit block, system-level modeling is concerned with determining what output results for a given input, not with the performance of the individual transistors inside. The inputs to a digital system-level model would not be detailed analog waveforms, but rather simple pulsed signals with "system-level" attributes such as delay time, rise time, fall time, minimum value, and maximum value. Although there is obviously some trade-off in accuracy, a simplistic description such as this saves tremendous amounts of simulation time, which is a critical issue in the simulation of systems that contain millions of transistors. When more detail is required, subsystems are often simulated at the transistor or circuit level using tools such as SPICE. The results of these simulations are then ported to a system-level simulator; in this fashion, system simulation can also be performed in a hierarchical manner. Existing digital system-level simulation tools are sophisticated enough to not only simulate systems of existing circuits, but to also generate new circuits automatically based on system-level specifications. This process is called synthesis and allows system-level tools to be of significant use in both design *and* simulation. Convincing equivalent tools and hardware description languages for the simulation of analog systems are rare, due to the difficulty of behaviorally describing systems that depend critically on continuous waveforms, specific device geometries, and layout considerations.

2.3.1 The VHDL modeling language

Verilog and VHDL are, at present, the primary modeling languages used in digital system simulation. Since an entire review of these languages is beyond the scope of this section, several brief aspects of VHDL will be highlighted in order to provide the reader with a feel for the considerable differences between circuit, device, and system simulation.

One of the main concepts in VHDL is the description of a device or circuit through **architecture** and **entity** statements. A device's **entity** specifies the input and output ports to the device and any input parameters required in the device description. For example, the **entity** description for a two-input NAND gate could take the form of

```
entity NAND2 is
    generic (T_RISE : TIME);
    port (A, B : in BIT; C : out BIT);
end NAND2;
```

In this statement, an **entity** NAND2 is defined with two input ports, A and B, and one output port, C. The ports are given type BIT, a predefined type in VHDL that can take on "0" or "1." Also defined in VHDL are traditional data types such as REAL, INTEGER, and BOOLEAN. Similar to many programming languages, VHDL allows for user-defined or enumerated types, as well. The **generic** statement is essentially a declaration of the input parameters to the NAND2 model. In this example, the input parameter is the rise time, assigned to type TIME, also a predefined type within VHDL. Notice that the **entity** statement contains absolutely no description of the device's behavior; it only describes the input and output ports and the input parameters. (The examples in this section follow the standard convention of using lowercase text to express VHDL keywords).

Once the **entity** for the device has been defined, an **architecture** for the device can be constructed. While the **entity** contained a description of how the device is seen to the outside world, the **architecture** describes the actual internal workings of the device. Since Boolean operations are defined within VHDL, the **architecture** could take on the form of

```
architecture NAND2_GATE of NAND2 is
begin
    C <= not (A and B) after RISE_TIME;
end NAND2_GATE;
```

This example shows the use of the keyword **after** to simulate rise time.

A similar description at the *circuit* level would involve constructing a netlist and a simulation input file, as described in previous sections. For a CMOS NAND gate, an input file could take the form of

```
vdd 1 0 5
p1 2 4 1 1 pmos
p2 2 5 1 1 pmos
n1 2 4 3 0 nmos
n2 3 5 0 0 nmos
va 4 0 .....
vb 5 0 .....
   .
   .
   .

.MODEL nmos
. . . . . .

.MODEL pmos
. . . . . .


   .
   .
   .
```

where the power supply is denoted as vdd and assigned a value of 5 V. The input voltages va and vb are connected to nodes 4 and 5 respectively; their exact descriptions are left unspecified since they depend on the type of simulation (DC, AC, transient). The NMOS and PMOS transistors are interconnected, as shown above, and the netlist is completed with .MODEL cards for both types of devices.

In the VHDL description, the basic "building block" was the NAND gate, whereas in the circuit description, it was the MOS transistor. Inspection of these two examples shows that the *methods* of model description in both circuit and digital system simulation are similar. If it were possible to look inside the actual circuit models for the NMOS and PMOS transistors, one would see a list of *external nodes* defined in the model (drain, gate, source, body); this is analogous to the **port** statement in the **entity** description of the VHDL NAND gate model. In addition, there would be a list of input parameters which the model would expect to receive from the .MODEL card in the input file; this is the analog of the **generic** statement shown previously. Finally, the circuit-level MOS model would contain detailed current-voltage and device equations, analogous to the simple Boolean NAND assignment in the VHDL model **architecture**. Thus, there are three main components common to both VHDL and circuit models: (1) description of the external nodes, (2) description of the external parameters, and (3) description of the device behavior.

It should be noted that the circuit netlist shown previously does not describe the circuit *models* (i.e., the NMOS and PMOS models); rather it describes the *connectivity* of the models. The VHDL **entity/architecture** pair was, on the other hand, a description of the model itself (i.e., NAND gate). The analogy of the circuit netlist for VHDL is yet another model (**entity/architecture** pair) which contains occurrences or *instances* of previously defined models; in this situation, the models are often referred to as *components*. Consider the example of Figure 2.12, which is a simple AOI (AND-OR-INVERT) circuit. Assuming that the AND, OR, and NOT gates are now predefined components in a library, the **component** statement is used in the **architecture** for AOI_CIRCUIT in order to declare that the AND2, OR2, and NOT1 gates are going to be used. The gates A1, A2, O1, and I1 are actually instantiated in the **architecture** body, with the **port map** command being used to specify the interconnectivity between the individual elements. Of interest also is the **signal** declaration which is used to define the "wires" or "nodes" E, F, and G. Thus, in VHDL (as in circuit simulation), a system can be described by (1) a set of device models and (2) the connectivity of the models.

The previous VHDL description of the AOI circuit is called a *structural* description. Structural descriptions essentially describe systems as hierarchical blocks constructed of smaller elements. Interestingly, VHDL also has other types of descriptive styles, such as the dataflow and behavioral descriptions [2.21]. In the dataflow method, the function of the circuit is described through a signal assignment. The system of Figure 2.12 using the dataflow method would be expressed as:

```
architecture AOI_CIRCUIT of AOI is
begin
   H <= not((A and B) or (C and D));
end AOI_CIRCUIT;
```

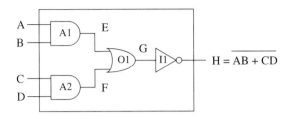

```
entity AOI is
    port (A, B, C, D : in BIT; H : out BIT);
end AOI;

architecture AOI_CIRCUIT of AOI is
    component AND2
      port (X1, X2 : in BIT; X3 : out BIT);
    end component;
    component OR2
      port (Y1, Y2 : in BIT; Y3 : out BIT);
    end component;
    component NOT1
      port (Z1 : in BIT; Z2 : out BIT);
    end component;
    signal E,F,G : BIT;
begin
    A1:   AND2 port map (A,B,E);
    A2:   AND2 port map (C,D,F);
    O2:   OR2 port map (E,F,G);
    I1:   NOT1 port map (G,H);
end AOI_CIRCUIT;
```

Figure 2.12: Gate-level schematic and VHDL description of
four-input AOI circuit.

This method is a very straightforward and simple way of describing the function of the AOI circuit. Also of interest is the behavioral method, in which circuit functionality is described through a set of sequentially executed statements. The AOI circuit expressed behaviorally would be

```
architecture AOI_CIRCUIT of AOI is
begin
    process (A, B, C, D)
       variable E, F, G;
       begin
          E := A and B;
          F := C and D;
          G := E or F;
          H <= not G;
       end process;
end AOI_CIRCUIT;
```

A VHDL **process** is a set of sequential statements that is executed every time there is a change in value on any of the listed signals (in this case, A, B, C, or D). Intermediate *variables* are defined within the **process** to facilitate the behavioral description of the AOI circuit. In contrast to VHDL, circuit simulators allow for only one type of description, namely, a list of the devices to be used and their interconnections (i.e., the netlist). Compare the VHDL description of the AOI circuit (Figure 2.12) with the transistor-level description of Figure 2.13. Clearly, the VHDL description is more straightforward; in view of the nonlinear DC and transient

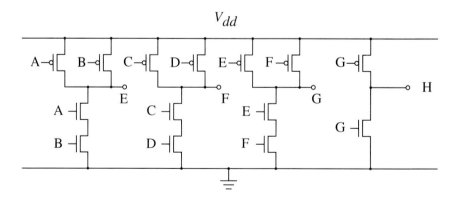

Figure 2.13: Implementation of AOI circuit with MOS transistors (CMOS) (NAND/NOT gates used instead of AND/OR gates to reduce transistor count). Bulk nodes omitted for simplicity: PMOS bulk → V_{dd}, NMOS bulk → GND.

numerical integration schemes presented in Section 2.1, the VHDL method should also be quicker. However, due to the simplistic (Boolean) nature of the VHDL circuit, obviously some degree of accuracy is lost.

In this section, the fundamental methods of system simulation using VHDL were outlined. In particular, the various similarities and differences between circuit-level and system-level descriptions were detailed. While a VHDL representation was shown to be clearly simpler than a circuit-level construction, no mention was made of the many other advantages that VHDL possesses, nor was any mention made of how to simulate a system once it has been described in VHDL. A complete overview would also include a discussion of the trade-offs between speed and accuracy when choosing between system and circuit simulation. As stated previously, the purpose of this section was solely to provide the reader with a feel for the differences in complexity between circuit- and system-level simulation.

2.4 Summary

In this chapter, an overview of the simulation methodologies used at the device, circuit, and system level was presented. While the focus of this book is *opto*electronic simulation, this chapter discussed the various techniques currently in place for conventional electronic simulation. Many of the tools and methodologies used in optoelectronic simulation rely heavily on those used for electronic simulation; in many cases, optoelectronic simulation is achieved by extending the capability of electronic simulators, rather than by creating entirely new tools. Thus, an understanding of conventional electronic simulation methods is useful for an understanding of the rest of this book.

The chapter began with circuit simulation and discussed the concept of the netlist as well as the various numerical methods involved in linear DC, nonlinear DC, AC, and transient circuit analyses. Clearly, SPICE was the pioneering work in this field, although many commercial circuit simulation tools now exist. Next, some of the fundamentals of device simulation were examined. While device simulation is very accurate, it requires a large amount of detail in the device description and consumes a proportionally large amount of time. Much of this can be attributed to the fact that the entire system of equations must be discretized and solved at each grid point in the mesh that is used to describe the device geometry. Finally, the system-level simulation of digital ICs was discussed using the VHDL language as an example. It was shown that a digital system can be described in three different ways: structurally, behaviorally, or through data flow. The choice of method is left to the designer and depends on the particular situation under consideration; in fact, a single system description can contain all three styles, each being used to describe different portions of the system.

One point that was not made was that a given simulation need not be restricted to only one of the three methodologies (device, circuit, or system). An entire field called "mixed-mode" simulation exists in which simulations at various levels are performed on various portions of a system. Most common is mixed-mode circuit-device simulation in which a complex device is simulated using device simulation while the circuit which contains the device is simulated using circuit simulation. Another example is mixed-mode circuit-system simulation, which is often used to simulate systems that contain both analog and digital blocks. As stated previously, since convincing system-level analog simulation tools are rare, designers often use circuit-level analysis to simulate complex analog blocks (e.g., PLL, VCO, I/O circuits) and system-level analysis to simulate digital blocks.

The discussions contained in this chapter highlighted those aspects that are directly pertinent to subsequent chapters; thus, this chapter should be considered as background material for the rest of the book, rather than a complete reference of electronic simulation. For more detailed information on individual topics, the reader is referred to the references.

2.5 References

[2.1]   A. Dharchoudhury and S. M. Kang, "Worst-case analysis and optimization of VLSI circuit performances," *IEEE Transactions on Computer-Aided Design*, vol. 14, no. 4, pp. 481-492, 1995.

[2.2]   S. R. Nassif, "Statistical worst-case analysis for integrated circuits," in *Statistical Approach to VLSI*, S. W. Director and W. Maly, editors. North Holland: Elsevier Science B.V., 1994.

[2.3]   S. W. Director, P. Feldmann, and K. Krishna, "Statistical integrated circuit design," *IEEE Journal of Solid-State Circuits*, vol. 28, no. 3, pp. 193-202, 1993.

[2.4]   T. L. Quarles, "SPICE3 version 3C1 user's guide," Electronics Research Laboratory Memorandum, University of California at Berkeley, no. UCB/ERL M89/46, 1989.

[2.5]   G. Massobrio and P. Antognetti, *Semiconductor Device Modeling with SPICE*. New York: McGraw-Hill, Inc., 1993.

[2.6]   R. Kielkowski, *Inside SPICE: Overcoming the Obstacles of Circuit Simulation*. New York: McGraw-Hill, Inc., 1994.

[2.7]   W. J. McCalla, *Fundamentals of Computer-Aided Circuit Simulation*. Boston, MA: Kluwer Academic Publishers, 1988.

[2.8]   J. Vlach and K. Singhal, *Computer Methods for Circuit Analysis and Design*. New York: Van Nostrand Reinhold Company, 1983.

[2.9]   L. O. Chua and P. M. Lin, *Computer-Aided Analysis of Electronic Circuits: Algorithms and Computational Techniques*. Englewood Cliffs, NJ: Prentice Hall, 1975.

[2.10] S. J. Leon, *Linear Algebra with Applications*. New York: Macmillan Publishing Co., 1986.

[2.11] A. S. Sedra and K. C. Smith, *Microelectronic Circuits*. New York: Holt, Rinehart and Winston, 1987.

[2.12] S. S. L. Chang, *Fundamentals Handbook of Electrical and Computer Engineering*, vol. III. New York, NY: John Wiley & Sons, Inc., 1983.

[2.13] C. M. Snowden, *Semiconductor Device Modeling*. London, U.K.: Peter Peregrinus Ltd., 1988.

[2.14] S. M. Sze, *Physics of Semiconductor Devices*. New York: John Wiley & Sons, Inc., 1981.

[2.15] K. Hess, *Advanced Theory of Semiconductor Devices*. Englewood Cliffs, NJ: Prentice Hall, 1988.

[2.16] C. M. Wolfe, N. Holonyak, Jr., and G. E. Stillman, *Physical Properties of Semiconductors*. Englewood Cliffs, NJ: Prentice Hall, 1989.

[2.17] M. R. Pinto, C. S. Rafferty, and R. W. Dutton, "PISCES-II — Poisson and continuity equation solver," Stanford Electronics Laboratory Technical Report, Stanford University, 1984.

[2.18] S. Selberherr, W. Fichtner, and H. W. Poetzl, "MINIMOS — a two-dimensional MOS transistor analyzer," *IEEE Transactions on Electron Devices*, vol. ED-27, pp. 1540-1550, 1980.

[2.19] E. M. Buturla, P. E. Cottrell, B. M. Grossman, and K. A. Salsburg, "Finite element analysis of semiconductor devices: The FIELDAY program," *IBM Journal of Research and Development*, vol. 25, pp. 218-231, 1981.

[2.20] J. M. Hammersley and D. C. Handscomb, *Monte Carlo Methods*. New York: Chapman and Hall, 1964.

[2.21] J. Bhasker, *A VHDL Primer*. Englewood Cliffs, NJ: Prentice Hall, 1995.

[2.22] D. R. Coelho, *The VHDL Handbook*. Boston, MA: Kluwer Academic Publishers, 1989.

[2.23] E. Sternheim, R. Singh, and Y. Trivedi, *Digital Design with Verilog HDL*. Cupertino, CA: Automata Publishing Company, 1990.

3

OPTOELECTRONIC CIRCUIT SIMULATION

As discussed in Chapter 1, when analyzing optoelectronic circuits and systems, simulation can be performed at a variety of different levels. Circuit simulation is most often chosen over more detailed methods because it is more time-efficient. While not as accurate as device simulation, the main advantage of circuit simulation is that it can be performed in minutes or seconds while device simulation often requires hours. In addition, while device simulation is used primarily to ascertain the properties of single devices, circuit simulation is used to analyze circuits or systems containing many such devices. In this chapter, the application of circuit simulation to optoelectronic integrated circuits will be investigated. This chapter will discuss the simulation of circuits that contain both electrical components, such as transistors, and optical elements, such as emitters and detectors.

3.1 Conventional Circuit Simulation

Conventional circuit simulators were created for electronic analysis; as such, they usually contain models for only standard electronic elements such as MOSFETs, BJTs, and diodes, as well as for simpler components such as resistors, capacitors, and inductors. To perform optoelectronic circuit simulation, models for optical elements, as well as their high-speed electronic counterparts, are required. Examples of such components are lasers, LEDs, and photodetectors, along with compound semiconductor transistors such as HEMTs and HBTs. The addition of such models to a standard simulation package often necessitates the coding and debugging of relevant device equations directly into the simulator source code, a process which can be not only complicated, but extremely time consuming, as well. Such an undertaking usually requires detailed knowledge of the program source code and structure. This is, indeed, a very difficult, if not impossible, task for the circuit designer, whose main job is to design circuits using established models, rather than to develop the models themselves.

Due to the lack of mature models for optoelectronic components, circuit designers are often left with no choice but to attempt to approximate the behavior of optoelectronic devices with the electronic models that are available. For example, to model the HBT, circuit designers often force the standard silicon BJT (Gummel-Poon) model to approximate HBT operation by assigning unrealistic values to the BJT model parameters. Another example is the photodetector which, for lack of an established model, is often represented in circuit simulation simply as a current source in parallel with a capacitor. The laser diode is often modeled as a capacitor in parallel with a resistor/voltage source series combination. An example of this will be shown in a subsequent section.

Circuit simulators operate by solving electrical networks of currents and voltages. It is natural to inquire, therefore, how a circuit simulator can be used to model devices which have not only electrical properties, but optical ones as well. The key is to realize that at the core of a circuit simulator is a differential equation solver. By properly interfacing to this solver, many physical systems described by differential equations can be modeled by a circuit simulator. The physical system needs not even be electronic. Consider, for example, a hypothetical system described by the following differential equation:

$$\frac{dy}{dx} + ay = b(x) \tag{3.1}$$

where a is a constant and b and y are functions of x. If both sides are multiplied by a constant C, the equation becomes

$$C\frac{dy}{dx} + Cay = Cb(x) \tag{3.2}$$

If the dependent variable y is then mapped into the variable v and the independent variable x into t, the result is

$$C\frac{dv}{dt} + Cav = Cb(t) \tag{3.3}$$

which can then be interpreted as a simple RC circuit with the resistor and capacitor expressed as in Figure 3.1. Performing transient circuit simulation on (3.3) will produce $v(t)$ from which $y(x)$ can be deduced.

If values for a and C of 10^3 and 10^{-6}, respectively, are assumed and for simplicity it is also assumed that b is a sinusoidal function with a 100 MHz frequency and a 1-μA amplitude, then the circuit can be implemented in SPICE format as in Figure 3.1 [3.1]. It is important to realize that no assumptions were placed on the function $y(x)$. It represents a generic function; it could represent optical power, it could represent mechanical stress, or it could even represent population growth. This example illustrates that it is possible to model a general system using a circuit simulator.

Since the circuit that resulted from this analysis is a combination of simple linear elements, it represents one example in which the laborious process of embedding new models into the program source code is not necessary. Indeed, the circuit that resulted from this analysis could be implemented directly into the circuit simulation input file as either a network of discrete elements or as a subcircuit. Unfortunately, however, the majority of devices of interest is considerably more complex than this example and requires more than simple linear elements to implement. While most standard circuit simulators do contain some nonlinear elements, implementation of variable assignments and program flow control (branching or conditional statements) is often not straightforward. Furthermore, simple model implementation through subcircuits in standard simulators does not

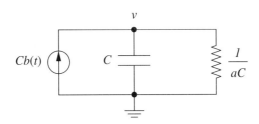

```
*   Simulation of hypothetical system
R1 1 0 1e3
C1 1 0 1e-6
Iin 0 1 SIN(0 1u 100e6 0 0)
.TRAN 1ns 100ns
.PRINT TRAN v(1)
.END
```

Figure 3.1: Equivalent circuit of hypothetical physical system.

offer the superior convergence or speed performance of a built-in model, nor does it allow for the detailed numerical procedures often required by complex devices.

3.2 The iSMILE Circuit Simulator

As discussed previously, the conventional approach to the circuit-level modeling of new devices involves developing and coding new model equations into an existing circuit simulator. This requires a significant amount of time not only for model development, but also for code debugging. While there are several circuit simulators that purport to overcome this bottleneck, for illustrative purposes, the iSMILE simulator, developed by Yang et al. [3.2], is chosen in this chapter. iSMILE (*i*llinois *S*imulator for the *M*odeling of *I*ntegrated-circuit *L*evel *E*lements) [3.2] is a versatile "SPICE-like" circuit simulator which allows for the easy implementation of models for new devices.   In contrast to the traditional method of model implementation, new models are added to iSMILE by specifying their equivalent-circuit topologies, their physical device equations, and their terminal current-voltage characteristics in a *model input file*.   From the model input file, iSMILE *automatically* generates new source code internally, which it then compiles and links to the original source code.   Once this is complete, the new model is added to iSMILE's standard library of devices and can be accessed at the input file level in the same manner as the built-in models. Through this process, the user is effectively shielded from the internal program source code. In view of the fact that it can be used to easily implement new, user-defined models, iSMILE can be considered to be a superset of SPICE.

Figure 3.2 shows the internal organization of iSMILE, which consists of four major components:

1) Model input processor
2) Model interface modules
3) Circuit simulator
4) Parameter extractor

The first step in developing a new model is to formulate a set of equations that accurately describes the device physics and the device's electrical terminal characteristics. Once a new device model has been developed, it is written in the form of a model input file (MIF). The MIF contains information about the device's equivalent-circuit topology, definitions of the variables that are to be accessed externally through the .MODEL statement, lists of variables that are for internal use only, and details about the device physics in the form of Fortran subroutines. The first of the four major components listed previously is the *model input processor* which takes this input file (MIF), parses it, and translates it into Fortran. The resulting sets of Fortran code are referred to as *model interface modules*. These modules, which form the second major component in the iSMILE program structure,

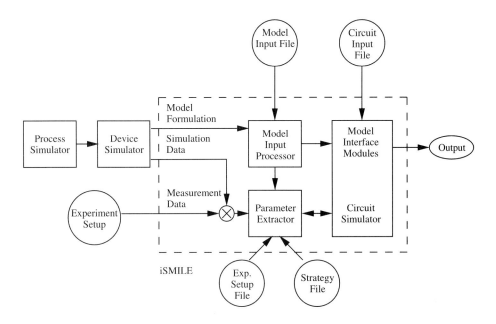

Figure 3.2: iSMILE internal system organization.

are automatically compiled and linked to the original iSMILE source code. Once this process is complete, the new model is added to iSMILE's library of devices and is accessible from the input file level in the same manner as any standard device. The third component in the iSMILE environment is the *circuit simulator*. The iSMILE circuit simulator has been structured to be compatible with that of SPICE. The circuit simulator takes the same input format as SPICE and can perform many of the same analyses (e.g., DC, AC, transient). The main difference is that when using iSMILE, the circuit designer has access to a larger library of devices that contains new, user-defined models. It is for this reason that it was previously stated that iSMILE can be essentially considered as a superset of SPICE. Finally, the fourth element in the iSMILE program structure is the *parameter extractor*. A parameter extractor is used to determine or extract the optimum device parameters required to fit the device model to measured device data, subject to particular ranges. When actual experimental data are not available, the results of device-level simulation are often used instead. This is a powerful method because it allows the circuit designer to complete his task without waiting for devices to be fabricated. Since the fabrication and characterization of new devices is often the bottleneck in the design process, this technique can be used to drastically reduce the time required for circuit and system analysis. Any variable that was defined as "external" in the model input file (i.e., those that appear in the device's `.MODEL` statement) can be optimized through the parameter extractor. It is important to note that most standard circuit simulators do not contain built-in parameter extraction. An example of new-model creation with iSMILE will be presented in Section 3.3, while an example of iSMILE's parameter extraction capability will be illustrated in Chapter 4.

3.3 Implementing New Models with iSMILE

In this section, a simple example of new-model implementation with Yang's iSMILE circuit simulator will be depicted. This will provide the necessary background for the remainder of the chapter, which will focus on optoelectronic models that have already been developed and added to iSMILE's device library.

As described in the previous section, at the heart of a new device model is the model input file, or MIF. While a model for the p-n junction diode is already built into iSMILE's permanent device library, in this section, a much simpler DC model will be used to illustrate new-model implementation. For this example, the basic diode equation will be used:

$$I_d = I_0 \left[\exp\left(\frac{qV}{nkT} \right) - 1 \right] \qquad (3.4)$$

In (3.4), q is the electronic charge, V is the voltage across the diode, n is the ideality factor, k is Boltzmann's constant, and T is the absolute temperature. The reverse saturation current I_0 is defined as

$$I_0 = I_s \cdot A = qA \left(\frac{D_p}{L_p} p_{no} + \frac{D_n}{L_n} n_{po} \right) \qquad (3.5)$$

where q is the electronic charge, A is the diode junction area, D_n and D_p are the electron and hole diffusion coefficients, L_n and L_p are the electron and hole diffusion lengths, and n_{po} and p_{no} are the electron and hole equilibrium minority carrier concentrations. While the individual diffusion parameters and carrier concentrations can be measured, calculated, or estimated, for the purposes of this example, it will be simpler to lump all of these quantities into the constant I_s.

The MIF used to implement this model is depicted in Figure 3.3, along with a description of each major MIF element. Also shown in Figure 3.3 is a schematic showing a representation of the diode in terms of circuit elements. Here the diode is modeled as a nonlinear resistor with a current-voltage characteristic as given by (3.4), with a linear resistor to represent the series resistance through the diode. As seen here, device physics are not the only things specified in the MIF; the current-voltage characteristic, or electrical terminal behavior, of the new device must be specified, as well. In addition, in order to aid in convergence, the first derivative of the current-voltage characteristic with respect to each voltage involved must be

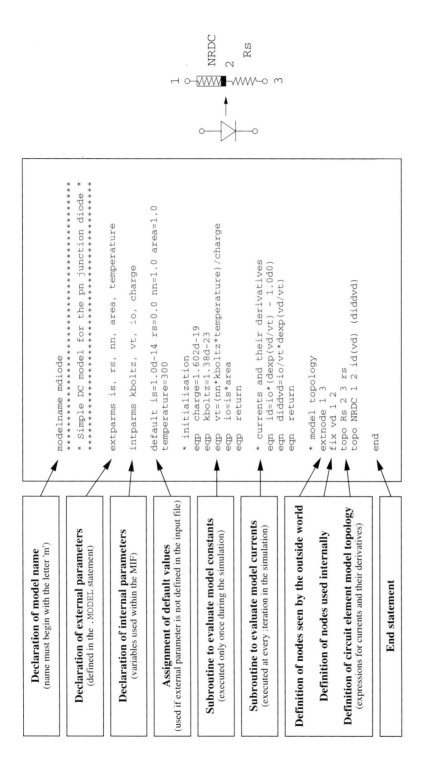

Declaration of model name
(name must begin with the letter 'm')

Declaration of external parameters
(defined in the .MODEL statement)

Declaration of internal parameters
(variables used within the MIF)

Assignment of default values
(used if external parameter is not defined in the input file)

Subroutine to evaluate model constants
(executed only once during the simulation)

Subroutine to evaluate model currents
(executed at every iteration in the simulation)

Definition of nodes seen by the outside world

Definition of nodes used internally

Definition of circuit element model topology
(expressions for currents and their derivatives)

End statement

```
modelname mdiode
*******************************************
* Simple DC model for the pn junction diode *
*******************************************

extparms is, rs, nn, area, temperature

intparms kboltz, vt, io, charge

default is=1.0d-14 rs=0.0 nn=1.0 area=1.0
temperature=300

* initialization
eqp  charge=1.602d-19
eqp  kboltz=1.38d-23
eqp  vt=(nn*kboltz*temperature)/charge
eqp  io=is*area
eqp  return

* currents and their derivatives
eqn  id=io*(dexp(vd/vt) - 1.0d0)
eqn  diddvd=io/vt*dexp(vd/vt)
eqn  return

* model topology
extnode 1 3
fix vd 1 2
topo Rs 2 3 rs
topo NRDC 1 2 id(vd)  (diddvd)

end
```

Figure 3.3: Model Input File (MIF) structure and equivalent-circuit topology for a p-n junction diode.

supplied analytically. While derivatives can always be computed numerically, iSMILE requires an analytical specification of the first derivative in order to increase computational efficiency. In the case of the diode example, the first derivative is easily expressed as

$$\frac{dI_d}{dV} = \frac{q}{nkT} \cdot I_o \exp\left(\frac{qV}{nkT}\right) \qquad . \qquad (3.6)$$

As Figure 3.3 shows, the name of the model must begin with the letter "m"; for the diode model, the name `mdiode` is chosen. The next statement is a declaration of the model's external parameters. These are the parameters that are to be accessible through the `.MODEL` card in the input file. The saturation current parameter, the series resistance, the ideality factor, the diode area, and the temperature have been made user definable. The next statement is a declaration of the model's internal parameters/variables. In the diode model, Boltzmann's constant and the electronic charge are defined as two internal variables. Of greater significance, however, is the declaration of the variables `vt` and `io`. As will be shown two blocks later, the assignments of the thermal voltage and the reverse saturation currents are made according to

$$V_T = \frac{nkT}{q} \qquad\qquad I_o = I_s A \qquad\qquad (3.7)$$

If the example of Figure 3.1 were implemented using iSMILE, a and C could have been made external parameters, and d could have been made an internal variable equal to $1/aC$. In the diode model, if the diffusion parameters and equilibrium minority carrier concentrations varied from device to device, they could have been made external parameters and I_s could have been made an internal parameter.

The next major block in the MIF is a list of default values that each of the external parameters is to be assigned in the event that the user does not specify them. The section commented as "initialization" in Figure 3.3 is also known as the `eqp` section. In this block, constants that are to be calculated only once in the simulation process are determined. This is in contrast to the next section, commented as "currents and their derivatives," or `eqn`. The `eqn` section, or subroutine, is executed multiple times, once during each iteration taken in the convergence process. At each iteration, the voltages are reevaluated, then the currents are recalculated based on these new voltages and on the new current-voltage derivatives. This process continues until relevant convergence criteria are met. The `eqn` and `eqp` subroutines are often so long and so involved that they are usually created in the format of separate Fortran files. In this case, instead of embedding the `eqn` and `eqp` statements within the MIF as in the diode example, calls are made to these subroutines from within the MIF via the Fortran `call` statement.

The last major part of the MIF is a declaration of the electrical topology for the new model. First, the external nodes are defined across the diode in a SPICE-like fashion as depicted in Figure 3.3. In this figure, the external nodes are labeled as "1" and "3." The diode voltage `vd` is assigned to the voltage across the junction by defining an "internal node" numbered as "2" in Figure 3.3. The next two statements describe the model topology, namely, a linear resistor to model the series resistance and a nonlinear resistor to model the current-voltage characteristic. A complete description of all the available components that can be used in the topology section can be found in [3.3]. Finally, the MIF is concluded with an `end` statement.

Once the diode MIF is completed and linked to iSMILE's source code, it is added to iSMILE's internal library of device models and is accessible from the input file. A sample input file using this new model is depicted in Figure 3.4. The circuit topology described by this netlist and the results of the DC simulation performed are also depicted in this figure.

The ease of new model implementation makes iSMILE ideal for simulating optoelectronic components for which new models have not yet been developed. In addition to models for conventional electronic devices (MOSFET, BJT, diode, etc.), models for multiple quantum-well laser diodes [3.4], MSM (Metal-Semiconductor-Metal) photodetectors [3.5], HEMTs/MODFETs [3.6], lossy transmission lines [3.7], and optical logic gates [3.8] have been implemented for use in optoelectronic circuit and system analysis. Each of these models is strongly based on device physics, enabling accurate determination of parameters such as circuit bandwidth, rise time, and gain. These models not only contain representations of DC behavior, like the diode example presented in this section, but are capable of predicting AC and transient behavior as well. Since, as previously discussed, circuit simulators operate on systems of branch currents and node voltages, all of the iSMILE optoelectronic models represent light or optical power as a node voltage. For the remainder of this chapter, three models that are of particular significance in the design of optical interconnect and optical communication systems will be highlighted: the quantum-well laser, the MSM photodetector, and the HEMT.

3.4 Multiple Quantum-Well Laser Diode

Multiple quantum-well semiconductor lasers are often used as sources in optoelectronic computing and communications systems. The electrical and optical properties of the active region of a quantum-well laser can be described by the well-known rate equations which describe the rates of change of the electron density, the photon density, and the optical phase in the quantum well in terms of physical parameters [3.9] – [3.13]:

$$\frac{dn}{dt} = \frac{J}{qNL_z} - (An + Bn^2 + Cn^3) - \Gamma gc'S \qquad (3.8)$$

```
*   Diode I-V characteristic

Vin 1 0 DC
Rin 1 2 1
mdiode 2 3 DIODE
Vtest 3 0 0

.MODEL DIODE mdiode is=1e-15   rs=1
+   area=10   temperature=300   nn=1

.DC Vin 0 1 .1
.PRINT DC I(Vtest)
.END
```

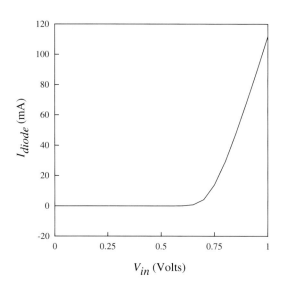

Figure 3.4: Input circuit topology and DC simulation results.

$$\frac{dS}{dt} = \beta B n^2 + \Gamma g c' S - \frac{S}{\tau_{ph}} \tag{3.9}$$

$$\frac{d\phi}{dt} = \frac{\alpha}{2}\left(\Gamma g c' - \frac{1}{\tau_{ph}}\right) \tag{3.10}$$

Definitions of the parameters in (3.8) – (3.10) are given in Table 3.1. The electron and photon *densities* n and S are defined to be the total number of electrons and photons, respectively, divided by the volume of the laser cavity.

The rate equations can be derived from Maxwell's equations operating on the electromagnetic field inside the laser cavity [3.9], [3.10]. Note that (3.8) – (3.10) are the rate equations for a *single-mode* laser. If multimode oscillation is to be considered, then a separate set of rate equations is required for each mode m:

$$\frac{dn}{dt} = \frac{J}{qNL_z} - (An + Bn^2 + Cn^3) - \sum_m \Gamma_m g_m c' S_m \qquad (3.11)$$

$$\frac{dS_m}{dt} = \beta_m Bn^2 + \Gamma_m g_m c' S_m - \frac{S_m}{\tau_{ph,m}} \qquad (3.12)$$

Parameter	Definition	Units
n	electron density	m^{-3}
S	photon density	m^{-3}
ϕ	phase of the optical field	—
t	time	s
J	injected current density	A/m^2
q	electronic charge	C
N	number of quantum wells	—
L_z	quantum-well thickness	m
A	nonradiative recombination coefficient	s^{-1}
B	radiative recombination coefficient	m^3/s
C	Auger recombination coefficient	m^6/s
Γ	optical confinement factor	—
g	optical gain	m^{-1}
c'	speed of light in the lasing medium	m/s
β	spontaneous emission coupling coefficient	—
τ_{ph}	photon lifetime	s
α	linewidth enhancement factor	—

Table 3.1: Definition of laser parameters.

$$\frac{d\phi_m}{dt} = \frac{\alpha_m}{2}\left(\Gamma_m g_m c' - \frac{1}{\tau_{ph,m}}\right)$$ (3.13)

A laser produces coherent optical power through the stimulated recombination of carriers. To generate these carriers, an injected current is required. The rate equation for the electron density (3.8) describes the various carrier mechanisms. The first term in (3.8) represents the carriers that enter the laser through the injected current J, while the second term models the electrons that are lost through spontaneous recombination. Of the three mechanisms in this term, only one (B) produces a photon as a result. The other two mechanisms are nonradiative, releasing energy into the lattice structure. The third term in (3.8) models the loss of carriers due to stimulated emission.

The rate equation for the photon density (3.9) describes the various mechanisms through which photons are added to the lasing mode. The spontaneously emitted photons in the first rate equation (Bn^2) are emitted randomly into all 4π steradians of space. The first term in (3.9) represents the fraction β of spontaneously generated photons that are coupled into the lasing mode. The second term in (3.9) models the increase in photons due to stimulated emission; notice that it is identical to the stimulated emission term in the electron rate equation, except for a sign change. Finally, the last term represents photon losses due to internal mechanisms within the laser cavity:

$$\tau_{ph} = \left[\Gamma c' \alpha_{int} - \frac{c'}{L}\ln(R)\right]^{-1}$$ (3.14)

The photon lifetime incorporates the effects of mirror losses due to the facet reflectivities R and losses due to intrinsic internal absorption mechanisms α_{int}. Note that the facet losses are distributed over the cavity length L.

The third rate equation (3.10) models the change in the optical phase with respect to time. The key component in this equation (3.10) is the linewidth enhancement factor α. As will be elaborated upon in a later section, fluctuations in intensity result directly in phase fluctuations. The degree to which this occurs is characterized by α. Since intensity is directly dependent on *net* gain, the expression in the parentheses of (3.10) represents gain in the cavity *minus* the losses.

It is important to consider the domain over which the rate equations are valid. Recall that n and S in (3.8) – (3.10) represent the total number of electrons and photons *averaged* over the entire cavity; thus, absolutely no information is contained about spatial distributions. In order to include such effects, the Poisson and continuity equations must be included to incorporate the carrier distributions and the Helmholtz eigenvalue equation should be included for the optical field [3.14]. Thus, inclusion of spatial effects is best performed through device or Monte Carlo simulation. More importantly, however, since the rate equations were derived from semiclassical considerations [3.9], [3.10] (i.e., using Maxwell's equations), they are incapable of describing quantum effects such as photon statistics, mode hopping [3.15], quantum amplitude and phase fluctuations, laser noise, and spectral width. In general, the rate equations are incapable of modeling phenomena that occur on very small (fs ~ ps) time scales. In order to treat these effects, it is necessary to use a fully quantum-mechanical description of the laser [3.16]. This type of treatment is *extremely* involved and requires the quantization of both the electromagnetic field and the atoms [3.17] – [3.19].

The standard rate equations are, however, quite capable of describing many properties of interest, such as laser turn-on transients, relaxation oscillations, output power, steady-state behavior, etc. In general, the rate equations accurately describe laser operation as long as the times involved are much greater than those required for atomic processes [3.15], [3.17]. One example of such an atomic process is intraband carrier relaxation, which takes place on the order of ~ 0.1 ps, (versus *inter*band relaxation which can require ~ 1 ns) [3.20]. Thus, the terminology "rate equation limit" or "rate equation approximation" is often used to emphasize the fact that the rate equations are to be used only on "relatively large" time scales. Although the standard rate equations are inadequate for modeling laser noise and spectral width, it is possible to introduce "noise-driven" rate equations which combine aspects of both the rate equation approximation as well as the fully quantum-mechanical theory. This will be explored further in Chapter 8.

The multiple quantum-well laser diode model of Gao et al. [3.4] that has been implemented in iSMILE, is based on the single-mode rate equations for the electron and photon densities (3.8) – (3.9). This model, however, does not include the second order effect of Auger (*A*) recombination in the rate equation for the electron density (3.8). The third recombination term Cn^3 is not *explicitly* included in (3.8), but nonradiative recombination is implicitly modeled through the inclusion of an electrical diode, as will be addressed shortly. While it is a simple matter to include the third (phase) rate equation, this model was developed mainly to investigate the circuit-level performance of laser diodes, rather than their spectral characteristics.

In a manner similar to that taken in the two examples of the previous sections, Gao maps each term in the rate equations into a current through an equivalent-circuit element (Figure 3.5). First, the electron and photon densities (*n* and *S*) are mapped into node voltages (V_n and V_S). The two nonlinear resistors on the electrical node V_n represent spontaneous (I_{sp}) and stimulated (I_{st}) emission, which are modeled by the last two terms of the electron density rate equation (3.8). The time change in electron density is represented by the nonlinear capacitor C_n, while the injected current term is modeled as the input current I_{in}. Attached to the optical node V_S are elements corresponding to the second rate equation (3.9). Specifically, there are two dependent current sources, $\alpha_{sp}I_{sp}$ and $\alpha_{st}I_{st}$, to model the spontaneous and stimulated emission contributions to the photon density. Also included are a resistor R_l to model photon losses in the laser cavity, as well as a capacitor C_{op} to model the

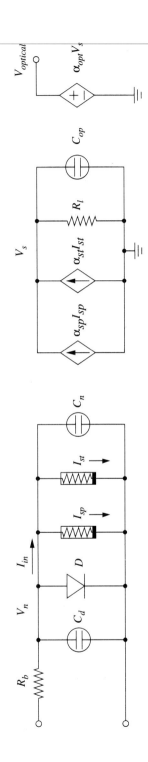

Figure 3.5: The iSMILE multiple quantum-well laser diode equivalent circuit.

change in photon density. Finally, since power is proportional to photon density through $P = (0.25c'WNL_z)S$ [3.4], an output node representing the laser's output power is included with a dependent voltage source. It should be noted that, as previously mentioned, not only must the current-voltage characteristic be defined for each element, but the first derivative of the current with respect to all voltages involved must be included as well.

The rate equations model the laser's optical properties *inside* the quantum well but they do not provide any information about the laser's electrical properties *outside* the active region. For example, the rate equations simply assume that a current density of J is injected into the quantum well. They cannot be used to determine how much voltage is generated as a result. Or, in the reverse scenario, they cannot be used to determine how much voltage must be applied to the laser in order to generate the current density J. The ability to predict the laser's current-voltage (I-V) characteristic is crucial for modeling circuits that contain both optical and electrical devices. For example, consider the simulation of a transmitter which consists of a transistor-based driver circuit and a laser diode. Without knowledge of the laser diode's I-V behavior, it would not be possible to predict the laser diode's effect on the electrical bias points of the transistors. Nor would it be possible to predict how much voltage the transistors must provide to generate sufficient current to drive the laser diode. To address these issues, three elements are added to the electrical node of Figure 3.5. Since the laser diode is a p-n junction, a diode D and a nonlinear capacitor C_d are added to model the electrical characteristics, including the diffusion and depletion capacitances as well as the nonradiative recombination current. In addition, a series resistor R_b is used to model the voltage drop outside the active region.

The gain is perhaps one of the most crucial parameters of interest in a semiconductor quantum-well laser. Since the gain is a function of the electron density (which is, in turn, a function of the injected current), it is often represented empirically for a bulk semiconductor laser as

$$g = a(n - n_o) \qquad (3.15)$$

where a is an experimentally determined gain coefficient and n_o is the electron density required for optical transparency [3.9]. For a quantum-well laser, the gain can be expressed through an empirical logarithm fit:

$$g = g_o \ln\left(\eta_i \frac{J}{J_o} \right) \qquad (3.16)$$

In this equation, g_o is a gain coefficient, η_i is the internal quantum efficiency, J is the current density, and J_o is the threshold current density [3.21].

Despite the applicability of (3.16), a more rigorous expression for gain in a quantum-well semiconductor laser can be developed from quantum-mechanical considerations. Indeed, one of the major contributions of Gao's laser diode model is the implementation of a quantum-mechanical model for the gain. While several laser diode circuit models exist, most use (3.15), or a similar form, as the basis for their gain formulation [3.11]. By relating the gain to the imaginary part of the susceptibility, the gain in a semiconductor quantum-well laser can be represented quantum mechanically as

$$g(\omega) = \frac{q^2 m_r^*}{\varepsilon_o m_o^2 c \bar{n} \hbar E L_z} |M|^2 [f_c(\omega) - f_v(\omega)] \qquad (3.17)$$

where q is the electronic charge, m_r^* is the reduced mass, ε_o is the permittivity, m_o is the electron rest mass, c is the speed of light, $\bar{n}$ is the index of refraction, $\hbar$ is Planck's constant divided by 2π, E is the photon energy, L_z is the thickness of the quantum well, M is the momentum matrix element, f_c and f_v are the Fermi distributions in the conduction and valence bands, respectively, and ω is the optical frequency. Derivation of this quantum-mechanical gain function can be found in [3.22] and [3.23]. It should be noted that (3.17) considers transitions between only the highest valence band state and the lowest conduction band state. If all transitions were to be considered, then (3.17) would contain a summation over all quantum states.

In addition to the quantum-mechanical description of the gain, Gao's model contains an expression for the electron density in the quantum well:

$$n = \frac{m_c^* kT}{\pi \hbar^2 L_z} \ln \left[1 + \exp\left(\frac{E_{fc} - E_{c1}}{kT} \right) \right] \qquad (3.18)$$

Here, m_c^* is the electron's effective mass in the conduction band and E_{c1} is the energy of the first conduction subband. Both the electron density and the gain are functions of the quasi-Fermi levels; however, Fermi levels are not physically observable quantities. In view of this fact, and of the fact that circuit simulators operate on networks of voltages, Gao's iSMILE model assumes that the potential difference between the quasi-Fermi levels is equal to the applied voltage. The quasi-Fermi levels are then computed internally within the laser diode model [3.4]. It should be noted that through the gain, the equivalent-circuit element representing stimulated emission depends indirectly on the Fermi levels and the applied voltage; thus, the derivative of the gain with respect to the applied voltage must be included in the model.

The gain equations in the previous analysis were derived on the assumption that the gain was independent of the intensity of light within the cavity. This is true under many sets of operating conditions, but in reality, the electromagnetic susceptibility, upon which the gain derivation is based, is inversely proportional to intensity [3.22]. While the effects of this inverse proportionality usually do not manifest themselves under normal operating conditions, they have been included in the iSMILE laser model for completeness. Because of the inverse dependence of the gain on the intensity, at high powers, the gain will decrease or saturate. The functional dependence of this gain saturation is often represented as

$$g = \frac{g_o}{1 + \dfrac{S}{S_{sat}}} \qquad (3.19)$$

where g_o is the normal, unsaturated gain as given by (3.17); it is to be recalled that the photon density is directly proportional to the intensity. In (3.19), the parameter S_{sat} is referred to as the saturation photon density; in effect, it provides a measure of how intense the optical power within the cavity must become before saturation effects occur. Typical values for S_{sat} are usually on the order of $10^{22} \sim 10^{24}$ m^{-3}. It is clear from (3.19) that for moderate intensities, the second term in the denominator will be negligibly small, and the saturated gain will be equal to the unsaturated gain. Since the photon density S is modeled by the node voltage V_S, the functional dependence of the gain on the photon density means that the first derivative of the gain with respect to the photon density must also be provided within the model.

As stated in Chapter 2, circuit simulators use the Newton-Raphson, or similar, methods to solve nonlinear circuit equations. Unfortunately, the solution of the rate equations consists of several different solution regimes; that is, multiple steady-state solutions exist for the rate equations, some more numerically stable than others. In order to prevent the circuit simulator from converging on a physically incorrect solution, it is necessary to "guide" the solver to the correct solution. One of the more numerically stable solutions of the rate equations will result in a negative value for the laser's photon density S, leading to a negative value for the output power. Indeed, this is one of the most common difficulties in the implementation of a circuit-level laser diode model. Gao's iSMILE laser model contains a heuristic algorithm to force the solver to converge to the correct solution [3.4]. This algorithm constantly samples the intermediate Newton-Raphson solution steps and, when necessary, forces the solver away from the "negative power" regimes. This sampling is accomplished through branching and conditional statements, the implementation of which is difficult at the input-file level of a conventional circuit simulator. Alternative methods to alleviate the problem of multiple rate equation solution regimes will be discussed in Chapter 5.

Using the values in Table 3.2, the DC and transient characteristics of the laser as simulated by iSMILE are depicted in Figures 3.6 and 3.7, respectively. As seen in Figure 3.6, the laser depicted in Table 3.2 has a threshold current of about 10 mA and a differential quantum (slope) efficiency of about 0.4 mW/mA. The transient characteristic of Figure 3.7 clearly shows the ringing, or relaxation oscillation, effect

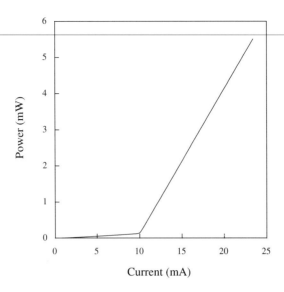

Figure 3.6: Simulated DC (L-I) characteristic.

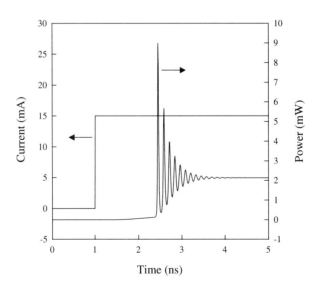

Figure 3.7: Simulated transient (turn-on) characteristic.

Model Parameter	Symbol	Value
Length	L	500 µm
Width	W	4 µm
Quantum-Well Thickness	L_z	70 Å
Number of Quantum Wells	N	1
Facet Reflectivity	R	0.32
Photon Energy	E_{photon}	1.556 eV
Temperature	T	300 K
Optical Confinement Factor	Γ	0.3
Dielectric Constant	$\bar{n}$	3.55
Radiative Recombination Coefficient	B	1.6×10^{-11} cm^3/s
Spontaneous Emission Coupling Coefficient	β	5×10^{-4}
Intrinsic Cavity Loss	α_i	3 cm^{-1}
Saturation Photon Density	S_{sat}	5×10^{23} m^{-3}
Series Resistance	R_b	1 Ω
Diode Ideality Factor	nn	2.11
Nonradiative Recombination Current	I_o	0.84 µA
Depletion Capacitance	C_o	3.17 pF

Table 3.2: Laser diode parameter values.

present during laser turn-on. This effect is usually caused by the dynamic equilibration of the photon and electron densities during laser transients. Figure 3.7 also shows the greater than 1-ns turn-on delay incurred while the electron population builds up. While Figures 3.6 and 3.7 depict simulations of the laser by itself, simulations of transmitters containing both laser diodes and compound semiconductor transistors will be illustrated in Chapter 4.

A comparison between Gao's laser diode model [3.4] and the model made popular by Tucker [3.11] is useful. There are many similarities between the models. For example, both are based on the single-mode rate equations and both provide large-signal equivalent circuits. However, Tucker's model was derived for (bulk) heterojunction lasers, as evidenced by the linear gain term used (3.15) in [3.11]. On the other hand, Gao's iSMILE laser model was specifically created to address the simulation of quantum-well lasers; as previously stated, an accurate description of the quantum-mechanical gain (3.17) is one of its major contributions. Thus, while the two models are similar in many respects, they are also quite different.

3.5 Metal-Semiconductor-Metal (MSM) Photodetector

The function of a photodetector is to detect incoming optical signals and to convert them to electrical signals. The MSM detector consists of a set of interdigitated metal fingers on a semiconductor surface which collects electron-hole pairs generated from incident light. Since the metal fingers directly contact the semiconductor, each finger forms a metal-semiconductor, or Schottky, contact. Figure 3.8 shows a schematic representation of the MSM, in which the metal fingers are indicated by the solid areas and the semiconductor active layer is shaded. Figure 3.9 is a photomicrograph of a fabricated MSM detector [3.24], while Figure 3.10 is a cross-section of the MSM. To fabricate an MSM, all that is necessary is to deposit metal on a semiconductor surface. As will be shown later, this step can be performed in the same step as the gate metal deposition in a FET process. Thus, one of the major advantages of using an MSM is its ease of integration into existing FET processes.

In Figure 3.10, incident photons strike the MSM from the top, penetrating the semiconductor active layer for a distance which is characterized by the absorption coefficient α. Once in the semiconductor, the photon is absorbed and, thus, generates an electron-hole pair. As seen in Figure 3.8, a voltage is applied across the MSM; as a result, the photogenerated electrons will migrate toward the positive contact while the photogenerated holes migrate toward the negative contact. This carrier motion constitutes the photocurrent.

Since each metal-semiconductor contact forms a Schottky barrier, a popular method of representing the electrical structure of the MSM is as a set of back-to-back Schottky diodes (Figure 3.11) [3.25]. Strictly speaking, however, this is not a valid representation. The metal contacts in conventional MSM detectors do not face each other; rather, they lie parallel to each other on a plane (Figure 3.12). This situation is somewhat analogous to early device structures for field-effect transistors. The JFET, for example, has a structure similar to Figure 3.11 with lateral source and drain contacts and vertical gate contacts. The MOSFET, on the other hand, is a completely planar structure, with drain, gate, and source contacts positioned on the same surface. The calculation of the electric field distribution, and the associated current-voltage characteristics, for the structure of Figure 3.12 is considerably more difficult than for the simple back-to-back Schottky diode case [3.26]. Essentially, the device of Figure 3.11 can be treated as one dimensional while the MSM of Figure 3.12 must be treated as two dimensional or quasi-two dimensional. The one-dimensional treatment of Figure 3.11 does, however, have a surprisingly wide range of applicability. In fact, with a few exceptions, nearly all analyses of the MSM have been based on this simple one-dimensional treatment.

The reverse bias applied across the MSM is used to deplete the semiconductor region between (Figure 3.11) and beneath (Figure 3.12) the metal contacts. As electron-hole pairs are generated in the semiconductor material, they migrate toward the metal contacts. If the reverse bias is not sufficient to completely deplete this

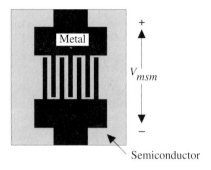

Figure 3.8: Schematic of MSM.

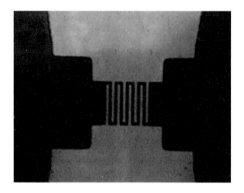

Figure 3.9: Fabricated MSM.

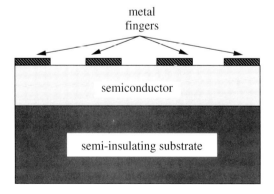

Figure 3.10: MSM cross-section.

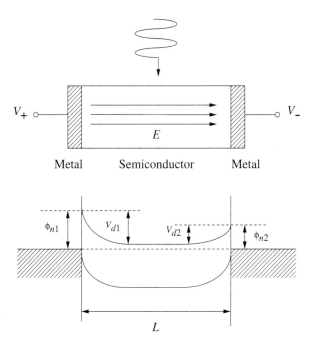

Figure 3.11: MSM electrical structure.

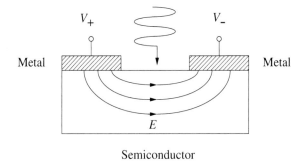

Figure 3.12: Planar MSM structure.

area, the electron-hole pairs must first diffuse to the depletion region before they drift through it. Since the diffusion process is slower than the drift process, this slows the response of the detector considerably. Thus, in normal operation, sufficient voltage is applied to the absorption region to completely deplete it, making the dominant transport mechanism drift. This, however, presents problems of its own. Since carriers move by drift, their motion is governed by their transit times. The electron and hole velocities (v_e and v_h) are proportional to their mobilities (μ_e and μ_h) and to the electric field E:

$$v_e = \mu_e E \qquad (3.20)$$

$$v_h = \mu_h E \qquad (3.21)$$

Thus, the time required for the carriers to drift to the metal contacts is inversely proportional to the mobility [3.27]. Since the hole mobility in most III-V compound semiconductors is lower than the electron mobility, holes require a longer time than electrons to reach the metal contacts. This behavior is often referred to as the "slow-hole tail" and is a fundamental speed limitation in MSMs (Figure 3.13). On the other hand, reports of 105-GHz MSMs [3.28] and 300-GHz MSMs [3.29] that were limited by parasitic capacitance and bond inductance indicate that with proper fabrication, the transit-time limit may not even be reached. It should be noted that these measurements were made with optical excitations of 532 nm and 620 nm, respectively, rather than the 850-nm wavelength of interest for short-distance optical interconnections. This is significant because the absorption coefficient is wavelength dependent, being higher for shorter wavelengths. The consequence of this is that shorter wavelength light will be absorbed closer to the surface and will thus be collected by the metal fingers more quickly. Since the velocity is also proportional to the electric field, it is possible to reduce the transit time by increasing the reverse

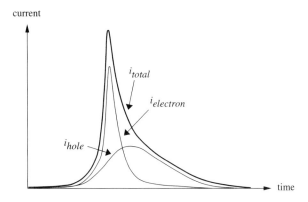

Figure 3.13: MSM transient photocurrents.

bias voltage; however, for practical system implementations, it is desirable to keep this voltage low.

Often, researchers characterize the speed of the MSM by measuring the FWHM (full width at half-maximum) of the impulse response. If the measurement exhibits a slow-hole tail, this is not accurate. In addition to the slow-hole tail, there is often a small amount of ringing or "ripple" at the end of the pulse response caused by impedance mismatches at the MSM output. This further reinforces the fact that the measurement of the FWHM is not always appropriate for speed characterization. A better figure of merit is the rate at which pulse can be reliably *repeated*; if optical input signal pulses are separated only by the FWHM, the resulting photocurrents will "smear" together. From a communications standpoint, this would cause nonnegligible intersymbol interference.

Another significant advantage of the MSM detector is its low capacitance, which is made possible by the very planar structure of the metal fingers (Figures 3.8 – 3.10). While calculation of the interdigitated capacitance for the planar MSM is very involved [3.30] – [3.32], very low values have been measured. For example, for the 100×100 μm^2 MSM of [3.33], the capacitance was only 100 fF. While this value can obviously be reduced by making the MSM smaller, if the detector is to be used in a practical system, coupling of light into the device must be considered. While a single-mode fiber has a core of less than 5 μm, in optical interconnects, multimode fiber (50 μm or 62.5 μm) is commonly used. Not only is it extremely difficult to couple light *into* single-mode fiber, but for short distances, the reduction in dispersion that results from using single-mode fiber is not required. Because multimode fiber is considered the transmission medium of choice for optical interconnects, the size of the detector should be made relatively large in these applications. Sometimes the MSM is even made with a round active (photosensitive) area to tailor it to the shape of an optical fiber [3.34].

It should also be noted that the MSM dark current can be made almost negligibly small (pA ~ nA) through proper processing [3.35] – [3.37], making the MSM competitive with the PIN, which exhibits low dark currents as well [3.38]. This raises the signal-to-noise ratio at the input, which is important since incoming optical signals are routinely on the order of 1 μW – 10 μW.

Perhaps the only drawback of the MSM is its low responsivity. Responsivity is defined to be the amount of photocurrent that is generated per unit amount of incident light power:

$$\Re = \frac{I_{ph}}{P_{in}}$$

(3.22)

In (3.22), I_{ph} is the generated photocurrent, P_{in} is the input optical power, and the responsivity $\Re$ is measured in units of mA/mW. Because of the metal fingers, one-half of the MSM active area is covered. In addition, there is a shadowing effect caused by the nonzero thickness of the fingers when the light strikes the detector at an angle. Once all factors are considered, responsivities on the order of 0.2 ~ 0.4 mA/mW for MSM detectors are not uncommon. This is to be contrasted with the ~ 1 mA/mW responsivities available with PIN detectors [2.26] and the >> 1 responsivities available with APDs (Avalanche PhotoDetectors). The low MSM responsivity is a serious problem since optical input powers on the μW scale are very common in lightwave systems. However, if the dark current can be lowered by the techniques of [3.35] – [3.37], a very low-level optical signal can be successfully detected and recovered with the aid of a low-noise, high-gain preamplifier.

As shown in (3.22), the photocurrent-optical power relationship is described by the detector responsivity. The next step is to determine the MSM's current-voltage relationship which, in the absence of optical excitation, is used to predict the MSM dark current. As shown in [3.25], the main current transport mechanism in the MSM is thermionic emission. The iSMILE MSM model of Bianchi et al. is essentially a one-dimensional model (Figure 3.11) based mainly on the formulation of Sze et al. [3.25], but it does include some two-dimensional effects, as will be mentioned later. This model assumes three regions of operation for current transport, each corresponding to a different level of bias voltage applied across the MSM fingers [3.5]. The first region is referred to as the *low-bias* region. In this region, the MSM bias voltage is insufficient to completely deplete the active absorption layer under the metal electrodes. While the MSM is not usually operated in this condition, it is, nevertheless, important to model this region to form a complete current-voltage characteristic. Unlike many conventional circuit models, which model the MSM simply as a current source in parallel with a capacitor, Bianchi's iSMILE MSM model is unique in that it separately models the electron and hole currents. In this region, the iSMILE model represents the electron and hole current densities as (3.23) and (3.24) [3.5], [3.25].

$$J_n = A_n^* T^2 \exp\left[\frac{q\left(\Delta\phi_n - \phi_n + \alpha_b E_m\right)}{kT}\right]\left[1 - \exp\left(\frac{-qV_-}{kT}\right)\right] \qquad (3.23)$$

$$J_p = \frac{qD_p p_{no} \tanh\left(xL_p\right)}{L_p}\left[1 - \exp\left(\frac{-qV_-}{kT}\right)\right] + \qquad (3.24)$$

$$\frac{A_p^* T^2 \exp\left[\left(\frac{-q}{kT}\right)\left(\phi_p + V_{bi}\right)\right]}{\cosh\left(x/L_p\right)}\left[\exp\left(\frac{qV_+}{kT}\right) - 1\right]$$

Recall that adjacent MSM fingers are biased with voltages of opposite polarity; V_+ is the voltage drop across the forward-biased metal-semiconductor junction while V_- is the voltage drop across the reverse-biased junction. The sum is thus equal to the total voltage applied across the detector, $V_{bias} = V_+ + V_-$. In (3.23) and (3.24), A_n^* and A_p^* are the effective Richardson constants for electrons and holes, respectively, T is the temperature, q is the electronic charge, k is Boltzmann's constant, ϕ_n and ϕ_p are the Schottky barrier heights for electrons and holes, respectively, α_b is the intrinsic barrier lowering coefficient, E_m is the maximum electric field at the negative junction, $\Delta\phi_n$ is the amount of Schottky barrier height lowering, D_p is the diffusion coefficient, p_{no} is the equilibrium hole density, L_p is the hole diffusion length, and x is the length of the *undepleted* active region. The total current through the detector is, of course, the sum of (3.23) and (3.24)

The second region of operation contained in Bianchi's model is the *reach-through* region, so called because in this region, just enough voltage is applied to cause the depletion regions underneath the individual electrodes to reach through to each other. In this region, the electron and hole currents are given by

$$J_n = A_n^* T^2 \exp\left[\frac{q(\Delta\phi_n - \phi_n + \alpha_b E_m)}{kT}\right]\left[1 - \exp\left(\frac{-qV_-}{kT}\right)\right] \tag{3.25}$$

$$J_p = A_p^* T^2 \exp\left[\frac{-q(\phi_p + V_{bi})}{kT}\right]\left[\exp\left(\frac{qV_+}{kT}\right) - 1\right] \tag{3.26}$$

where all terms are as previously defined.

The third, and final, region is termed the *flat-band* region because sufficient bias voltage is applied to make the electric field at the positive contact zero, resulting in a flat energy band. In this region, the currents are implemented as

$$J_n = A_n^* T^2 \exp\left[\frac{q(\Delta\phi_n - \phi_n + \alpha_b E_m)}{kT}\right]\left[1 - \exp\left(\frac{-qV_-}{kT}\right)\right] \tag{3.27}$$

$$J_p = A_p^* T^2 \exp\left[\frac{q(\Delta\phi_p - \phi_p)}{kT}\right] \tag{3.28}$$

where $\Delta\phi_p$ is the amount of Schottky barrier lowering. As stated previously, (3.23) – (3.28) were derived from the theory of thermionic emission; the detailed physical analysis can be found in [3.5] and [3.25].

There is often a photocurrent gain found in the MSM response. While many speculate that this is due to charge trapping at the interface [3.26], [3.37], the actual mechanism involved is still not well understood. Although gain enhances sensitivity due to increased photocurrent, because of its random nature, gain is also a source of noise which *degrades* the sensitivity [3.39]. As described in [3.5], the DC gain is usually a very slow effect which can ultimately degrade the MSM's high-speed response [3.40]. The photocurrent gain, when constant, can be modeled as [3.41]

$$G = \frac{I_{ph}}{q\,(1-R)\,[\,1 - \exp\,(-\alpha t_{abs})\,]} \qquad (3.29)$$

where R is the reflectivity of the semiconductor surface, t_{abs} is the thickness of the absorption layer, and α is, in this case, the absorption coefficient. However, this gain has also been observed to be voltage dependent. Since no suitable models for voltage-dependent gain exist, Bianchi represents the gain empirically as (Figure 3.14)

$$G = 1 + \frac{G_o}{1 + \exp\!\left(\dfrac{V_o - V_{bias}}{slope}\right)} \qquad (3.30)$$

The quantity V_{bias} is the voltage applied across the MSM and the three parameters G_o, V_o, and *slope* are varied until a suitable fit can be obtained. The voltage-

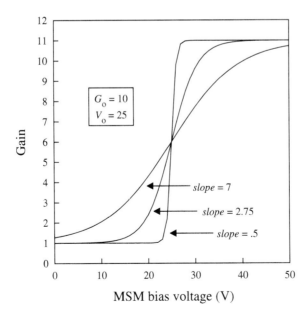

Figure 3.14: iSMILE MSM voltage-dependent gain function.

dependent gain term was taken from the Fermi-Dirac occupation probability distribution and was chosen because it varies smoothly between 0 and 1.

While Bianchi's model is based heavily on one-dimensional theory, it does include some quasi two-dimensional effects by modeling the extent of the electrical field not by the physical distance between electrodes, but by the actual length of the lines of force of the electric field, as seen in (3.31).

$$L = s + \pi \sqrt{\frac{1}{2}\left[\left(\frac{t_{abs}}{2}\right)^2 + \left(\frac{w}{4}\right)^2\right]} \tag{3.31}$$

In this equation, L is an approximation of the length of the electric field lines in the middle of the absorption layer, s is the finger spacing, w is the finger width, and t_{abs} is the thickness of the absorption layer. The term added to the finger spacing s is an approximation of a second-kind elliptic integral and gives the length of two arcs of an ellipse with half-axes of $t_{abs}/2$ and $w/4$ [3.5].

The transient response of the MSM is modeled by Bianchi through (3.32) [3.42].

$$I_{ph}(t) = \Re P\left[\frac{\tau_e - t}{\tau_e^2} + \frac{\tau_h - t}{\tau_h^2}\right] \tag{3.32}$$

In this equation, the electron and hole transit times are given by

$$\tau_e = \frac{L}{v_e} \qquad\qquad \tau_h = \frac{L}{v_h} \tag{3.33}$$

where v_e and v_h are the electron and hole saturated drift velocities, respectively. The parameter L represents the extent of the electric field; in the strictly one-dimensional treatment, L would coincide with the spacing between electrodes. In the quasi two-dimensional approach taken in Bianchi's model, however, L is an approximation of the true length of the electrical field lines in two dimensions (3.31).

Finally, the MSM capacitance C_{pd} in Bianchi's model is based on the standard two-dimensional conformal mapping approach introduced in [3.31] and [3.32]:

$$C_{pd} = \varepsilon_0 l (1 + \varepsilon_r) (N - 1) \frac{K(k)}{K(k')} \tag{3.34}$$

$$K(k) = \int_0^{\pi/2} \frac{d\phi}{\sqrt{1 - (k\sin\phi)^2}} \tag{3.35}$$

$$k = \tan^2\left(\frac{\pi}{4} \cdot \frac{w}{w+s}\right) \qquad (3.36)$$

$$k' = 1 - k^2 \qquad (3.37)$$

In these equations, ε_0 and ε_r are the absolute and relative permittivities, respectively, N is the number of fingers, l is the finger length, and all other parameters are as previously defined.

The equivalent-circuit representation of Bianchi's iSMILE MSM model is depicted in Figure 3.15, while the parameters used in the model are defined in Table 3.3. The first node in the equivalent circuit V_{op} models the optical input as a node voltage, much in the same way that the laser diode model represents optical output power. The next node V_x has five elements attached to it. The first is a dependent current source which is used to convert the optical input to a photocurrent via the responsivity relationship (3.22). The remaining elements form two sets of RC pairs whose RC time constants are used to model the MSM's transient response (3.32), (3.33). Notice that the electron (RC_e) and hole (RC_h) time constants are modeled separately. The resulting photocurrent is transferred to the output node through another dependent current source I_{ph}. The function defining this dependent source accounts for any current transport effects that occur during carrier collection. Also attached to the output node are nonlinear resistors to model the MSM electron and hole dark currents (3.23) – (3.28). The capacitance between fingers C_{pd} is also modeled through this node according to (3.34) – (3.37). Accurate modeling of this quantity is important since it often dominates the MSM response. The resistors R_s and R_d account for the resistance of the MSM fingers and the bulk resistance, respectively. The bulk resistance is often so large ($\sim 100\,\text{M}\Omega$) that it can be neglected. Similarly, the capacitance between the substrate and the semiconductor C_p is usually negligible if the substrate is semi-insulating.

As previously mentioned, the most common approach taken to model the MSM in a conventional simulator is to use a current source (to model the photo and dark currents) in parallel with a capacitor (to model the MSM capacitance). However,

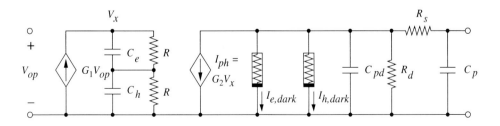

Figure 3.15: The iSMILE MSM equivalent circuit.

Parameter	Definition	Units
s	finger spacing	m
w	finger width	m
l	finger length	m
t	finger thickness	m
N	number of fingers	—
ρ	finger resistivity	Ω-m
λ	optical wavelength	m
N_d	doping in the absorption layer	m^{-3}
n_i	intrinsic carrier concentration of semiconductor	m^{-3}
ρ_{bulk}	bulk semiconductor resistivity	Ω-m
t_{abs}	thickness of absorption layer	m
α	absorption coefficient	m^3/s
V_{bi}	built-in potential	V
R	reflectivity of semiconductor material	—
L_p	hole diffusion length	m
A_n^*, A_p^*	effective Richardson constants of electrons, holes	$A/m^2/K^2$
v_e, v_h	saturation drift velocities of electrons and holes	m/s
ϕ_n, ϕ_p	Schottky barrier heights of electrons and holes	V
α_b	intrinsic barrier-lowering coefficient	—
ε_r	relative dielectric constant	—
$G_o, V_o, slope$	gain function fitting parameters	—

Table 3.3: Definition of MSM parameters.

this approach cannot model the separate electron and hole current effects, the I-V response, the dependence of capacitance on device dimensions, nor any of the other detailed effects presented in this section. For completeness, however, an example of this approach will be presented in Chapter 4.

Figure 3.16 shows a plot of the MSM's iSMILE DC I-V response taken from [3.5]. Both the dark current as well as the photocurrent due to an approximately 3-μW optical input are shown to be functions of the MSM bias voltage. Figure 3.17 shows the simulated MSM impulse response. This was generated using an "optical"

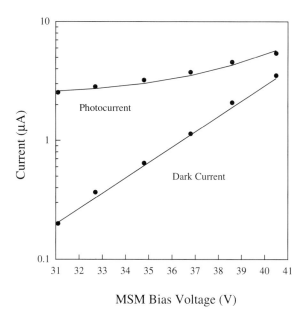

Figure 3.16: Log plot of the MSM DC I-V relation.
Markers represent experimental values; curves represent simulation results.

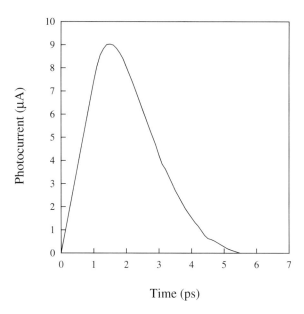

Figure 3.17: Simulated MSM impulse response.

voltage pulse at time zero with a very short pulse width (~ 1 ps) and a very high magnitude (~ 10^6 V) to replicate the behavior of a Dirac delta function. Notice the lagging tail of the MSM response; the photocurrent peaks after about 1.5 ps but does not end until more than 5 ps after the impulse. Examples of using the MSM model within photoreceiver simulations will be presented in Chapter 4.

3.6 Compound Semiconductor Field-Effect Transistors

Optoelectronic integration often requires optical and electrical elements to be fabricated either on a common substrate or, at least, to be composed of similar materials. Since most optical elements are made from compound semiconductors, this requires the ability to fabricate compound semiconductor transistors. The monolithic integration of both the electrical and the optical devices onto a single substrate has been viewed as a method to significantly lower the cost of optoelectronic subsystems while simultaneously improving their performance. While compound semiconductor transistors are, thus, clearly necessary for monolithic optoelectronic integrated circuits (OEICs), they are often used even in hybrid integrated systems for their high-speed and low-noise performance.

The most common compound semiconductor transistors used in optoelectronic integrated systems are the HEMT (High Electron Mobility Transistor), the MESFET (MEtal Semiconductor Field Effect Transistor), and the HBT (Heterojunction Bipolar Transistor). Because of similarities in device structure, field-effect transistors (FETs) are often monolithically integrated with MSM detectors to form photoreceiver front ends. Similarly, monolithic photoreceivers are often composed of HBTs and PIN detectors.

Due to the lack of suitably accurate models for the HEMT and the HBT, device researchers have traditionally used reverse modeling in order to curve fit HEMT data to conventional MESFET models and HBT data to conventional BJT models. In this section, the iSMILE HEMT model of Cioffi et al. will be presented [3.6], as will the iSMILE implementation of the popular MESFET model of Statz et al. [3.43], [3.44].

It should be mentioned, in passing, that the speeds of silicon transistors have reached impressive levels in recent years; in [3.45], a 25 Gb/s silicon decision circuit and a 40 Gb/s silicon demultiplexer were presented based on silicon bipolar technology. These results have direct relevance because of the critical role that these subsystems play in photoreceiver design. In [3.46] – [3.47], 10-Gb/s hybrid integrated photoreceivers were presented, again based on silicon bipolar technology. These speed benefits must, however, be weighed against the additional power requirements brought on by bipolar designs. A low-power monolithic CMOS photoreceiver was presented in [3.48]; however the speeds attained in that work were well below 1 Gb/s. Circuit simulation of these silicon-based systems can be performed with models for the bipolar junction transistor (BJT) and the metal-oxide-semiconductor field-effect transistor (MOSFET) that already exist in the iSMILE device libraries.

3.6.1 High-Electron Mobility Transistor (HEMT)

The HEMT, which is also referred to as the MODFET (MOdulation Doped Field-Effect Transistor), is based on the principle of a two-dimensional conducting channel, commonly referred to as the two-dimensional electron gas. By putting a donor layer, which is of different material than the channel itself, between the gate and the channel, a heterojunction is formed. As these layers are epitaxially grown, a discontinuity occurs in the energy bands at the heterojunction interface [3.49]; the resulting space and donor charges equilibrate, bending the bands as in Figure 3.18, which depicts a GaAs-based HEMT. In this figure, the donor layer is implemented in AlGaAs and the channel layer is implemented in GaAs. The band bending is such that a triangular potential well is formed at the channel. The thickness of this well is small enough to induce quantum effects; carriers are confined in the lateral direction and can propagate only in the direction perpendicular to the page in Figure 3.18. It is because of this one-dimensional carrier confinement that the channel has come to be known as the two-dimensional electron gas.

In addition, an intrinsic AlGaAs layer is added between the donor layer and the channel in Figure 3.18. This is called a spacer layer and it is used to physically separate the electrons in the channel from their ionized donors in the n-AlGaAs region; such a separation is not present in conventional FETs, such as the MOSFET or MESFET. Without such a separation, carriers in the conducting channel must flow near the impurities (dopants). This causes Coulombic scattering, the net result

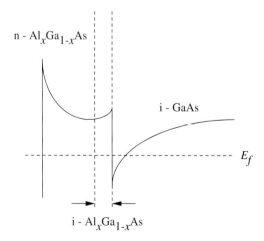

Figure 3.18: Energy-band diagram of conventional heterostructure FET.

being lower mobility and lower effective electron velocity [3.49]. Although using separate donor and channel layers already physically separates the donors and the carriers, the additional spacer layer further shields the carriers in the channel, providing an even greater mobility enhancement.

Because of the higher electron mobility in GaAs, such devices should inherently operate faster than Si FETs. To carry this argument one step further, since the electron mobility in many other compound semiconductors is even higher, FETs made of those materials should operate even faster. For example, the electron mobility in Si is about 800 cm^2/V-s, in GaAs it is about 4600 cm^2/V-s [3.49]. In InSb, the electron mobility is about 80,000 cm^2/V-s [3.50]. However, for heterojunctions to be properly grown, the lattice constants of the two materials must be very close; if they are not close, defects will form at the interface rendering the heterojunction useless. In the case of the AlGaAs/GaAs heterojunction, the lattice constant for GaAs is about 5.65 Å and that of AlAs is about 5.66 Å. The lattice constant for InSb, however, is about 6.48 Å [3.50], making its integration with GaAs-based structures very difficult. Thus, while other compound semiconductors have very high mobility, practical materials issues make fabrication of devices using these compounds difficult, if not impossible.

On the other hand, when $x{\sim}0.20$, $In_xGa_{1-x}As$ has a mobility of over 6000 cm^2/V-s, as well as a lattice constant of about 5.73 Å. While this lattice constant does not exactly match the 5.65 Å value for GaAs or AlGaAs, it is close enough for suitably thin layers to be grown without defects. Because of the lattice mismatch, *strain* will be induced at the heterojunction interface as the atoms of one material shift in an attempt to accommodate atoms of the other material. In such structures, the thickness of the strained layer must be kept below the *critical thickness* to avoid cracking or defects. While strained-layer heterostructures have applications in many different areas, HEMTs that employ strained-layer channels are commonly known as *pseudomorphic* HEMTs or P-HEMTs. Another benefit of using an InGaAs channel is that it has a lower band gap than GaAs. This results in larger heterojunction energy-band discontinuities at the AlGaAs interface which, in turn, increases the sheet-carrier concentration and improves carrier transport. Finally, it has been shown that devices with some indium content have inherently higher reliability [3.51]. An example of this type of structure is depicted in Figure 3.19.

Cioffi's iSMILE HEMT model is depicted in Figure 3.20. As a field-effect device, the HEMT has many properties in common with the MOSFET. However, the two-dimensional nature of the HEMT channel necessitates a unique analysis. This is especially evident in Cioffi's derivation of the I-V characteristic, which is represented in Figure 3.20 as the voltage-dependent current source I_{ds}. In Cioffi's analysis of [3.52], the derivation of the DC I-V relation is broken up into three steps: (1) derivation of the channel charge versus gate voltage relationship, (2) derivation

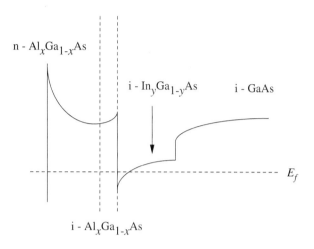

Figure 3.19: Energy-band diagram of pseudomorphic HEMT.

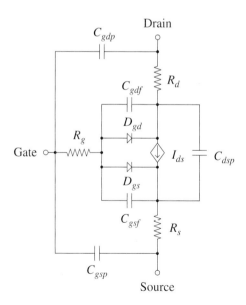

Figure 3.20: HEMT equivalent circuit.

of the velocity versus electric field relationship, and (3) combination of (1) and (2) to generate the current versus voltage relationship.

In the first step, the charge n at a point along the channel y versus the gate voltage V_g is modeled as

$$n(y) = \frac{n_{so} X^3(y)}{[1 + X^4(y)]^{3/4}} \tag{3.38}$$

$$X(y) = \frac{V_g - V(y) - V_{To} - V_1}{V_p} \tag{3.39}$$

$$V_p = q\frac{d}{\varepsilon}n_{so} \tag{3.40}$$

where n_{so} is the maximum charge concentration, V_{To} is the nominal threshold voltage, $V(y)$ is the channel potential, q is the electronic charge, d is the gate-to-channel spacing, ε is the dielectric constant, and V_1 is a constant equal to -0.401 V_p.

Of the many popular HEMT models in existence, two in particular are appropriate for circuit simulation since they describe the channel charge with smooth transitions between the linear charge region and the maximum well charge [3.52]. The first, the Yeager-Dutton model [3.53], uses a quadratic spline function over this region. However, according to [3.52], incorporation of this function in the current equations requires the assumption of a constant electric field which leads to a significant overestimation of the channel charge. The second model is based on the work of Rohdin and Roblin [3.54]. However, this model can result in negative charge under low gate voltages. In order to correct this deficiency, an artificial constraint must be imposed on the value of the channel charge which results in discontinuous derivatives and, thus, convergence problems. Cioffi's HEMT model, however, overcomes these problems through well-behaved, accurate expressions throughout the channel (3.38) – (3.40).

The second step in the derivation of the DC I-V characteristic is the determination of the velocity versus electric field relationship. This is expressed in [3.52] as

$$v(y) = \frac{\mu_o E(y)}{1 + E(y)/E_c} \tag{3.41}$$

$$E_c = \frac{v_s}{\mu_o} \tag{3.42}$$

where v and v_s are the velocity and the saturation velocity, respectively, μ_o is the mobility, and E and E_c are the electric field and the critical field, respectively.

The third, and final, step is the combination of these effects into, and the integration of, the one-dimensional drift equation for electrons

$$I = qWn(y)v(y) \qquad (3.43)$$

where W is the channel width. The diffusion current in this model, as in many well-accepted models [3.52], [3.55], is neglected. Using the fact that the electric field is the negative spatial derivative of the potential, (3.38), (3.41), and (3.43) can be expressed as

$$I = q\mu_o W n_{so} V_p \cdot \frac{X^3(y)}{[1 + X^4 y]^{3/4}} \cdot \frac{dX}{dy} \cdot \frac{1}{1 + \dfrac{V_p}{E_c} \cdot \dfrac{dX}{dy}} \qquad (3.44)$$

The derivation is completed by integrating over the channel length y. Omitting the lengthy calculations carried out in [3.52], the DC I-V characteristic is expressed as

$$I = k_p M_r \left[\left(1 + X_{gs}^4\right)^{1/4} - \left(1 + X_{gp}^4\right)^{1/4} \right] (1 + \lambda V_{ds}) \qquad (3.45)$$

where k_p, M_r, and the normalized voltages (X_{gs}, X_{gd}, and X_c) are defined in (3.46), (3.47), and λ is an empirical parameter used to model the HEMT output conductance in saturation. In (3.44) – (3.47), all other symbols are as previously defined.

$$k_p = \frac{q\mu_o n_{so} V_p W}{L_g} \qquad M_r = \frac{1}{1 + \dfrac{X_{gs} - X_{gd}}{X_c}} \qquad (3.46)$$

$$X_{gs} = \frac{V_{gs} - V_{To} - V_1}{V_p} \qquad X_{gd} = \frac{V_{gd} - V_{To} - V_1}{V_p} \qquad X_c = \frac{E_c L_g}{V_p} \qquad (3.47)$$

Unlike the MOS equations, where completely different equations are used to describe the DC I-V characteristic over different regimes of operation, this model can be used to describe HEMT operation over all regimes of operation provided that (3.48) is used to define X_{gp} in (3.45).

$$X_{gp} = \begin{cases} X_{gd} & \text{if } X_{gd} < X_{gd,sat} & \text{(linear regime)} \\ X_{gd,sat} & \text{if } X_{gd} > X_{gd,sat} & \text{(saturation regime)} \end{cases} \qquad (3.48)$$

Cioffi determines the saturation voltage $X_{gd,sat}$ through Newton iteration of (3.49).

$$(X_c + X_{gs})\, X^3_{gd,\,sat} + 1 - (1 + X^4_{gs})^{1/4}\, (1 + X^4_{gd,\,sat})^{3/4} = 0 \qquad (3.49)$$

In Figure 3.21, a simulation of the HEMT "family" of curves is depicted. In this simulation, the amount of drain current generated for a given drain-source voltage bias is calculated for various gate-source voltages.

Another distinguishing feature of the HEMT (versus the MOSFET) is that the gate directly contacts semiconducting material. In Figure 3.18, the gate metal would appear directly to the left of the n-$Al_xGa_{1-x}As$ layer; this direct contact forms a metal-semiconductor Schottky contact. Cioffi's HEMT model represents these Schottky barriers as simple diodes, as seen in Figure 3.20 (D_{gd}, D_{gs}). The current-voltage relationships of these diodes are expressed as

$$I_{gs} = \frac{J_o}{2} WL_g \left[\exp\left(\frac{qV_{gs}}{nkT}\right) - 1 \right] \qquad (3.50)$$

$$I_{gd} = \frac{J_o}{2} WL_g \left[\exp\left(\frac{qV_{gd}}{nkT}\right) - 1 \right] \qquad (3.51)$$

In these equations, J_o is the diode saturation current, W and L_g are the gate width and length, respectively, and n is the diode ideality factor. A simulation of the gate current characteristic is depicted in Figure 3.22.

The two intrinsic capacitances in Figure 3.20 (C_{gdf} and C_{gsf}) are defined as

$$C_{gsf} = \frac{\partial Q_g}{\partial V_{gs}} \qquad\qquad C_{gdf} = \frac{\partial Q_g}{\partial V_{gd}} \qquad (3.52)$$

The gate, drain, and source charges are expressed in [3.52] as

$$Q_g = -(Q_s + Q_d) \qquad (3.53)$$

$$Q_d = -qW \int_0^{L_g} \frac{y}{L} n(y)\, dy \qquad (3.54)$$

$$Q_s = -qW \int_0^{L_g} \left(1 - \frac{y}{L}\right) n(y)\, dy \qquad (3.55)$$

where L is the channel length, L_g is the gate length, and n is the channel charge. The numerical evaluation of these charge integrals is treated in [3.52]. Once expressions for the charges have been developed, a charge-based description of the drain current can be developed for AC and transient analysis (3.56).

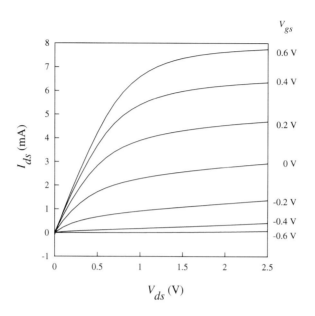

Figure 3.21: Simulation of HEMT DC I-V characteristic.

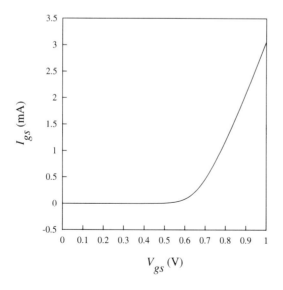

Figure 3.22: Simulation of HEMT gate current.

$$I_{ds} = \frac{\partial Q_s}{\partial V_{gd}} \cdot \frac{dV_{gd}}{dt} - \frac{\partial Q_d}{\partial V_{gs}} \cdot \frac{dV_{gs}}{dt} \qquad (3.56)$$

The three nonchannel parasitic capacitances in Figure 3.20 (C_{gsp}, C_{gdp}, and C_{dsp}) are due to parasitic interelectrode and fringing interactions and are derived in Cioffi's HEMT model through a two-dimensional analysis. Several unique issues are involved in the modeling of the HEMT interelectrode capacitances. First, most other HEMT capacitance models assume infinitely thin gate, source, and drain regions with no nonfringing capacitances. Second, these models neglect the shielding effect that the gate electrode has on the drain-to-source capacitance. A third issue is that HEMT source and drain electrodes usually extend below the surface in order to contact the two-dimensional conducting channel, an effect normally not represented in other HEMT models [3.52].

By solving the integral form of Green's function to Poisson's equation through the method of moments, expressions for the parasitic capacitances can be derived. The capacitance-per-gate width equations implemented in Cioffi's model for GaAs-based HEMTs are

$$C_{gsp} = C_{gdp} = 0.82\varepsilon_o \left(\varepsilon_{GaAs} + 1 \right) f(n) + \varepsilon_o \frac{d_{as} + \varepsilon_{GaAs} d_{sg}}{L_{gs}} \qquad (3.57)$$

$$C_{dsp} = \varepsilon_o \left(\varepsilon_{GaAs} + 1 \right) f(m) f_g(p) \qquad (3.58)$$

$$f(n) = \begin{cases} \pi \left\{ \ln \left[2 \cdot \frac{1 + (1-n^2)^{0.25}}{1 - (1-n^2)^{0.25}} \right] \right\}^{-1} & 0 \leq n \leq 0.707 \\[2em] \frac{1}{\pi} \ln \left[2 \cdot \frac{1 + \sqrt{n}}{1 - \sqrt{n}} \right] & 0.707 \leq n \leq 1 \end{cases} \qquad (3.59)$$

$$n = \frac{L_{gs}}{L_{gs} + 2L_g} \qquad\qquad m = \frac{L_{ds}}{L_{ds} + 2L_s} \qquad (3.60)$$

$$f_g(p) = \frac{0.235}{\sqrt{p + 0.156}} \qquad\qquad p = \frac{L_g}{L_{gs} + L_s} \qquad (3.61)$$

The various parameters in (3.57) – (3.61) are defined in Figure 3.23, where $L_{ds} = L_{gs} + L_g + L_{gd}$. The first term in (3.57) accounts for the fringe capacitance, while the second term accounts for the parallel-plate component of the capacitance. The term $f_g(p)$ in (3.58) is an empirical function that accounts for the shielding effect of the gate electrode on the drain-source capacitance. This function should increase as the gate length decreases, corresponding to a greater shielding effect.

While the gate, drain, and source resistors (R_g, R_d, and R_s, respectively) in Figure 3.20 are usually defined as external .MODEL parameters, the drain and source resistors can also be calculated from fundamental device dimensions and parameters. For a GaAs-based HEMT, if the contact sheet resistance is denoted as r_c, the sheet resistance of the cap layer as r_{cap}, the resistance across the AlGaAs as r_e, and the sheet resistance of the electron gas as r_{2deg}, the source and drain resistances are calculated by Cioffi [3.52] from the model of Feuer [3.56] as

$$R_{s,d} = \frac{1}{W} \left[\frac{r_{s1} r_{s2}}{r_{s1} + r_{s2}} L_{gs} + \frac{1}{r_{s1} + r_{s2}} \cdot \frac{\alpha + \beta C + \gamma k S}{(r_{s1} + r_{s2}) C + (R_{c1} + R_{c2}) k S} \right] \tag{3.62}$$

where

$$\alpha = 2 r_{s2} (r_{s1} R_{c2} - r_{s2} R_{c1}) \tag{3.63}$$

$$\beta = 2 r_{s2}^2 R_{c1} + (r_{s1}^2 + r_{s2}^2) r_{c2} \tag{3.64}$$

$$\gamma = (r_{s1} + r_{s2}) R_{c1} R_{c2} + r_{s2}^2 \rho_{12} \tag{3.65}$$

$$C = \cosh (k L_{gs}) \tag{3.66}$$

$$S = \sinh (k L_{gs}) \tag{3.67}$$

$$k = \sqrt{(r_{s1} + r_{s2})/\rho_{12}} \tag{3.68}$$

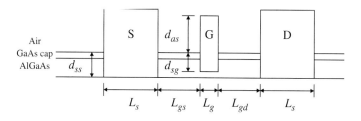

Figure 3.23: Cross-section of HEMT for use in capacitance calculations.

$$r_{s1} = \frac{1}{q\mu_2 N_{dcap} d_{cap}} \qquad\qquad r_{s2} = \frac{1}{q\mu_o n_{so}} \qquad (3.69)$$

In (3.69), μ_2 is the mobility of doped GaAs, N_{dcap} is the doping of the cap layer, and d_{cap} is the thickness of the cap layer. According to [3.52], Fueur's model does not account for variations of the contact resistance R_{c1}, R_{c2}, or the barrier resistivity ρ_{12} on the doping densities, so they are assumed constant and equal to 0.255 Ω·mm, 1.2 Ω·mm, and 1.5×10^{-5} Ω·cm^2, respectively.

Cioffi's iSMILE HEMT model input parameters are listed in Table 3.4. Examples of the use of this model in circuit and subsystem simulations will be presented in Chapter 4.

3.6.2 MEtal-Semiconductor Field-Effect Transistor (MESFET)

For the simulation of MESFETs, the established Statz model has been implemented into iSMILE. However, several other popular MESFET models deserve mention; excellent reviews of these models are presented in [3.43] and [3.57]. One of the first modeling approaches was to use the JFET model [3.57] since both the MESFET and JFET operate on similar principles. The current-voltage relationship for the JFET model is given by (3.70), where all of the parameters take on their usual meanings.

$$I_{ds} = \begin{cases} 0 & V_{gs} < V_T \\ \beta[2(V_{gs}-V_T)V_{ds}-V_{ds}^2](1+\lambda V_{ds}) & V_{ds} < V_{gs}-V_T \\ \beta(V_{gs}-V_T)^2(1+\lambda V_{ds}) & V_{ds} \geq V_{gs}-V_T \end{cases} \qquad (3.70)$$

The first major improvement, which is known as the Curtice model [3.57], [3.58], incorporated an empirical hyperbolic tangent function in the drain current above threshold:

$$I_{ds} = \beta(V_{gs}-V_T)^2(1+\lambda V_{ds})\tanh(\alpha V_{ds}) \qquad (3.71)$$

While this does not add any physical insight to the model, it does provide a good fit to measured data over *all* voltage ranges. Thus, only one equation is needed to formulate the I-V characteristic. The empirical parameter α is extracted from measured data. As stated in [3.57], while computation of the hyperbolic tangent increases simulation times, this speed loss is compensated by improved convergence speed due to the well-behaved nature of the tanh function (smooth, continuous derivatives).

Parameter	Definition	Units
μ_0	low-field mobility	$cm^2/V{\cdot}s$
V_{To}	threshold voltage	V
n_{so}	maximum electron concentration	cm^{-2}
v_s	saturation velocity	cm/s
λ	output conductance parameter	—
R_g	gate resistance	Ω
R_s	source resistance	$\Omega{\cdot}mm$
R_d	drain resistance	$\Omega{\cdot}mm$
n	diode ideality factor	—
j_0	diode saturation current	A/cm^2
C_{gsp}	extrinsic gate-to-source parasitic capacitance	F/cm
C_{gdp}	extrinsic gate-to-drain parasitic capacitance	F/cm
C_{dsp}	extrinsic drain-to-source parasitic capacitance	F/cm
N_d	doping density of $Al_xGa_{1-x}As$	cm^{-3}
x	Al mole fraction in $Al_xGa_{1-x}As$	—
Φ_b	Schottky barrier height	V
N_{dcap}	doping of cap layer	cm^{-3}
N_{bi}	interface state density	cm^{-2}
L	channel length	m
W	channel width	m
d_d	n-$Al_xGa_{1-x}As$ thickness	Å
d_i	i-$Al_xGa_{1-x}As$ thickness	Å
Δd	two-dimensional electron gas thickness	Å
L_{sg}	source-to-gate spacing	m
L_{dg}	drain-to-gate spacing	m
L_s	length of source or drain	m
d_{cap}	thickness of cap layer	Å
d_{as}	height of source or drain electrode above surface	Å
d_{ss}	height of source or drain electrode below surface	Å

Table 3.4: Input parameters for Cioffi's iSMILE HEMT model.

A follow-up to the Curtice model was the Curtice cubic (or modified Curtice-Ettenberg) model [3.59] which modeled the current voltage relationship as a cubic, rather than quadratic, function:

$$I_{ds} = [A_0 + A_1 V_{gs} + A_2 V_{gs}^2 + A_3 V_{gs}^3] \tanh(\gamma V_{ds}) \qquad (3.72)$$

In this formulation, the A coefficients are completely empirical and are extracted in [3.59] through least-squares fitting.

One of the most widely accepted models is a modification of the Curtice model by Statz et al. [3.44]. In the Statz, or Raytheon, model an empirical parameter b was added to the Curtice model to better model velocity saturation effects. In addition, the hyperbolic tangent function in the Curtice model was replaced by a polynomial expansion to improve computational efficiency. The current-voltage relationship is expressed as (3.73).

$$I_{ds} = \begin{cases} \left[\dfrac{\beta(V_{gs} - V_T)^2}{1 + b(V_{gs} - V_T)} \right](1 + \lambda V_{ds})\left[1 - \left(1 - \alpha \dfrac{V_{ds}}{3}\right)^3 \right] & 0 < V_{ds} < \dfrac{3}{\alpha} \\[4mm] \left[\dfrac{\beta(V_{gs} - V_T)^2}{1 + b(V_{gs} - V_T)} \right](1 + \lambda V_{ds}) & V_{ds} \geq \dfrac{3}{\alpha} \end{cases} \qquad (3.73)$$

It is the Statz model that has been implemented in both iSMILE and SPICE. The various parameters for the Statz model are listed in Table 3.5 [3.44], [3.57]. Examples of iSMILE simulations using the Statz model are depicted in Figures 3.24 and 3.25. The intent of this section, thus far, has been to review some of the more common MESFET models in use. Detailed explanations of each are beyond the scope of this text; the reader is referred to [3.44] and [3.57] – [3.59] for in-depth analyses.

It should be emphasized that the three MESFET models presented in this section are only the most common ones. Of the many, many other reported models, a few have gained general acceptance. Among these is the Triquint model [3.60], [3.61] which, among other things, addresses the fact that the threshold voltage has a dependence on the drain-source voltage:

$$V_T = V_{To} - \gamma V_{ds} \qquad (3.74)$$

In (3.74), γ is a fitting parameter. The Triquint model, which is based on the Statz model, also contains a factor δ for modeling variations in the output current with the drain current and bias point (3.75), (3.76).

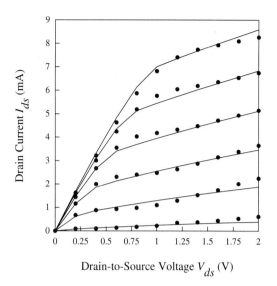

Figure 3.24: MESFET I-V characteristic (family of curves)
using iSMILE Statz model.
Curves represent simulation results; markers represent experimental data.

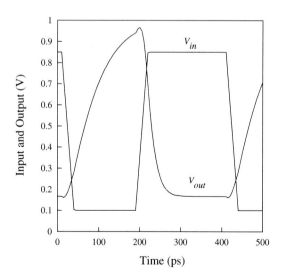

Figure 3.25: Transient simulation of MESFET inverter
using iSMILE Statz model.

Parameter	Definition	Units
β	transconductance parameter	A/V^2
α	saturation/hyperbolic tangent factor	1/V
λ	channel-length modulation parameter	1/V
b	doping tail extending parameter	1/V
V_{To}	threshold voltage	V
R_s	source resistance	Ω
R_d	drain resistance	Ω
C_{gs}	zero-bias gate-to-source parasitic capacitance	F
C_{gd}	zero-bias gate-to-drain parasitic capacitance	F
ϕ_b	gate junction potential	V
I_s	gate junction saturation current	A

Table 3.5: Definition of Statz MESFET model parameters.

$$I_{ds} = \frac{I_{dso}}{1 + \delta V_{ds} I_{dso}} \tag{3.75}$$

$$I_{dso} = \beta (V_{gs} - V_T)^Q \left[1 - \left(1 - \alpha \frac{V_{ds}}{3} \right)^3 \right] \tag{3.76}$$

It is emphasized in [3.60] that γ and δ replace λ and b in the Statz model (3.74). In (3.76), the exponent Q is used to generate an arbitrary, rather than a quadratic, dependence on the gate voltage.

3.7 Summary

In this chapter, an overview of circuit simulation as it pertains to optoelectronics was presented. As an illustrative vehicle, the iSMILE simulator was used to demonstrate the implementation of models for various optoelectronic devices, such as the quantum-well laser, the MSM detector, the HEMT, and the MESFET. While circuit simulation requires an equivalent-circuit representation of the device physics, the level of detail of the models shown was too complex to implement simply through the circuit-simulation input file. While the traditional approach is to embed such detailed models directly into the simulator source code, iSMILE facilitates new model development through a user interface which automatically translates, compiles, and links the equivalent-circuit topology and device physics, then appends the resulting code onto the simulator engine. While this chapter focused on the models themselves, Chapter 4 contains several examples of their use in realistic circuits and systems.

3.8 References

[3.1]   L. W. Nagel, "SPICE2: A computer program to simulate semiconductor circuits," Electronics Research Laboratory Memorandum, University of California, Berkeley, no. ERL-M520, 1975.

[3.2]   A. T. Yang and S. M. Kang, "iSMILE: A novel circuit simulation program with emphasis on new device model development," *Proceedings of the 26th ACM/IEEE Design Automation Conference*, pp. 630-633, 1989.

[3.3]   A. T. Yang and S. M. Kang, *iSMILE User's Manual*, University of Illinois at Urbana-Champaign, 1989.

[3.4]   D. S. Gao, S. M. Kang, R. P. Bryan, and J. J. Coleman, "Modeling of quantum-well lasers for computer-aided analysis of optoelectronic integrated circuits," *IEEE Journal of Quantum Electronics*, vol. 26, no. 7, pp. 1206-1216, 1990.

[3.5]   J.-P. Bianchi, "A new model for the MSM photodetector and its application to MSM-HEMT photoreceiver design," M. S. thesis, University of Illinois at Urbana-Champaign, 1991.

[3.6]   K. R. Cioffi, S. M. Kang, and T. N. Trick, "Circuit simulation models for the high electron mobility transistor," *Proceedings of the IEEE International Symposium on Circuits and Systems*, pp. 405-409, 1988.

[3.7]   D. S. Gao, A. T. Yang, and S. M. Kang, "Accurate modeling and simulation of interconnection delays and crosstalks in high-speed integrated circuits," *IEEE Transactions on Circuits and Systems*, vol. 37, no. 1, pp. 1-9, 1990.

[3.8]   J. J. Morikuni, D. S. Gao, and S. M. Kang, "Modelling of optical logic gates for computer simulation," *IEE Proceedings*, vol. 139, pt. J, no. 2, pp. 105-116, 1992.

[3.9]   G. P. Agrawal and N. K. Dutta, *Long-Wavelength Semiconductor Lasers*. New York: Van Nostrand Reinhold, 1986.

[3.10] D. Marcuse, "Classical derivation of the laser rate equation," *IEEE Journal of Quantum Electronics*, vol. QE-19, no. 8, pp. 1228-1231, 1983.

[3.11] R. S. Tucker, "Large-signal circuit model for simulation of injection-laser modulation dynamics," *IEE Proceedings*, vol. 128, pt. I, no. 5, pp. 180-184, 1981.

[3.12]   J. Buss, "Principles of semiconductor laser modeling," *IEE Proceedings*, vol. 132, pt. J, no. 1, pp. 42-51, 1985.

[3.13]   Y. Arakawa and A. Yariv, "Quantum well lasers — gain, spectra, dynamics," *IEEE Journal of Quantum Electronics*, vol. QE-22, no. 9, pp. 1887-1899, 1986.

[3.14]   G. H. Song, K. Hess, T. Kerkhoven, and U. Ravaioli, "Two dimensional simulation of quantum well lasers," *European Transactions on Telecommunications and Related Technologies*, vol. I, no. 4, pp. 375-381, 1990.

[3.15]   D. Marcuse, "Computer simulation of laser photon fluctuations: Theory of single-cavity laser," *IEEE Journal of Quantum Electronics*, vol. QE-20, no. 10, pp. 1139-1148, 1984.

[3.16]   K. Shimoda, *Introduction to Laser Physics*. Berlin: Springer-Verlag, 1986.

[3.17]   M. Sargent III, M. O. Scully, and W. E. Lamb, *Laser Physics*. Reading, MA: Addison-Wesley Publishing Company, 1974.

[3.18]   D. E. McCumber, "Intensity fluctuations in the output of cw laser oscillators. I," *Physical Review*, vol. 141, no. 1, pp. 306-322, 1966.

[3.19]   H. Haug, "Quantum-mechanical rate equations for semiconductor lasers," *Physical Review*, vol. 184, no. 2, pp. 338-348, 1969.

[3.20]   G. P. Agrawal and C. M. Bowden, "Concept of linewidth enhancement factor in semiconductor lasers: Its usefulness and limitations," *IEEE Photonics Technology Letters*, vol. 5, no. 6, pp. 640-642, 1993.

[3.21]   T. A. DeTemple and C. M. Herzinger," On the semiconductor laser logarithmic gain-current density relation," *IEEE Journal of Quantum Electronics*, vol. 29, no. 5, pp. 1246-1252, 1993.

[3.22]   A. Yariv, *Quantum Electronics*. New York: John Wiley & Sons, Inc., 1989.

[3.23]   R. H. Yan, S. W. Corzine, L. A. Coldren, and I. Suemune, "Corrections to the expression for gain in GaAs," *IEEE Journal of Quantum Electronics*, vol. 26, no. 2, pp. 213-216, 1990.

[3.24]   M. H. Tong, "An enhancement/depletion process for III-V compound semiconductor heterostructure field-effect transistors and optoelectronic integrated circuits," Ph.D. dissertation, University of Illinois at Urbana-Champaign, 1992.

[3.25]   S. M. Sze, D. J. Coleman, Jr., and A. Loya, "Current transport in metal-semiconductor-metal (MSM) structures," *Solid-State Electronics*, vol. 14, pp. 1209-1218, 1971.

[3.26] J. B. D. Soole and H. Schumacher, "InGaAs metal-semiconductor-metal photodetectors for long wavelength optical communications," *IEEE Journal of Quantum Electronics*, vol. 27, no. 3, pp. 737-752, 1991.

[3.27] B. E. A. Saleh and M. C. Teich, *Fundamentals of Photonics*. New York: John Wiley & Sons, Inc., 1991.

[3.28] B. J. Van Zeghbroeck, W. Patrick, M.-M. Halbout, and P. Vettiger, "105-GHz bandwidth metal-semiconductor-metal photodiode," *IEEE Electron Device Letters*, vol. 9, no. 10, pp. 527-529, 1988.

[3.29] S. Y. Chou and M. Y. Liu, "Nanoscale tera-hertz metal-semiconductor-metal photodetectors," *IEEE Journal of Quantum Electronics*, vol. 28, no. 10, pp. 2358-2368, 1992.

[3.30] J. B. D. Soole and H. Schumacher, "Transit-time limited frequency response of InGaAs MSM photodetectors," *IEEE Transactions on Electron Devices*, vol. 27, no. 11, pp. 2285-2291, 1990.

[3.31] W. Hilberg, "From approximations to exact relations for characteristic impedances," *IEEE Transactions on Microwave Theory and Techniques*, vol. MTT-17, no. 5, pp. 259-265, 1969.

[3.32] Y. C. Lim and R. A. Moore, "Properties of alternately charged coplanar parallel strips by conformal mappings," *IEEE Transactions on Electron Devices*, vol. ED-15, no. 3, pp. 173-180, 1968.

[3.33] A. A. Ketterson, M. Tong, J.-W. Seo, K. Nummila, J. J. Morikuni, S. M. Kang, and I. Adesida, "A high-performance AlGaAs/InGaAs/GaAs pseudomorphic MODFET-based monolithic optoelectronic receiver," *IEEE Photonics Technology Letters*, vol. 4, no. 1, pp. 73-76, 1992.

[3.34] J. Burm, K. I. Litvin, W. J. Schaff, and L. F. Eastman, "Optimization of high-speed metal-semiconductor-metal photodetectors," *IEEE Photonics Technology Letters*, vol. 6, no. 6, pp. 722-724, 1994.

[3.35] D. H. Lee, S. S. Li, S. Lee, and R. V. Ramaswamy, "A study of surface passivation on GaAs and $In_{0.52}Ga_{0.47}As$ Schottky-barrier photodiodes using SiO_2, Si_2N_4 and polyimide," *IEEE Transactions on Electron Devices*, vol. 35, no. 10, pp. 1695-1696, 1988.

[3.36] A. Aboudou, J. P. Vilcot, D. Decoster, A. Chenoufi, E. Delhaye, P. Boissenot, C. Varin, F. Deschamps, and I. Lecuru, "Ultralow dark current GaAlAs/GaAs MSM photodetector," *Electronics Letters*, vol. 27, no. 10, pp. 793-795, 1991.

[3.37] M. Ito and O. Wada, "Low dark current GaAs metal-semiconductor-metal (MSM) photodiodes using WSi_x contacts," *IEEE Journal of Quantum Electronics*, vol. QE-22, no. 7, pp. 1073-1077, 1986.

[3.38] O. K. Kim, B. V. Dutt, R. J. McCoy, and J. R. Zuber, "A low dark-current, planar InGaAs p-i-n photodiode with a quaternary InGaAsP cap layer," *IEEE Journal of Quantum Electronics*, vol. QE-21, no. 2, pp. 138-143, 1985.

[3.39] O. Wada, H. Hamaguchi, L. Le Beller, and C. Y. Boisrobert, "Noise characteristics of GaAs metal-semiconductor-metal photodiodes," *Electronics Letters*, vol. 24, no. 25, pp. 1574-1575, 1988.

[3.40] M. Zirngibl, J. C. Bischoff, D. Theron, and M. Ilegems, "A superlattice GaAs/InGaAs-on-GaAs photodetector for 1.3 μm applications," *IEEE Electron Device Letters*, vol. 10, no. 9, pp. 336-338, 1989.

[3.41] T. Kikuchi, H. Ohno, and H. Hasegawa, "$Ga_{0.47}In_{0.53}As$ metal-semiconductor-metal photodiodes using a lattice mismatched $Al_{0.4}Ga_{0.6}As$ Schottky assist layer," *Electronics Letters*, vol. 24, no. 19, pp. 1208-1210, 1988.

[3.42] W. B. Jones, *Introduction to Optical Fiber Communication Systems*. New York: Holt, Rinehard and Winston, Inc., 1988.

[3.43] G. Massobrio and P. Antognetti, *Semiconductor Device Modeling with SPICE*. New York: McGraw-Hill, Inc., 1993.

[3.44] H. Statz, P. Newman, I. W. Smith, R. A. Pucel, and H. A. Haus, "GaAs FET device and circuit simulation in SPICE," *IEEE Transactions on Electron Devices*, vol. ED-34, no. 2, pp. 160-169, 1987.

[3.45] A. Felder, R. Stengl, J. Hauenschild, H.-M. Rein, and T. F. Meister, "25 to 40 Gb/s Si ICs in selective epitaxial bipolar technology," *IEEE International Solid-State Circuits Conference Digest of Technical Papers*, vol. 36, pp. 156-157, 1993.

[3.46] T. Suzaki, M. Soda, T. Morikawa, H. Tezuka, C. Ogawa, S. Fujita, H. Takemura, and T. Tashiro, "Si bipolar chip set for 10 Gb/s optical receiver," *IEEE Journal of Solid-State Circuits*, vol. 27, no. 12, pp. 1781-1786, 1992.

[3.47] H. Tezuka, M. Soda, T. Suzaki, H. Takemura, and S. Fujita, "All-silicon IC 10 Gb/s optical receiver," *IEEE Photonics Technology Letters*, vol. 5, no. 7, pp. 803-805, 1993.

[3.48] P. J.-W. Lim, A. Y. C. Tzeng, H. L. Chuang, and S. A. St. Onge, "A 3.3 V monolithic photodetector/CMOS preamplifier for 531 Mb/s optical data link applications," *IEEE International Solid-State Circuits Conference Digest of Technical Papers*, vol. 36, pp. 96-97, 1993.

[3.49] S. M. Sze, *High-Speed Semiconductor Devices*. New York: John Wiley & Sons, Inc., 1990.

[3.50] J. T. Verdeyen, *Laser Electronics*. Englewood Cliffs, NJ: Prentice Hall, 1989.

[3.51] P. A. Kirkby, "Dislocation pinning in GaAs by the deliberate introduction of impurities," *IEEE Journal of Quantum Electronics*, vol. QE-11, no. 7, pp. 562-568, 1975.

[3.52] K. R. Cioffi, "Circuit simulation models for the high electron mobility transistor," Ph.D. dissertation, University of Illinois at Urbana-Champaign, 1987.

[3.53] H. R. Yeager and R. W. Dutton, "Circuit simulation models for the high electron mobility transistor," *IEEE Transactions on Electron Devices,* vol. ED-33, no. 5, pp. 554-563, 1986.

[3.54] H. Rohdin and P. Roblin, "A MODFET dc model with improved pinchoff and saturation characteristics," *IEEE Transactions on Electron Devices*, vol. ED-33, no. 5, pp. 664-672, 1986.

[3.55] D. Delagebeaudeuf and N. T. Linh, "Metal-(n) AlGaAs - GaAs two-dimensional electron gas FET," *IEEE Transactions on Electron Devices*, vol. ED-29, no. 6, pp. 955-960, 1982.

[3.56] M. D. Feuer, "Two-layer model for source resistance in selectively doped heterojunction transistors," *IEEE Transactions on Electron Devices,* vol. ED-32, no. 1, pp. 7-11, 1985.

[3.57] S. I. Long and S. E. Butner, *Gallium Arsenide Digital Integrated Circuit Design.* New York: McGraw-Hill Publishing Company, 1990.

[3.58] W. R. Curtice, "A MESFET model for use in the design of GaAs integrated circuits," *IEEE Transactions on Microwave Theory and Techniques*, vol. MTT-28, no. 5, pp. 448-455, 1980.

[3.59] W. R. Curtice and M. Ettenberg, "A nonlinear GaAs FET model for use in the design of output circuits for power amplifiers," *IEEE Transactions on Microwave Theory and Techniques*, vol. MTT-33, no. 12, pp. 1383-1393, 1985.

[3.60] A. J. McCamant, G. D. McCormack, and D. H. Smith, "An improved GaAs MESFET model for SPICE," *IEEE Transactions on Microwave Theory and Techniques*, vol. 38, no. 6, pp. 822-824, 1990.

[3.61] *HSPICE User's Manual*, vol. 2, Meta-Software, Campbell, CA, 1992.

4

CIRCUIT AND SUBSYSTEM SIMULATION EXAMPLES

Yang's iSMILE circuit simulator was discussed in Chapter 3 as a means of simulating devices for which new models have not yet been created. The model development process was reviewed and the simulation results of individual optoelectronic devices were depicted. In this chapter, several applications of the iSMILE simulator and its related models will be discussed. Many simulation examples will be depicted, with the main focus on optoelectronic transmitters and receivers, subsystems critical to optical communication links. The examples presented in this chapter were purposely made very basic in order to make them understandable by engineers with various backgrounds, such as traditional (silicon) circuit designers who have no optoelectronic design experience, optoelectronic device designers who have no circuit design expertise, and electronic CAD tool designers who have no knowledge of optoelectronic modeling and simulation.

In addition to simulation examples using iSMILE, examples of optoelectronic simulation using other, more conventional circuit simulators will be depicted as well. As mentioned in Chapter 3, simulation using conventional CAD tools requires a crude representation of optoelectronic devices in terms of simplistic electronic circuit elements. Such rough approximations, while useful for first-order estimates, often cannot capture true device behavior. Nevertheless, several such examples will be shown for illustrative purposes.

4.1 Preamplifier Topologies

As a prelude to discussion about photoreceiver simulation, it is instructive to review the various types of popular electrical preamplifiers used in photoreceiver design. Once a low-level optical signal has been received by a photodetector, the resulting photocurrent must be converted to a usable output voltage by a preamplifier. In this section, the three most commonly used photoreceiver preamplifier topologies will be discussed, with an emphasis on the transimpedance amplifier, which will be depicted in the simulations of this chapter. It should be mentioned that since MSM and FET models were presented in Chapter 3, the simulation examples in this section will rely heavily on those technologies.

4.1.1 The resistor-terminated configuration

Probably the simplest configuration is the resistor-terminated (voltage) preamplifier [4.1]. This is shown in Figure 4.1 where the input capacitance, resistance, and voltage gain of the amplifier are denoted by C_{in}, R_{in}, and A_v. In this design, the photocurrent flows through the resistor and capacitor creating a voltage v_{in}. In a practical implementation, v_{in} is the gate voltage of a FET; since current does not flow through the FET gate in proper operation, the resistor should be sized so that the voltage created across it by the detected photocurrent is enough to either turn on or modulate the transistor. The biggest design constraint for this configuration is that the resistor must be made small enough to minimize the effect of the input capacitance C_{in}, which is usually an uncontrollable, parasitic capacitance. As will be shown in the next section, the effect of C_{in} is to "integrate" v_{out}; thus, its presence must be minimized as much as possible. To illustrate this fact, it is helpful to derive the output voltage using simple circuit techniques:

$$v_{out} = A_v v_{in} = \left(R_{in} \| \frac{1}{j\omega C_{in}} \right) i_{in} \cdot A_v = \left(\frac{1}{1 + j\omega R_{in} C_{in}} \right) R_{in} A_v \cdot i_{in} \qquad (4.1)$$

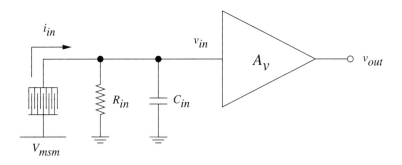

Figure 4.1: Resistor-terminated design.

If the input capacitance were eliminated, the output voltage would become dependent only on the photocurrent, the input resistance, and the gain:

$$v_{out} \approx R_{in} A_v \cdot i_{in} \tag{4.2}$$

To accomplish this effect (assuming that C_{in} is an uncontrollable parasitic capacitance), the input resistor must be made small enough so that in (4.1), $j\omega R_{in} C_{in} \ll 1$. Physically speaking, this means that the impedance presented by the resistor to the photocurrent is much less than that presented by the capacitor. From a practical standpoint, [4.1] states that the input resistor should be sized so that

$$R_{in} \leq \frac{1}{2\pi C_{in} B} \tag{4.3}$$

where B is approximately the bandwidth range of interest. However, the noise current that results from a resistor is given by

$$\langle i_R^2 \rangle = \frac{4kT\Delta f}{R} \tag{4.4}$$

Thus, small values of resistance lead to large amounts of noise and reduced sensitivity.

The speed of this configuration is limited by the $R_{in} C_{in}$ time constant; thus, small values of R_{in} will give correspondingly large bandwidths. Since the output voltage is given by $v_{out} \approx R_{in} A_v \cdot i_{in}$, this increased bandwidth is obtained at the expense of gain. Although the resistor-terminated topology is very easy to fabricate (only an input resistor, a FET, and a FET load resistor are necessary), because of its poor sensitivity, it is rarely used.

4.1.2 The high-impedance configuration

The high-impedance configuration has the same circuit topology as the resistor-terminated configuration (Figure 4.1). The difference between these two preamplifiers is in the size of the input resistor. For the resistor-terminated configuration, the input resistor was sized small, so that the effect of the input capacitance was minimized. In the high-impedance configuration, the input resistor is sized very large; thus, by (4.4) the input noise is low and the sensitivity is high.

With a large input resistor, (4.1) simplifies to (4.5) since $j\omega R_{in} C_{in} \gg 1$. By making use of the fact that division by $j\omega$ in the frequency domain results in integration in the time domain, it is clear from (4.5) that in the time domain, the input signal is integrated. To recover the desired input-output relation (4.2), this integrated signal must be differentiated (i.e., *multiplied* by $j\omega$). The only way to do this is to provide an equalizer at the output. Although the large resistor will degrade the bandwidth, with typical values of R_{in} and C_{in}, relatively high speeds can still be attained [4.2].

$$v_{out} \approx \left(\frac{1}{j\omega R_{in} C_{in}} \right) R_{in} A_v \cdot i_{in} \qquad (4.5)$$

The biggest drawback of the high-impedance design is the need to provide equalization. As stated in [4.1], a passive equalizer will increase the circuit noise and seriously degrade the sensitivity. Active equalizers, on the other hand, will have better noise performance but will have difficulty functioning at high frequencies. In addition, while the large input resistor increases sensitivity, it decreases dynamic range by increasing the probability of amplifier saturation [4.1].

4.1.3 The transimpedance configuration

The transimpedance preamplifier combines the advantages of both the resistor-terminated and the high-impedance configurations by providing both high sensitivity and high bandwidth. This is accomplished through negative feedback, as shown in Figure 4.2.

By examination of Figure 4.2, at low frequencies, the capacitors are open circuited and the transimpedance can be approximated as

$$Z_T = \frac{v_{out}}{i_{in}} = \frac{-R_f}{1 - \frac{1}{A_v}} \qquad (4.6)$$

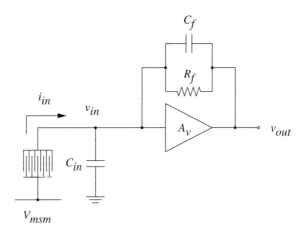

Figure 4.2: Transimpedance amplifier.

Thus, if the open-loop voltage gain A_v is sufficiently high, the magnitude of the DC transimpedance is approximately equal to the value of the feedback resistance. In Figure 4.2, a feedback capacitor is included to model the parasitic effects of the resistor. Through Miller's theorem [4.3], the feedback resistor and capacitor, as seen from the input, appear as a shunt combination (Figure 4.3).

The topology of Figure 4.3 is identical to that of the resistor-terminated and the high-impedance configurations (Figure 4.1); thus, some of the same design techniques can be applied. For example, the feedback resistor should be large enough to provide sufficient transimpedance gain (4.6) and low noise (4.4), but it should be small enough to prevent integration of the input signal (which would require output equalization). In the case of the transimpedance amplifier, the voltage gain provides an additional degree of freedom to these design criteria. Since the feedback resistor is divided by the voltage gain at the input, the sensitivity of the transimpedance amplifier is generally not as good as the high-impedance amplifier; however, with proper design, good noise performance can still be achieved. The smaller input resistance of the transimpedance amplifier improves its dynamic range performance since the amplifier will not tend to saturate as easily [4.1].

From Figure 4.3, the bandwidth of the transimpedance amplifier can be roughly approximated as

$$f_{3\text{ dB}} = \frac{1}{2\pi R_{in} C_{total}} \tag{4.7}$$

Thus, a figure of merit can be defined, the transimpedance bandwidth (TZBW), which is independent of the feedback resistance [4.4]:

$$\text{TZBW} = Z_T \cdot f_{3\text{ dB}} = \frac{A_v}{2\pi C_{total}} \tag{4.8}$$

The transimpedance gain of Figure 4.2 can be more rigorously defined than (4.6) not only by considering the contents of the voltage amplifier but by

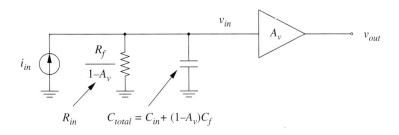

Figure 4.3: Transimpedance amplifier as viewed from the input.

incorporating frequency dependence as well. In a design which uses FETs, the voltage amplifier can be modeled as a simple common-source amplifier. This topology, along with its small-signal equivalent, is depicted in Figures 4.4 and 4.5. In these figures, C_{in} and C_{out} are, for the time being, described only as the input and output capacitances; they will be elaborated upon shortly. The parameter g_m is the FET transconductance and R_o is the output/load resistance of the common-source amplifier. The explicit feedback resistor and its associated parasitic capacitance are represented by R_f and C_f, respectively. While C_f is caused mainly by the feedback resistor, it contains other factors as well. This will be discussed later, in conjunction with the description of C_{in} and C_{out}. From Figures 4.4 and 4.5, the transimpedance can be rigorously expressed as

$$Z_T(\omega) = \frac{v_{out}(\omega)}{i_{in}(\omega)} = \frac{R_f\left(\dfrac{A'_v}{1-A'_v}\right)}{1+j\omega R_f\left(C_f+\dfrac{C_{in}}{1-A'_v}\right)} = \frac{A+j\omega B}{C+j\omega D+(j\omega)^2 E} \qquad (4.9)$$

In (4.9), A'_v is the open-loop voltage gain modified to take into account the effect of feedback loading and is expressed as

$$A'_v = \frac{-g_m+\dfrac{1}{R_f}+j\omega C_f}{\dfrac{1}{R_o}+\dfrac{1}{R_f}+j\omega(C_{out}+C_f)} \qquad (4.10)$$

Thus, the constants $A - E$ in (4.9) are given by

$$A = R_o(1-g_m R_f) \qquad (4.11)$$

$$B = R_o R_f C_f \qquad (4.12)$$

$$C = 1+g_m R_o \qquad (4.13)$$

$$D = R_o(C_{in}+C_{out}) + R_f(C_f+C_{in}) + g_m R_o R_f C_f \qquad (4.14)$$

$$E = R_o R_f[(C_{in}+C_{out})C_f+C_{in}C_{out}] \qquad (4.15)$$

Similar formulations have been presented in [4.5] – [4.7]. Since the forward transmission of the main (common-source) amplifier $|y_{21a}|$ is invariably greater than that of the feedback network $|y_{21f}|$ [4.6], [4.7], the transfer function of (4.9)

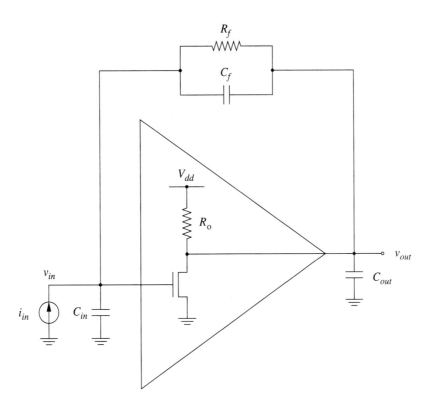

Figure 4.4: Transimpedance amplifier with common-source voltage amplifier.

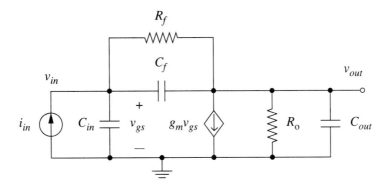

Figure 4.5: Small-signal equivalent circuit.

can be considerably simplified. That is, since (4.16) is often satisfied, the numerator of (4.9) can be represented by (4.17).

$$\left| g_m \right| \gg \left| \frac{1}{R_f} + j\omega C_f \right| \tag{4.16}$$

$$A + j\omega B \approx -g_m R_o R_f \tag{4.17}$$

Using (4.9) and (4.17), the 3-dB bandwidth of the transimpedance amplifier can also be more rigorously defined than (4.7) in terms of its individual circuit components:

$$f_{3\ dB} = \frac{1}{2\pi} \left\{ \frac{-(D^2-2CE) \pm \sqrt{(D^2-2CE)^2 + 4C^2 E^2}}{2E^2} \right\}^{1/2} \tag{4.18}$$

This is done by taking the magnitude of (4.9) and realizing that in units of dB-Ω, the transimpedance is defined to be

$$Z_T = 20 \log\left(\frac{v_{out}}{i_{in}}\right) \tag{4.19}$$

It should be noted, in passing, that the small-signal representation of Figure 4.5 and of (4.9) – (4.19) is usually justified by the extremely low optical input powers that are typically used in optical interconnect and communication systems. To reduce repeater spacing in optical communication systems, the receiver is usually made sensitive enough to detect signals in the microwatt range. Thus, it is unlikely that any optical inputs will perturb the amplifier bias points.

The feedback in the transimpedance amplifier is often implemented with a transistor rather than with a passive resistor. Active feedback offers several advantages over passive feedback [4.8] – [4.11]. For example, a feedback transistor generally requires less area than a feedback resistor. In addition, since it is a transistor, it can be made at the same time the other FETs are being fabricated, with no additional processing steps. Another significant advantage of active feedback is that it allows for tunability of the transimpedance gain, which allows the amplifier to function over a wide range of input signals. Tunability also makes it possible to assess the impact of the feedback resistance on various parameters such as gain, noise, and bandwidth [4.4].

A feedback transistor can also provide lower capacitance than a passive resistor due to the low C_{ds} of FETs. However, the gate-to-source and the gate-to-drain capacitances of the feedback transistor must also be accounted for in actual operation. At this point, the issue of the C_{in}, C_f, and C_{out} capacitances of the generic transimpedance amplifier of Figures 4.4 and 4.5 will be addressed. When using FETs, these capacitors can now be expressed as (4.20) – (4.22).

$$C_{in} = C_{gs} + C_{pd} + C_{gsf} \tag{4.20}$$

$$C_{out} = C_{ds} + C_{gdf} + C_{load} + C_{next} \tag{4.21}$$

$$C_f = C_{gd} + C_{dsf} \tag{4.22}$$

In these equations, the subscripts g, d, and s refer to the transistor gate, drain, and source, respectively, and the subscript f denotes those quantities belonging to the feedback transistor. The capacitors C_{pd}, C_{load}, and C_{next} are used to model the capacitances of the photodetector, the common-source amplifier load, and the next stage, respectively. Examination of (4.20) shows that the gate-to-source capacitor of the feedback transistor increases the input capacitance, while (4.21) shows that the feedback gate-to-drain capacitor adds to the output capacitance. The feedback capacitance (4.22) has two components: the first is from the drain-to-source capacitance of the feedback transistor itself, while the second is from the gate-to-drain capacitance of the common-source amplifier driver.

The transimpedance amplifier has gained widespread acceptance in photoreceiver design [4.4], [4.12] – [4.21] because of its high bandwidth, good noise performance, and good dynamic range. For these reasons, the receiver simulations depicted in this chapter will be based on the transimpedance topology.

4.2 Depletion-Mode MSM-HEMT Photoreceiver

The popular transimpedance-based photoreceiver is often implemented with an MSM detector and depletion-mode HEMT transistors. This section will depict a simulation of this topology using the tools and models discussed in Chapter 3.

Photoreceiver design using only depletion-mode HEMTs (versus enhancement- *and* depletion-mode HEMTs) is attractive because growth and processing of only one type of transistor is required. However, design with only depletion-mode transistors is made difficult because the negative threshold voltage V_T of these devices requires level shifting between stages to maintain compatibility between voltage levels. Consider, for example, a typical inverter circuit (Figure 4.6). By tying the gate to the source of the load in Figure 4.6, a gate-to-source voltage (V_{gs}) of 0 V is created. Since the threshold voltage of depletion-mode transistors is negative, this transistor is always on. In fact, because of the channel-length modulation factor λ, if the transistor is biased correctly, the current through it varies linearly with the voltage across it (V_{ds}); thus, it acts as a resistor. This configuration is very similar to a depletion-mode NMOS inverter with a resistive load [4.22].

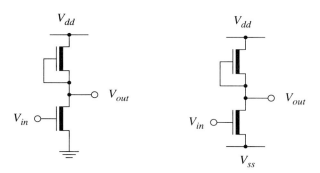

Figure 4.6: Depletion-only inverter with negative power supply equal to
(a) ground and (b) V_{ss}.

The problem with this depletion-only configuration is as follows. Assume, for argument's sake, that the output high voltage V_{OH} is equal to the positive voltage supply, and that the output low voltage V_{OL} is equal to the negative supply voltage. (In reality, V_{OH} and V_{OL} will not extend all the way to the supply voltages.) Consider the case of a logic high voltage applied at the input. Because this is an inverter, the logic high voltage at the input will produce V_{OL} at the output; that is, the output should be interpreted as a "logic 0." It was assumed that V_{OL} was roughly equal to V_{ss}; thus, the next stage will see V_{ss} at its input (gate). However, this means that for the next stage, $V_{gs} = V_{OL} - V_{ss} = 0$. Because of the negative threshold voltage V_T of depletion-mode transistors, $V_{gs} = 0 > V_T$, and the driver of the next stage will turn on. Thus, what was meant to be interpreted as a logic low was not. This is equally true for $V_{ss} = 0$, as in Figure 4.6a.

To have the next stage successfully interpret the inverter output as "low," the output voltage level must somehow be "shifted" down so that V_{gs} of the next stage is less than V_T (Figure 4.7). This can be accomplished through a level shifter, which is simply a source-follower amplifier with some Schottky diodes. Since the turn-on voltage of Schottky diodes V_D in GaAs is about $0.6 - 0.7$ V, the example of Figure 4.7 shifts the inverter output down by about 1.2 V. Thus, in Figure 4.7, $V_{in,2}$ will be roughly -1.2 V which is sufficiently low to keep the second inverter turned off. Because the voltage has to be pulled down below zero, a V_{ss} supply is needed even if ground is used for the inverter's second supply voltage. Logic gates that are designed in this fashion belong to the BFL family (Buffered FET Logic) [3.57]. Design with BFL requires level shifting between *each* stage.

Using only depletion-mode transistors, the common-source amplifier of Figure 4.4 can be implemented with the inverter structure of Figure 4.6. By adding an explicit feedback path, this can be expanded into a transimpedance amplifier. Active feedback can be implemented for the reasons discussed previously. However, since the output of the common-source amplifier is fed back to its input (which can be

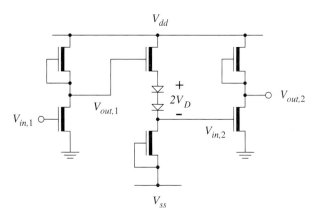

Figure 4.7: Cascaded inverters using depletion-mode transistors.

considered as the "next stage"), level shifting is required. In this case, since the inverter is now being used as a small-signal amplifier, the level shifter is used to maintain a stable and optimum DC bias point, rather than to perform digital voltage level shifting. Finally, the addition of a photodetector (in this case an MSM) makes the photoreceiver complete (Figure 4.8).

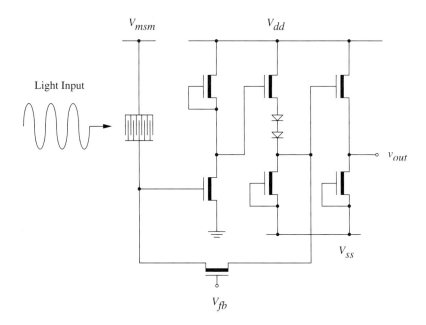

Figure 4.8: Depletion-mode MSM-HEMT transimpedance photoreceiver.

In this design, a source-follower amplifier has been added at the output to provide impedance matching to 50 Ω. This ensures that ringing and reflections do not occur as the photoreceiver output is connected to a transmission line. While the source-follower output resistance is given by the middle term of (4.23), where g_m is the transconductance, g_o is the FET output conductance, and R_L is the effective resistance of the source-follower load, the quantities g_o and $1/R_L$ are usually small enough to approximate the source-follower output resistance by the right-hand side of this expression.

$$R_{out} = \frac{1}{g_m + g_o + \dfrac{1}{R_L}} \approx \frac{1}{g_m} \tag{4.23}$$

The load resistance R_L can be made arbitrarily large by decreasing the width of the source-follower load (within the allowable minimum feature sizes) and typical HEMT values for g_m and g_o are 400 mS/mm and 15 mS/mm, respectively, where the units of conductance are given in mS (milliSiemens) and mm refers to millimeters of gate width [4.4].

The photoreceiver design was simulated with iSMILE using the Cioffi HEMT model of [3.6] and the Bianchi MSM model of [3.5] (Figure 4.9). The photoreceiver was fabricated, and the transimpedance gain was calculated from S-parameter measurements (4.24) according to [4.23].

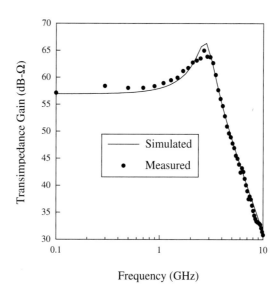

Figure 4.9: Depletion-mode MSM-HEMT photoreceiver frequency response.

$$G_T = \frac{S_{21}}{1 - S_{11}} R_L \qquad (4.24)$$

In (4.24), R_L is, in this case, the load impedance at both the input and output (50 Ω). Figure 4.9 depicts a simulation of the frequency response with measured data superimposed. From this figure, the low-frequency gain is seen to be 56.8 dB-Ω and the 3-dB frequency was measured to be slightly over 4.0 GHz. Both of these are in good agreement with their simulated values. The electrical measurements were made in response to 0.85-μm light input.

The photoreceiver layers were MBE-grown and the circuits were fabricated with a seven-level all-electron-beam lithography process. A cross-sectional layer diagram and the fabricated photoreceiver are depicted in Figures 4.10 and 4.11, respectively. The HEMTs were fabricated with quarter-micron T-gates ($L = 0.25$ μm) for low resistance. As stated earlier, the MSM can be made in the same step as the gate metal deposition (as indicated by the equal shading of the MSM and the gate in Figure 4.10).

The MSM was made with an area of 100×100 μm^2 (3-μm finger width, 6-μm pitch) and was passivated with SiN$_x$, which also served as an antireflective (AR) coating. While the resulting capacitance ($\sim$ 100 fF) could have been lowered by using a smaller detector, as discussed in Section 3.5, a large area is desirable to facilitate coupling of the optical input signal. The common-source driver and load transistors were 20 μm and 6 μm wide, respectively, the level shifter FETs were both 20 μm and the output buffer transistors were both 75 μm [4.4]. The feedback transistor was sized with a W/L ratio of 10 μm/2.5 μm. The HEMTs had an average threshold voltage of about -0.7 V and the positive and negative voltage supplies were + 2.5 V and - 2.0 V, respectively.

The measured transistor transconductance was about 500 mS/mm and the output conductance was about 30 mS/mm. The high transconductance of these devices resulted in a high transit frequency f_T of 66 GHz. The MSM dark current was measured to be about 10 nA. The responsivity of the MSM was measured to be about 0.25 mA/mW and the static power dissipation was estimated to be about 90 mW.

4.3 Enhancement-Depletion Photoreceiver

While the speed of the depletion-only photoreceiver presented in the previous section was quite good, there are several shortcomings inherent to BFL (depletion-only) designs. Due to the negative threshold voltage of depletion-mode devices, level shifting between each stage is required in order to maintain voltage compatibility. As shown in the previous section, a simple two-transistor amplifier in a depletion-only process will automatically require two additional transistors, as well as several Schottky diodes. From a systems perspective, this wasted area creates a

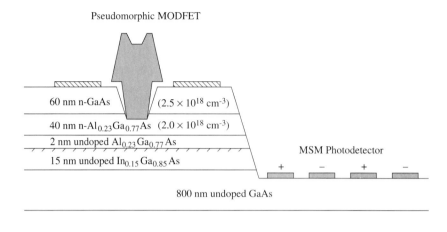

Pseudomorphic MODFET

60 nm n-GaAs $(2.5 \times 10^{18}$ cm$^{-3})$

40 nm n-Al$_{0.23}$Ga$_{0.77}$As $(2.0 \times 10^{18}$ cm$^{-3})$

2 nm undoped Al$_{0.23}$Ga$_{0.77}$As

15 nm undoped In$_{0.15}$Ga$_{0.85}$As

MSM Photodetector

$+$ $-$ $+$ $-$

800 nm undoped GaAs

S. I. GaAs Substrate

Figure 4.10: MSM-HEMT pseudomorphic AlGaAs/InGaAs/GaAs
layer structure.

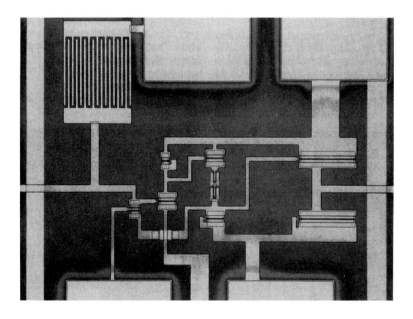

Figure 4.11: Photomicrograph of the fabricated photoreceiver.

huge amount of overhead; this area requirement makes large-scale integration of depletion-only circuits extremely difficult. Since the level shifter maintains a constant current path between the positive and negative power supplies, it dissipates a large amount of static power. As mentioned in Section 4.2, the depletion-mode photoreceiver was estimated to dissipate over 90 mW of static power. If photoreceivers are to be used as a substitute for electronic input stages in a computer network or interconnect scheme, it is important that this power dissipation be minimized. As with the area constraint, the static power dissipation incurred by level shifting would make any form of large-scale integration difficult.

To address the deficiencies of the depletion-only photoreceiver, many researchers choose to utilize a design which uses both enhancement-mode and depletion-mode (E-D) transistors. As a review, a highly conductive channel under the gate of an enhancement-mode transistor is created when a *positive* (threshold) voltage is applied to the gate. Following the example in Section 4.2, an inverter similar to that shown in Figure 4.6, but with an enhancement-mode driver and a depletion-mode load, will produce high and low outputs near the positive voltage supply and ground, respectively. Because of the positive threshold voltage of enhancement-mode transistors, a 0-V output from one stage will be successfully interpreted as a logic low by the next stage. That is, the driver of the next stage will not turn on erroneously as was the case with a depletion-only design; thus, level shifting is not required and successive stages can simply be cascaded. Digital circuits which are fabricated with E-D transistors in this manner are categorized under the DCFL (Direct Coupled FET Logic) family [3.57]. DCFL design in GaAs is very similar to enhancement-driver depletion-load NMOS design in silicon [4.22].

The system-level advantages of this type of design are obvious. Since there is no need for level shifting, the E-D design requires one less stage than the D-only design. The level shifter in the photoreceiver of Figures 4.8 and 4.11 consumed roughly 4,250 μm^2 of area; thus, an E-D design can produce an area savings of about 20%. Considering the fact that level shifting must be performed between *each* stage, when mapped into VLSI or even LSI, this savings is tremendous. As an example, under these conditions, after only 235 logic gates, the amount of *wasted* area will exceed 1 mm^2. The circuit topology of an E-D photoreceiver is depicted in Figure 4.12.

In this section, computer-aided design of the E-D photoreceiver will be depicted, again using the iSMILE simulator and the MSM and HEMT models. It should be noted that Cioffi's HEMT model can be used to simulate both depletion- and enhancement-mode transistors, simply by appropriate assignment of the threshold voltage V_T parameter in the .MODEL portion of the input file.

The first step in any electronic (or optoelectronic) design is establishing appropriate transistor sizes and DC bias points. In the design of the amplifier/inverter first stage of Figure 4.12, both the size of the enhancement-mode transistor, which determines the amplifier's drive capability, and the size of the

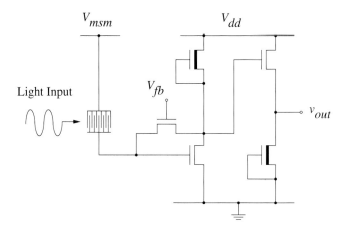

Figure 4.12: Enhancement-depletion (E-D) photoreceiver.

depletion-mode transistor, which determines the load resistance of the amplifier must be determined. Iterative optimization of these two factors is often necessary to ensure good operation. For reasons which will become apparent, a 40 μm width is first chosen for the enhancement-mode drive transistor. As with the depletion-only design, quarter-micron transistor lengths are chosen. A DC analysis of the drive transistor is depicted in Figure 4.13. In this simulation, the 40-μm wide, 0.25-μm long enhancement-mode transistor had a 0.5-V bias at its drain and a grounded source. The gate voltage was allowed to vary between 0 and 1 V. The channel current I_{ds} was then monitored as a function of varying gate voltage V_{gs}. This gives a good measure of the gate voltage to channel current conversion efficiency, that is, the transconductance, g_m. The transconductance was arrived at by taking the iSMILE DC simulation output and mathematically calculating according to

$$g_m = \left. \frac{\partial I_{ds}}{\partial V_{gs}} \right|_{V_{ds}} \qquad (4.25)$$

It should be noted that the transconductance is directly proportional to the transistor width; that is, the wider the transistor, the higher the transconductance. For this reason, g_m is often expressed in the mS/mm units used on the last section. (The 0.5-V drain bias (as well as the 40-μm width) was arrived at through an iterative design process in which both the driver and load transistors were considered. More about this will be mentioned shortly.)

The transconductance is a good figure of merit in photoreceiver design, since the AC voltage gain of the common-source amplifier and, thus, the transimpedance-bandwidth (4.8) are directly proportional to g_m, and the input noise of the

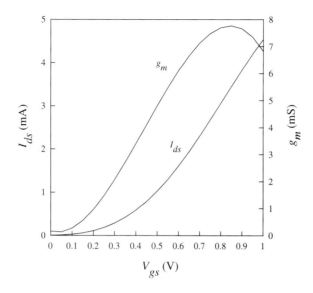

Figure 4.13: DC analysis of first-stage driver.

photoreceiver is inversely proportional to g_m [3.33]. Thus, it would seem that the higher the g_m, the better. Why then was a 40-μm width chosen? The answer is that other factors must be considered, as well. For example, while increasing the width would create a larger g_m, it would also create a larger transistor capacitance which, when speed and noise [3.33] are considered, would serve to counteract the effects of a large g_m. Many factors must be considered in determining optimum values.

Since photocurrents in the μA range and feedback resistances in the kΩ range are common, the DC voltage drop across the feedback resistor/transistor is often negligible; hence, the driver gate and drain voltages are roughly equal. If this is the case, from Figure 4.13, it would seem that an approximately 0.85-V gate (and hence drain) bias would be more effective than the 0.5-V bias chosen in the simulation. This choice of drain bias again represents a trade-off, since the resistance of the load must be considered, as well. If a 0.85 V bias were chosen, the corresponding drain current would be about 3.5 mA (versus $\approx$ 1 mA @ 0.5 V). To sink this larger current, the depletion load would have to be sized so that its DC resistance were smaller. A smaller DC resistance requires a larger transistor width resulting in a larger load capacitance. Correspondingly, a larger drain current would result in a smaller AC driver output resistance since $r_o = (\lambda I_{ds})^{-1}$ [4.6]. This would then result in a smaller common-source amplifier open-loop gain

$$A_v = \frac{v_{out}}{v_{in}} = -g_m (r_o \| R_L) \qquad (4.26)$$

At a bias of 0.5 V, the open-loop voltage gain is about twice that at 0.85 V. Thus, for this particular design example, a 40-μm driver width and a 0.5-V drain bias are chosen as a good compromise between resistance, capacitance, transconductance, and gain.

The second design criterion mentioned previously is the size of the depletion load transistor. A simulation of the load transistor's I-V characteristic as a function of transistor width W_D is depicted in Figure 4.14. With a supply voltage of 2 V and a driver drain bias of 0.5 V, the drain-source voltage of the load must be 1.5 V. From this figure, it is clear that to obtain a current of 1 mA as prescribed by Figure 4.13, a 20-μm-wide load transistor must be chosen. Multiple plots of Figure 4.13 along with Figure 4.14 can be combined to depict the DC voltage transfer characteristic for various load sizes (Figure 4.15). For successful AC operation as a common-source amplifier, the inverter must be DC biased at or about the transition point between logic high and logic low. As seen in Figure 4.15, the transition takes place about V_{in} = 0.5 V when a 20-μm-wide load is used.

The next step in the photoreceiver design is the addition of a feedback transistor. This converts the common-source amplifier into a transimpedance amplifier. A minimally sized transistor (5 μm) was adopted for this purpose not only because of the need for moderately large resistances, but also to reduce any extraneous parasitic capacitances. (A good review of the capacitive trade-offs involved between active and passive feedback is contained in [4.4]). A DC simulation of the transimpedance amplifier closed-loop transfer characteristic is

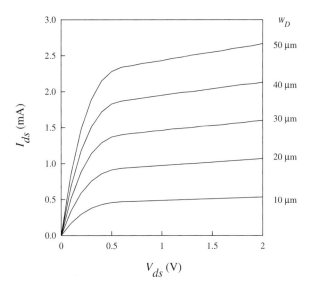

Figure 4.14: Current-voltage characteristic of depletion load.

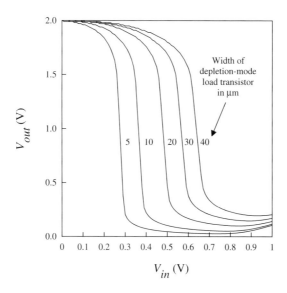

Figure 4.15: Inverter voltage transfer characteristic.

depicted in Figure 4.16. In this figure, I_{in} is the input photocurrent, V_{out} is the voltage at the output of the transimpedance amplifier and V_{fb} is the voltage applied to the gate of the feedback transistor. From Figure 4.16, it is clear that by varying V_{fb}, different amounts of feedback resistance are created. It should be noted that when the input current is zero, the feedback path is nearly open-circuited (negligible current flow); in this case, the output voltage is 0.5 V as expected from the design steps outlined for the common-source amplifier.

The final step in the design of the E-D photoreceiver is the addition of the source-follower output buffer. As mentioned previously, the function of this stage is to provide 50-Ω output impedance matching to any probes or transmission lines used to connect to the photoreceiver output. While analytical techniques can be used to approximate the output resistance of the buffer, circuit simulation can provide considerable insight. However, since this method requires AC analysis, it will be discussed after the simulation of the photoreceiver frequency response.

Figures 4.17 and 4.18 depict the circuit schematic and the input file, respectively, for the simulation of the receiver's frequency response. In this design, a 10×10 μm^2 MSM detector was used, although, as mentioned previously, larger detector sizes are usually desirable to facilitate optical coupling. The detector has six 10-μm-long fingers that are 1 μm wide and are spaced 1 μm apart. It should be noted that the notation for the MSM model parameters in Figure 4.18 differs somewhat from the definitions given in Table 3.3. This is because [3.5] uses nonstandard notation for many of the MSM parameters. For a correlation between

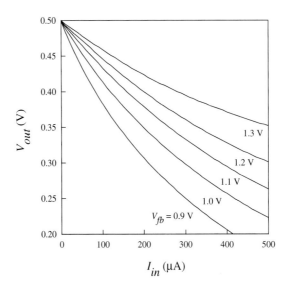

Figure 4.16: Closed-loop transfer characteristic of transimpedance amplifier.

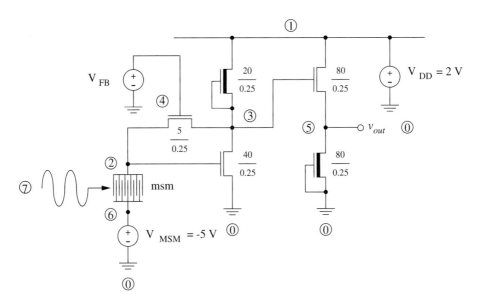

Figure 4.17: Circuit schematic for iSMILE simulation of E-D photoreceiver.

```
*E-D Receiver Design

msm 7 6 2 6 M1
Vop 7 6 AC 5.3e-3
Vmsm 6 0 DC -5

Vdd 1 0 2

mfet2 1 3 3 HD w=20u l=.25u
mfet2 3 2 0 HE w=40u l=.25u

mfet2 3 4 2 HE w=5u l=.25u
Vfb 4 0 1.1

*Source Follower
mfet2 1 3 5 HE w=80u l=0.25u
mfet2 5 0 0 HD w=80u l=0.25u

.MODEL   HE mfet2
+        vt=+0.328 lambda=0.15 vsat=4.18e7 u0=800
+        j0=0.4e-5 ns0=0.28e13 nd=0.2e19 x=0.25 di=40
+        rss=1.662 rdd=0.01 rgg=15.61 dd=240
+        lsg=0.5e-4 ldg=0.125e-4 dcap=100 ndcap=1e18
+        cgsp=.78e-12 cgdp=.72e-12 nid=1.55

.MODEL   HD mfet2
+        vt=-0.328 lambda=0.15 vsat=4.18e7 u0=800
+        j0=0.4e-5 ns0=0.28e13 nd=0.2e19 x=0.25 di=40
+        rss=1.662 rdd=0.01 rgg=15.61 dd=240
+        lsg=0.5e-4 ldg=0.125e-4 dcap=100 ndcap=1e18
+        cgsp=.78e-12 cgdp=.72e-12 nid=1.55

.MODEL   M1 msm(lf=1d-6 fw=1d-6 sf=10d-6 nbfin=6
+        rfin=2.824d-8 thf=3.0d-7 ndop=.965d18 fn=2.470
+        fp=1.900 vbi=0.408 idcfac=1 areafac=1.000
+        al1=2.095d-8 er=14.60 wlay=1.5e-6 an=8.160e4
+        ap=7.440e5 listvar=0 nint=2.25d12
+        abso=1.2e6 opga=0.0d0 vthr=169.7 slop=1 mexp=1
+        ve=50d6 vh=34d6 capfac=1.0 robulk=3.33d6)

.AC DEC 10 10e6 100e9
.PRINT AC VM(5)
.END
```

Figure 4.18: E-D MSM-HEMT photoreceiver iSMILE
AC simulation input file.

Figure 4.18 and Table 3.3, the reader is referred to [3.5]. The frequency response of the E-D photoreceiver is depicted in Figure 4.19. This figure was generated from an AC sweep of the circuit of Figures 4.17 and 4.18 with an incident optical input of 5.3 mW. From Figure 4.19, it is apparent that the output voltage that results from this particular set of operating conditions is approximately 1 V. Furthermore, it is clear that the 3-dB frequency attained is about 10.5 GHz.

The final stage is the design of the output buffer; as stated previously, the main consideration is output impedance matching to 50 Ω. To this end, iSMILE simulations are used to choose the proper transistor sizing to ensure a good impedance match. Figure 4.20 depicts the topology required for a small-signal simulation of the source-follower output resistance, while Figure 4.21 contains the actual iSMILE AC simulation. Although it is the AC resistance that must be determined, the first step in this process is actually a DC simulation of the *entire* photoreceiver. This is necessary to determine the DC bias points at nodes 3 and 5 in Figure 4.17, which correspond to the source-follower input and output, respectively. A determination of the DC bias points from Figure 4.20 rather than Figure 4.18 would be incorrect, since it would not take into account the DC effects of the transimpedance amplifier on nodes 3 and 5. Once these DC bias points are determined, the source-follower can be linearized and simulated in the AC domain, as detailed in Chapter 2. The voltage sources in Figure 4.20 represent, then, the DC bias points determined from a DC analysis of Figure 4.17; this ensures proper linearization. The AC resistance is then determined by placing a 1-mV AC voltage source at the output, sweeping the frequency over a wide range, and monitoring the

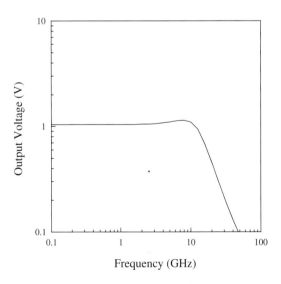

Figure 4.19: iSMILE frequency response of the E-D photoreceiver.

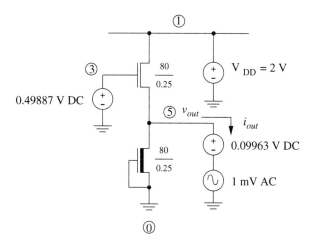

Figure 4.20: Circuit for determining source-follower output
resistance with iSMILE.

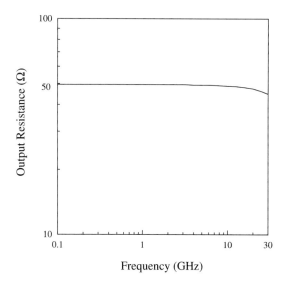

Figure 4.21: Simulation of source-follower output resistance.

current that flows through the 1-mV source. The output resistance is 1 mV divided by the monitored current. This approach is essentially the Thévenin method for determining the equivalent resistance at a circuit output [4.24]. From Figure 4.21, it is clear that over the frequency ranges of interest, the output resistance of the source follower is approximately 50 Ω. Relatively large 80-μm transistors were chosen mainly to provide sufficient output drive current.

4.4 Alternate Photoreceiver Simulation Methods

The simulations of previous sections showed the usefulness of an analysis in which optical and electrical elements were treated in the same input file. However, it is not always necessary to have models for optoelectronic devices in order to simulate optoelectronic systems. As alluded to previously, a crude, commonly used approach is to model optoelectronic elements through simple linear circuit elements. For example, the MSM detector is often modeled as a constant current source in parallel with a capacitor (Figure 4.22).

The use of the current source is somewhat justified by the fact that the output of a photodetector is a current (the photocurrent). However, this approach is unsound since the actual photocurrent is *not* constant. As depicted in Figure 3.16, the photocurrent (and the dark current) is actually a function of the voltage applied to the detector.   This approach also fails to account for the different photocurrent components due to the different electron and hole mobilities, as depicted in Figure 3.13. Furthermore, modeling the detector as a constant current source removes all physical, geometrical, and material dependencies of the photocurrent. It is not possible to assess the effect of, for example, lengthening the fingers, or modifying the finger pitch. It is furthermore not possible to predict the detector responsivity.

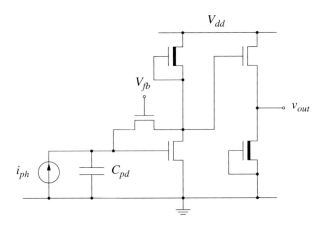

Figure 4.22: Conventional method for photoreceiver simulation.

Nevertheless, this continues to be an extremely popular approach due mainly to the lack of widely available models.

When using this method, the electronic preamplifier and the optoelectronic detector are usually optimized separately. Indeed, the most common situation is one in which the preamplifier and detector are designed and fabricated by completely separate groups or even companies, then integrated hybridly. The party responsible for the design of the preamplifier often has no control over the design of the detector except, for example, to specify that the detector capacitance must be below some specified limit, or that the detector bandwidth must be above some specified limit. Care must be taken, however, to choose values for the current source and capacitor that are appropriate not only for the choice of physical structure, but also for the particular voltage bias points in use.

4.5 Power Simulation

Along with speed, power dissipation of optoelectronic components is also an important consideration, especially if optoelectronic photoreceivers and transmitters are to be used as replacements for their electronic counterparts. Despite their many advantages, optoelectronic components will be unsuitable for integration into existing electronic systems if they dissipate too much power.

The simplest estimation of power dissipation is static power. This is defined as the total amount of DC current drawn from the voltage supplies multiplied by the voltage values themselves, as seen in (4.27), where the subscript n is used to sum over all voltage sources.

$$P_{static} = \sum_n I_{DC,n} V_{DC,n} \tag{4.27}$$

In the case of the depletion-only receiver, $n = 4$ (positive and negative power supplies for the electronic preamplifier, tunable voltage for the feedback transistor, and bias voltage for the detector); for the E-D receiver, $n = 3$. Since the feedback voltage is applied to the gate of the feedback transistor (a FET), the current flow (gate current) through this source is usually negligible. It was this technique that was used to arrive at the 90-mW static power estimation of the depletion-only receiver. Reduction of static power dissipation is of paramount importance, as was seen in the wasted power in the level-shifting stage of the depletion-only receiver. Very low *static* power dissipation is one of the main advantages of CMOS, a technology of importance even in optoelectronics, as evidenced by [3.48].

While static power is an important consideration, at high frequencies and speeds, dynamic power dissipation plays a critical role, as well. Power is dynamically expended during transient and switching events; more concretely, dynamic power dissipation is the rate at which energy is consumed during the charge and discharge of capacitors. One popular method for estimating the dynamic power dissipation of CMOS circuits is through (4.28), where C_T is the total effective capacitance affected by a switching event, V_{DD} is the voltage supply, and f is the switching frequency [4.25].

$$P_{dynamic} = C_T V_{DD}^2 f \qquad (4.28)$$

It should be noted that since there is negligible static power dissipation in CMOS circuits, the dynamic power dissipation is essentially the same as the total power dissipation. This method, however, is difficult to apply since it is difficult to estimate the total effective capacitance for even moderately sized circuits [4.26].

A much simpler method, introduced in [4.26], uses simulation to estimate the total power dissipation. In this approach, a "noninvasive power meter" is inserted into the circuit input file to avoid disturbing or loading the circuit or device under test. As mentioned previously, the total power dissipated can be calculated by multiplying the total current flow through each voltage source by the voltage values, themselves. This approach is easy to use when computing static power dissipation (4.27), but is difficult to apply when multiple, high-speed switching events are occurring, and the corresponding currents through each voltage source are changing. The concept of average power dissipation, as defined in (4.29), is often used to circumvent this issue.

$$P_{avg} = \frac{1}{T} \int_0^T p(t)\, dt = \frac{1}{T} \int_0^T \sum_n V_{DC,n} i_n(t)\, dt \qquad (4.29)$$

In this equation, the instantaneous power $p(t)$ is integrated over one period T, then divided by the period to arrive at the average power P_{avg}. The instantaneous power $p(t)$ is defined as the product of the instantaneous current and the instantaneous voltage. However, since the value of the voltage supply ideally does not change, it is indicated as a constant V_{DC} in (4.29), while the instantaneous current is indicated as $i(t)$. It should be noted that $i(t)$ is the total instantaneous current (i.e., $i(t) = I_{DC} + i_{transient}$); thus, (4.29) is an expression for the *total* average power dissipation.

The "noninvasive power meter" approach is illustrated in Figure 4.23. In this figure, a zero-valued voltage source V_s is inserted directly after the voltage supply V_{DD}. This is the often-used current meter technique — since most circuit simulators use node voltages as the state variables, currents are often not readily available as simulation outputs. The only method to obtain a current measurement in most circuit simulators is to request the simulator to output the current that is flowing through a given voltage source. The current through V_s can thus be monitored in a completely noninvasive manner since $V_s = 0$ corresponds to a short circuit. Using SPICE, iSMILE, or any other circuit simulator, the current through V_s is assigned the

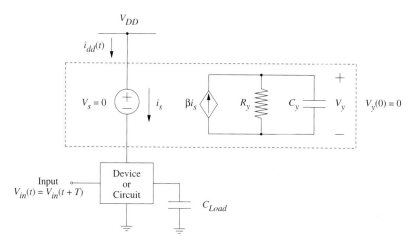

Figure 4.23: Noninvasive power meter technique.

name i_s. (If a circuit simulator that allows the monitoring of current through any branch is available, then this step is unnecessary.)

The element values β, R_y, and C_y are chosen so that the voltage V_y corresponds to the average power dissipation at time $t = T$. From Figure 4.23, it is clear that

$$C_y \frac{dV_y}{dt} = \beta i_s - \frac{V_y}{R_y} \tag{4.30}$$

where it is noted, for clarity, that β is the multiplying factor for the current-controlled current source. It should also be noted that the voltage V_y must be set so that $V_y(0) = 0$. This can usually be accomplished through a .IC or "Initial Conditions" command. Continuing the analysis of [4.26], the solution of (4.30) leads to

$$V_y(t) = \frac{\beta}{C_y} \int_0^t \exp\left[\frac{-(t-\tau)}{R_y C_y}\right] i_s(\tau)\,d\tau \tag{4.31}$$

If R_y and C_y are chosen such that $R_y C_y \gg T$, then at $t = T$,

$$V_y(T) = \frac{\beta}{C_y} \int_0^T \exp\left[\frac{-(T-\tau)}{R_y C_y}\right] i_s(\tau)\,d\tau \approx \frac{\beta}{C_y} \int_0^T i_s(\tau)\,d\tau \tag{4.32}$$

Thus, if (4.33) is met, then $V_y(T) \approx P_{avg}$.

$$\beta = \frac{V_{DD}C_y}{T} \qquad\qquad (4.33)$$

Because this method is merely a simplified way of measuring the average current drain from each voltage source, this technique is general, and can be used to measure the average power dissipation of any circuit, electronic or optoelectronic, silicon or compound semiconductor. In the case of many types of compound semiconductor FETs (e.g., MESFETs), however, static power dominates dynamic power dissipation. While (4.27) can be more useful in these situations, the method depicted in Figure 4.23 still offers a convenient method for total power measurement.

4.6 Monolithic Optoelectronic Transmitter

While the simulation examples of the previous sections concentrated on photoreceivers, the examples in this section will focus on monolithic transmitter design, analysis, and simulation. As with the photoreceiver simulations, the transmitter examples will use Cioffi's HEMT model; however, instead of Bianchi's MSM photodetector model, the exercises in this section will use Gao's multiple quantum-well laser diode model as the optoelectronic element. The section will begin with some background material about transmitters, continuing with a description of both the electrical and optical devices to be considered. Also presented in this section is an example of parameter extraction, as performed by iSMILE. Finally, a design example of a laser driver consisting of a differential topology will be presented.

4.6.1 Transmitter design issues

The success of optical communications can be significantly enhanced by the availability of low-cost, monolithic optoelectronic components; mature monolithic integration can make large-scale implementation more economical. Monolithic photoreceiver implementations are quite common, due to the compatibility between the epitaxial structures used for the transistors and detectors [4.13], [4.27]. However, because of the dissimilarity between the laser and transistor structures, transmitter integration is usually done through hybrid bonding, a process which not only has poorer performance than monolithic integration but is often expensive and unreliable, as well. It is quantitatively shown in [4.28] and [4.29] that the inductance induced by the bonding wire in traditional hybrid implementations can cause excessive peaking in the transmitter frequency response as well as ringing in the transmitter pulse (transient) response.

It is natural to inquire why a laser *driver* is required; indeed, because of the considerable difficulty in integrating a laser diode with transistors, it would be desirable to design a transmitter that does not require a laser driver. In this case, the laser inputs could be connected directly to the output of a digital circuit. When applied across the contacts of a laser, the "logic 0" output of the digital circuit would result in no light output, or OFF, and the "logic 1" state would result in a nonzero light output, or ON. The problem is that switching a laser on (light) and off (no light) is an extremely slow process. The buildup of electron population in a laser is very time consuming and can easily approach the nanosecond scale. Once the laser exceeds threshold, however, high-speed modulation is easily achieved. That is, it is much quicker to modulate a laser which is already above threshold than it is to switch it on and off. Thus, the logic low and high states are more efficiently achieved by forcing the laser to output "some" light and "more" light, respectively, rather than turning it completely on and then off again.

One way to achieve high-speed modulation is by sinking an amount of current from the laser that is equal to, or slightly greater than, the threshold current of the laser. This DC current is referred to as a *bias current* and serves to maintain the laser at or slightly above threshold. A signal current is then superimposed on the bias current to achieve high-speed modulation (Figure 4.24), since as the total current through the laser is modulated, so is the light output.

To sink a bias current and to superimpose a signal current, an external circuit is obviously required. This circuit is referred to as a laser *driver*, while the word *transmitter* denotes a system consisting of a laser integrated with its driver. The scenario just described is called direct modulation since the laser output is modulated by modulating the amount of injected current. In contrast, the light can be modulated by injecting a DC current into the laser, then modulating the output light via an externally placed modulator (Figures 4.25 and 4.26).

The principal advantage of external modulation is that it can achieve higher speeds than direct modulation. Another significant advantage is low-chirp modulation. However, several challenges exist in external modulator fabrication and processing, particularly when the modulator and the laser are integrated monolithically. When hybrid integration is performed, much more effort is required since sometimes only a fraction of the laser output can be successfully coupled into the external modulator. Alignment, reflections, and mode matching play significant roles in this problem.

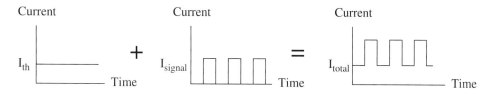

Figure 4.24: Current flow through laser.

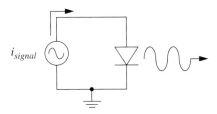

Figure 4.25: Direct modulation.

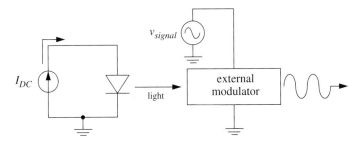

Figure 4.26: External modulation.

Another reason for using a driver circuit is that lasers operate on the principle of current injection. While a voltage applied across a laser *will* generate current flow through the laser, the output voltage of a digital circuit may not produce sufficient current to drive a laser. Furthermore, the output levels of digital circuits may not be directly compatible with the laser structure; for example, high-speed digital circuits are often configured differentially. As mentioned earlier, it is desirable to maintain a constant (static) current flow through the laser which is slightly greater than the threshold current. With threshold currents commonly in the 10–20 mA range, special precautions must be taken with the driver circuits, as such high currents can potentially damage the transistors. Typically, the bias transistor which sinks the threshold current is made very wide (large) to avoid overheating. Thus, the driver circuit also acts as an interface between the external electronic circuit and the laser, both for level shifting/current adjustment and for protection of the external circuit.

One final application of a laser driver is to perform *photomonitoring*. There are several factors which can cause a laser to deviate from its design specifications. For example, the threshold current has been shown to vary with temperature according to

$$I_{th}(T) = I_{o} \exp\left(\frac{T}{T_{o}}\right) \tag{4.34}$$

where T is the absolute temperature and I_{o} and T_{o} are constants which depend on the laser material and structure. Thus, a change in temperature can radically affect laser performance. To avoid such operation, a photomonitor circuit samples the light output of a laser and constantly adjusts the injection current so that the output conforms to specifications. Laser performance is also subject to aging and degradation; this is particularly true for the threshold current and the slope efficiency. By providing constant feedback from the laser to the driver circuit, the photomonitor can be used to adjust the injection current to maintain the light output at a desired level. Since a laser is essentially a p-n junction, a reverse-biased laser can be used as a photodetector; thus, the photomonitor device is often created simply by etching a trench or gap in the laser, separating it into two segments as depicted in Figure 4.27. Light emanates from the front facet of the laser while the rear facet is treated so that only a very small amount of light can escape. This light then strikes the photodetector, generating a photocurrent whose magnitude is used to determine the level of the output light. Once this feedback loop is completed, the injection current is adjusted by the drive circuit to regulate the laser output power. The properties of lasers used as photodetectors are described in [4.30].

The fluctuations in laser output characteristics are so great that commercially available transmitters almost invariably contain some degree of photomonitoring to ensure reliable, long-term operation. The creation of etched, rather than cleaved, facets requires an etching technology such as CAIBE (Chemically Assisted Ion Beam Etching) or RIE (Reactive Ion Etching). If etching, rather than cleaving, is used for both facets, then all mirror processing can be carried out on the wafer level. (Recall that cleaved-facet lasers can be placed only on the edge of the wafer; actually, cleaving *creates* a wafer edge along the plane of the cleave.) Most commercial transmitters, however, use completely separate lasers and detectors and bond them together hybridly.

A survey of the literature shows that several different institutions have successfully fabricated monolithic integrated transmitters (Table 4.1). The most work has been done in the GaAs materials system at the 0.85-μm wavelength due to

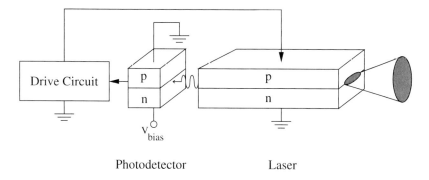

Figure 4.27: Typical photomonitoring circuit.

	Year	Institution	Speed	I_{th} (mA)	Transistor	Material
[4.31]	1980	Hitachi	—	100	MESFET	GaAs
[4.32]	1983	Honeywell	160 MHz	30–40	MESFET	GaAs
[4.33]	1984	NEC	2 Gb/s	30	MISFET	InP
[4.34]	1984	Cal. Tech	25 MHz	15–20	MESFET	GaAs
[4.35]	1984	Hitachi	4.5 GHz	—	MESFET	GaAs
[4.36]	1984	Rockwell	2 GHz	30	MESFET	GaAs
[4.37]	1985	Fujitsu	1 Gb/s	15	MESFET	GaAs
[4.38]	1986	CNET	—	22	MESFET	GaAs
[4.39]	1986	Hitachi	2 Gb/s	31	MESFET	GaAs
[4.40]	1989	Fujitsu	1.5 Gb/s	15–20	MESFET	GaAs
[4.41]	1990	Bellcore	5 Gb/s	8	MODFET	InP
[4.42]	1991	AT&T Bell	5 Gb/s	18–25	HBT	InP
[4.43]	1993	RTI	10 Gb/s	27	HBT	GaAs
[4.44]	1993	Lockheed	—	6	MESFET	GaAs

Table 4.1: Summary of reported OEIC transmitters.

the maturity of GaAs processing technology [4.31], [4.32], [4.34]–[4.40], [4.43], [4.44]. However, there has also been some work done in the more difficult InP materials system due to the 1.3-μm–1.5-μm wavelength window of low dispersion and loss in optical fibers [4.33], [4.41], [4.42]. While most of the reported transmitters are simple drive circuit/laser combinations, several are very sophisticated, including the presence of complete photomonitoring functionality. It should be emphasized that reports of laser drivers (integrated drive circuits *without* a monolithically integrated laser) are not included in Table 4.1. It is of interest to note that while most of the integrated transmitters in Table 4.1 contained traditional, edge-emitting lasers, in [4.44], a VCSEL (Vertical Cavity Surface Emitting Laser) was integrated with a MESFET. The low currents required by VCSELs can significantly relax demands on the transistors in the laser driver circuit.

4.6.2 Device description

The device chosen for the electronic driver circuit for the examples in this section is the doped-channel pseudomorphic Metal-Insulator-Semiconductor Field-Effect Transistor (MISFET). Figure 4.28 illustrates typical device structures for three popular compound semiconductor field-effect transistors. The MISFET was chosen over the more mature MESFET technology due to its higher

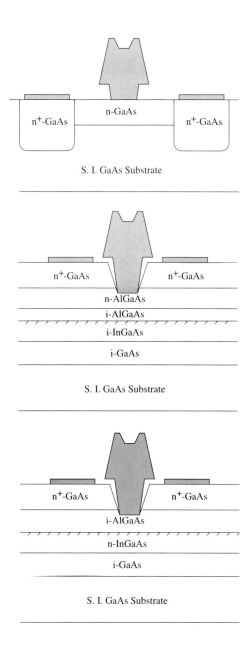

Figure 4.28: Typical device structures for MESFET, HEMT, and MISFET, from top to bottom, respectively.

transconductance (g_m) and better threshold voltage control.   Furthermore, the Schottky barrier height of AlGaAs is higher than that of GaAs. Since the gate metal directly contacts GaAs in a MESFET and AlGaAs in a MISFET, higher input voltages can be tolerated in a MISFET before excessive gate conduction occurs. The Schottky barrier heights of materials such as InGaAs, which have much higher electron mobilities than GaAs, are even lower than GaAs, precluding their use in conventional MESFET structures.

The MISFET was chosen over the HEMT for several reasons. The prime advantage of the MISFET over the HEMT is better threshold voltage control. While the threshold voltage of a HEMT is a very sensitive function of the thickness of the AlGaAs layer as well as its doping concentration, the threshold voltage of a MISFET depends only on the type of material and composition [4.45]. The doped channel of the MISFET makes it less sensitive to background doping levels than the HEMT, as well as giving it higher sheet carrier concentrations. Finally, since the gate contacts undoped AlGaAs, the MISFET exhibits wider drain current swings as a result of its higher gate-drain breakdown voltage [4.45]. This makes the MISFET ideal for high current applications such as the laser drivers under consideration.

As seen in [4.42] and [4.43], Heterojunction Bipolar Transistors (HBTs) are often used in laser driver circuits because of their excellent high-speed performance and current driving capability. However, since an analytical HBT model has not yet been developed for iSMILE, the examples in this section will use the MISFET, instead. (A numerical, table-based HBT model will be presented in Chapter 7.) The actual MISFET structure that will be used in the examples of this section is depicted in Figure 4.29.

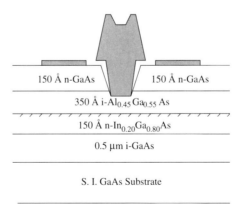

Figure 4.29: MISFET layer structure.

4.6.3 Parameter extraction

The first step of a circuit design is to determine the device parameters of the transistors to be used (device parameters for a FET, for example, would include threshold voltage V_T, mobility μ_o, and the channel-length modulation factor λ). Typically, this process involves several steps. First, a sample of the wafer that will be used for fabrication is obtained. Next, several test devices are processed on this sample, and their terminal (I-V) characteristics are measured. Finally, the device parameters are extracted from the measured data. To initiate extraction, the designer must usually specify a range (upper and lower bounds) for each parameter to be extracted. The extractor then attempts to minimize the difference between measurement and calculation by adjusting the parameters, subject to their respective ranges. By iteratively modifying the parameters, recalculating the device equations with these modified parameters, and comparing these calculations with the measured data, the parameter extractor arrives at optimum values for the parameters. Thus, parameter extraction is basically an optimization problem with the parameter ranges as the constraints.

As an example, consider the equation for a diode:

$$I = I_o \left[\exp\left(\frac{qV}{nkT}\right) - 1 \right] \tag{4.35}$$

The parameter extraction algorithm iteratively modifies the device parameters (I_o and n) and recalculates I until it finally arrives at a good match between the measured and calculated values. This example is somewhat overly simplistic in that all of the parameters could be extracted using one set of data. In general, several different sets of data (I-V curves) are necessary to determine all of the model parameters. This is, in fact, the case for the MISFET, which requires three I-V curves just to determine the DC parameters (I_{gs} versus V_{gs}, I_{ds} versus V_{gs}, and I_{ds} versus V_{ds}).

Some parameter extraction software packages are linked to test or measurement equipment. The advantage of this is that the parameters can be extracted immediately when the test devices are measured, with little or no work for the designer. The disadvantage is that parameters can be extracted only for the device models and equations that are stored in the software. This method is difficult to use if a completely new device is to be modeled. When such software packages are not available, another method is to use circuit simulation in conjunction with either a numerical optimizer or a stand-alone parameter extraction program.

Another approach is to use a circuit simulation program, such as iSMILE, which contains integrated parameter extraction capability. For illustrative purposes, iSMILE will again be used in this section, though it should be emphasized that the techniques used are general and can be applied regardless of the extraction tool. Three files are required as inputs: a set of measured data, a list of the parameters to be extracted along with their associated ranges, and a netlist describing the circuit setup from which the data were measured. iSMILE optimizes the parameters to achieve the best fit between measured data and simulation, subject to the specified parameter ranges. Once the parameters have been determined, their values are inserted in the corresponding `.MODEL` card in the iSMILE input deck and simulation proceeds in a manner identical to that of previous sections.

Examination of Figure 4.28 will show that the device structures for the HEMT and MISFET are virtually identical. Since there is no dedicated iSMILE model for the MISFET, rather than develop one, parameter extraction can be used to fit MISFET measured data to the HEMT model. To extract parameters, a 25-µm-wide, 0.2-µm-long MISFET was fabricated on the layer structure shown in Figure 4.29, and several I-V characteristics were measured.

The first set of extracted parameters models the forward conduction, or gate current, of the MISFET. These parameters can be extracted from a measurement of gate current I_{gs} as a function of gate voltage V_{gs} with the source grounded (essentially a Schottky diode). From this I-V curve, the diode ideality factor n and the reverse saturation current density J_o were determined to be 1.525 and 5.0×10^{-4} A/cm^2, respectively, while the gate resistance R_g was determined to be 20 Ω (Figure 4.30). While these parameters were extracted from the gate-to-source Schottky diode, in reality, there are two Schottky diodes in the HEMT model, one between the gate and the source and another between the gate and the drain. The HEMT model assumes that these diodes are identical and uses the same set of parameters to describe both.

The second set of parameters is extracted from a measurement of drain current I_{ds} as a function of gate voltage V_{gs} with the drain held at 2.5 V. From these data, the threshold voltage V_T was determined to be -0.3785 V (Figure 4.31). The channel-length modulation factor λ was extracted from a third set of data which consisted of the drain current I_{ds} measured as a function of both the drain voltage V_{ds} and the gate voltage V_{gs}. The optimum value for λ was 0.6167 V^{-1}. The I_{ds} versus V_{gs} and I_{ds} versus V_{ds} curves are both used iteratively to determine the following parameters: mobility μ_o, maximum electron concentration n_{so}, saturation velocity v_{sat}, source resistance R_s, and drain resistance R_d. Multiple sets of data are required because in some cases, minimizing the difference between measurement and calculation can be achieved in multiple ways. For example, in Figure 4.31, the drain current can be increased by increasing μ_o, n_{so}, v_{sat} or all three. If only one set of data is available, it would be difficult to determine which parameter(s) to adjust. On the other hand, parameters such as V_T and λ can clearly be determined from only one set of data. By running multiple extraction runs iteratively, these parameters are estimated to be: $\mu_o \approx 1500$ cm^2/V·s, $n_{so} \approx 2.386 \times 10^{12}$ cm^{-2}, $v_{sat} \approx 2.266 \times 10^7$ cm/s, $R_s \approx 1.600$ Ω·mm and $R_d \approx 3.760$ Ω·mm. It is important to note that these parameters describe only the DC behavior of the HEMT. To predict transient and frequency-domain properties, capacitors (C_{gsp}, C_{dsp}, C_{gdp}) and other frequency-dependent parameters must be extracted, as well. Unfortunately, AC parameter extraction is not supported in

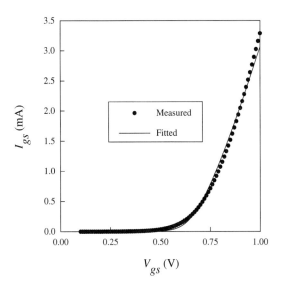

Figure 4.30: Extraction of Schottky diode parameters.

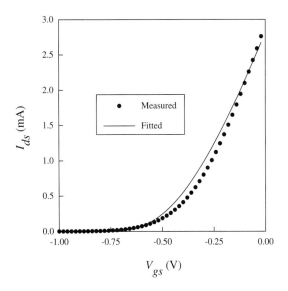

Figure 4.31: Extraction using measured drain current
as a function of gate voltage.

iSMILE. However, the methods detailed in this section are general enough to be used for AC extraction with the appropriate parameter extraction or optimization tools.

The layer structure for the quantum-well laser to be used in this transmitter example is depicted in Figure 4.32. This structure is an 85% Graded Index Separate-Confinement Heterostructure (GRIN-SCH) ridge laser and has been designed for MOCVD growth. The graded-index structure results in a parabolic index profile which improves the optical confinement. The GRIN-SCH structure also increases the efficiency at which carriers are collected since the energy bands are sloped (on either side of the quantum well) toward the active region. This has the added benefit of reducing the threshold current. Since a GaAs quantum well is used as the active layer, lasing occurs at about $\lambda = 0.8$ μm. The length of the laser cavity is 400 μm, while the width is 40 μm. The facet reflectivity is estimated to be about 0.32.

Parameter extraction, measurements, and simulation suggest a threshold current of about 20 mA, a slope efficiency of 0.40 mW/mA, and a series resistance of 1 Ω. Based on these attributes, the iSMILE multiple quantum-well laser model can be used for DC simulation. To achieve a threshold current of 20 mA and a slope efficiency of 0.40 mW/mA, several internal model parameters must be properly set: spontaneous emission coupling coefficient $\beta \approx 1.5 \times 10^{-4}$, radiative recombination coefficient $B \approx 7.0 \times 10^{-16}$ cm^3/s, diode ideality factor $n \approx 2.11$, and diode reverse saturation current $I_o \approx 8.40 \times 10^{-7}$ A. From these parameters, iSMILE simulations result in I-V and L-I (light-current) characteristics as shown in Figures 4.33 and 4.34.

0.2 μm	GaAs	p^+
1.0 μm	$Al_{0.85}Ga_{0.15}As$	$p = 10^{18}$ cm^{-3}
0.12 μm	$Al_{0.85}Ga_{0.15}As$ to $Al_{0.20}Ga_{0.80}As$ Linear Grade	i
100 Å	GaAs QW	i
0.12 μm	$Al_{0.20}Ga_{0.80}As$ to $Al_{0.85}Ga_{0.15}As$ Linear Grade	i
1.0 μm	$Al_{0.85}Ga_{0.15}As$	$n = 10^{18}$ cm^{-3}
1000 Å	GaAs contact	$n > 3 \times 10^{18}$ cm^{-3}

Figure 4.32: Laser layer structure.

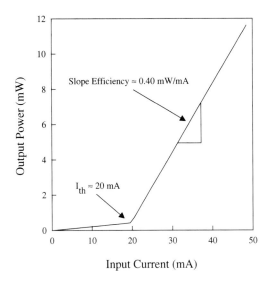

Figure 4.33: DC light-current characteristic.

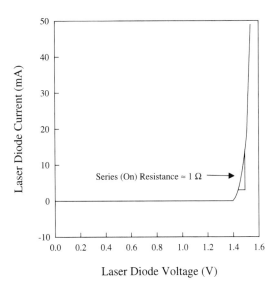

Figure 4.34: DC current-voltage characteristic.

4.6.4 Device integration

While fabrication of the MISFET and laser *individually* is fairly straightforward, it is not clear how to integrate the two devices monolithically. Such integration is particularly difficult because the MISFET and laser layer structures are completely different. One possible approach is to use regrowth. In this method, growth and processing are first performed for one type of device. Once this is completed, the layers on the unused portions of the die are stripped down to the substrate and a second growth is performed for the other type of device. Once the second type of device is processed, the two types are connected by metal lines. The advantage of regrowth schemes is that almost no compatibility is required between the two device structures; thus, a wide range of OEICs can be fabricated. However, regrowth can be very complex and time consuming; thus, yields are typically low. Since regrowth involves many process steps, device reliability is often low, as well. For these reasons, regrowth is often considered to be uneconomical for practical implementation.

Another approach is to grow both the FET layers and the laser layers at the same time, one on top of the other. After laser processing is complete, the laser layers on the unused portions of the die are stripped down to the FET layers to allow for FET processing. This method is very cost effective since is requires only one pass through the growth chamber. It is not, however, without problems. For effective processing, a die should be as planar as possible. A nonplanar surface with protruding structures, such as the one just described (the laser protrudes by more than two microns), can present serious processing challenges.

To planarize this structure, the laser/FET layers can be grown on a patterned substrate. A patterned substrate (usually S. I. GaAs) is formed by creating grooves, or patterns, in the substrate before growth. When the wafer is placed in the growth chamber, the deposited epilayers follow the contour of the pattern. The resulting layers are initially highly nonplanar; however, planarity can be achieved by placing the laser in the groove and the FETs outside the groove. This can be accomplished simply by etching away the laser layers outside the groove which separates the structure of Figure 4.35 into three parts, one part with laser layers on top of FET layers, and two parts with only FET layers. Once planarity has been achieved, the laser (in the groove) and the FETs (outside the groove) can be processed separately.

Once processing of both types of devices has been completed, the devices must be interconnected. Since the laser is fundamentally a p-n junction, contact must be made to both its top and bottom. To contact the bottom of the laser, use is made of the heavily doped, 1000 Å n-GaAs contact which is the last layer of the laser structure (Figure 4.32). Since this layer is heavily doped, it can be effectively used as an interconnection line (Figure 4.36). However, since this layer is GaAs and not metal, it will have a nonnegligible amount of series resistance. Preliminary measurements and calculations indicate that this resistance is about 0.6 $\Omega/\mu m$. Thus, if a 15-μm line is assumed, the series resistance will be about 9 Ω. It is important to

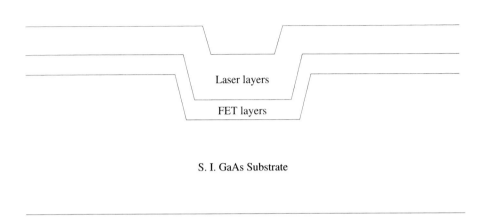

Figure 4.35: Growth on patterned substrate.

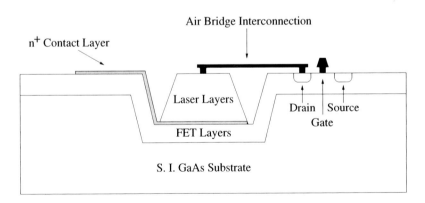

Figure 4.36: Integration of laser and FET structures.

realize that this series resistance must be added to the series resistance of the laser (1 Ω) to determine the total series resistance. To contact the top of the laser, either a layer of metal or an air bridge can be used. The device structure using an air bridge is depicted in Figure 4.36.

4.6.5 Circuit design and simulation

As mentioned previously, one of the main functions of a laser driver is to maintain a constant current flow so that the laser is biased slightly above its threshold. Another function is to act as an interface between the laser and the external electronics. Since many high-speed circuits are configured differentially, the design example presented here will be differential. In addition, use of a differential pair is advantageous since it lessens the circuit dependence on several transistor parameters such as threshold voltage, temperature, and process variations. This is due mainly to the fact that it is the *relative* rather than the absolute voltage that is important. Thus, as long as each side, or leg, of the differential pair is fabricated under the same conditions and process variations, the overall amplifier performance is more robust than that of a single-ended amplifier. For this example, it is assumed that the differential inputs V_{in}^+ and V_{in}^- both range from - 0.5 V to + 0.5 V. Thus, the total differential swing ($\Delta V_{in} \equiv V_{in}^+ - V_{in}^-$) is -1 V < ΔV_{in} < 1 V, although the procedure demonstrated here can be used to design over an arbitrary input range. For this example, a design goal of at least a 1-mW relative swing in output light power between the low and high states into a fiber will be set. The *absolute* light level is not crucial since the photoreceiver can be designed to interpret any amount of light as "logic 0"; however, excessive amounts of output power should be avoided because they can produce safety hazards and induce nonlinear effects in fibers [4.46], [4.47]. Since the optical output specification was for light into a fiber, losses in the laser-to-fiber coupling must be taken into account. While several methods for coupling lasers to fibers exist, the coupling efficiency can range anywhere from about 10% to over 80%, with typical efficiencies around 25% to 50% in systems using lenses [4.48]–[4.50]. Here, coupling efficiency is defined to be the percentage of the laser output which successfully enters the fiber.

The schematic of the differential transmitter is depicted in Figure 4.37, where the node numbers corresponding to the simulation input file of Figure 4.38 are present. A bias transistor has been provided to sink a constant threshold current through the laser. The width of this transistor was chosen large enough to sink at least 20 mA of (threshold) current without excessive heat dissipation. From experimental measurements, the current-carrying capability of the MISFET of Figure 4.29 has been estimated to be about 180 mA per mm of gate width. With a transistor width of 150 μm, the bias transistor is able to sink about 20.4 mA from the laser. In this particular example, the gate voltage on the bias transistor was made adjustable so that the bias current could be adjusted after fabrication; however, it

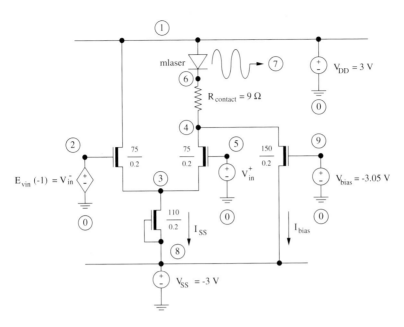

Figure 4.37: Transmitter circuit schematic.

would be desirable to tie the gate to the source of this device, use it as a resistor, and, thus, sink a more or less constant bias current from the negative power supply. This can be accomplished with proper sizing of the bias transistor.

The differential amplifier operates on the principle of current switching. First, a constant current source is provided at the base of the amplifier. This current is referred to as I_{ss} and can be switched almost entirely to the left (reference) leg or right (laser) leg by varying the input voltages. The size of the transistor providing this current was determined by two conditions. In operation, I_{ss} will be switched to the reference leg when the difference between V_{in}^+ and V_{in}^- is at its minimum value (-1 V), and to the laser leg when ΔV_{in} is at its maximum value (1 V). When $\Delta V_{in} = 0$, I_{ss} will be split equally between the two legs. Since the laser has a slope efficiency of 0.40 mW/mA, when $\Delta V_{in} = +1$ V, 4 mW of output light can be attained if I_{ss} is set to about 10 mA. This is sufficient optical output to account for a coupling efficiency as low as 25%. To accomplish this with a -3 V V_{ss}, the width of this transistor was set to 110 μm.

The positive and negative input transistors are sized based on the 180-mA/mm condition; each must be capable of sinking $I_{ss} = 10$ mA. In addition, their sizes must be set in a manner that allows them to always remain in saturation, given the predetermined V_{ss} and V_{dd}. For these reasons, the input transistors were assigned a width of 75 μm.

```
* Laser Transmitter

Vdd 1 0 3
Vss 8 0 -3
mlaser 1 6 7 0 lasmod
Rcontact 6 4 9

* Input transistor with laser
mfet2 4 5 3   HD w=75u l=.2u
Vin 5 0   DC

* Input transistor without laser
mfet2 1 2 3   HD w=75u l=.2u
Eref 2 0 5 0 -1

* Current source for electronic circuit
mfet2 3 8 8   HD w=110u l=.2u

* Current source for laser bias
mfet2 4 9 8 HD w=150u l=.2u
Vbias 9 0 -3.05

.MODEL  HD mfet2 vt=-.3785
+       lambda=.6167 vsat=.2266e8 ns0=.2386e13
+       rgg=20 rss=1.6 rdd=3.76 phib=.5 nd=.2e19
+       dd=.24e3 di=40 j0=0.5000e-3 nid=1.525 u0=1500
+       cgsp=.78e-12 cgdp=.72e-12 cdsp=.12e-11 x=.25
+       xvp=1 nbi=.5e10 lsg=.5e-4 ldg=.125e-4
+       dcap=100 ndcap=1e18

.MODEL  lasmod mlaser length=400E-6 width=4E-6
+       lz=100E-10 lambda=8350E-10 reflect=0.32
+       beta=5e-4 i01=8.40E-7 nn=2.11 alphai=300
+       c0=1.5835E-3 vj=1.15 m=0.26 fc=0.74 kco=0.3
+       gamma2=7.0D-16 ktau=1.2 keq=1.0
+       eqg=1.482 rbs=1 betavg=0.05 km=1.0 kn=2.5

.DC Vin -1 1 0.05
.PRINT DC V(7)
.END
```

Figure 4.38: Input file for transmitter simulation.

In most textbooks, differential amplifiers are completely symmetric. The transistor sizes on both sides of the input are identical as are their resistive loads. In most applications, differential pairs are used as voltage amplifiers; that is, through current switching and resistive loads, a differential output voltage is created according to $\Delta V_{out} = -(g_m R_L) \Delta V_{in}$, where ΔV_{in} and ΔV_{out} are the differential input and output voltages, g_m is the transconductance, and R_L is the load resistor. The circuit used in this driver, however, has a laser (and GaAs interconnect resistor) on one leg, and no load on the other. The reason is that for this application, *current* rather than voltage switching is required; that is, modulation in the output power occurs by modulating *current* through the laser, not voltage across it. Thus, it is not necessary to put a "dummy" resistor on the reference leg to "balance" the amplifier. Furthermore, the resistance on the laser leg which is the sum of the laser series resistance with the GaAs interconnect resistance, is quite small ($\approx 10\ \Omega$). When current is switched to the laser leg, a voltage of magnitude $V_{laser} = I_{ss} R_{laser}$ will be formed. As long as this voltage is small enough to keep the input transistor in saturation, the circuit will operate properly [4.6]. If V_{laser} is too large, however, the bias point of the transistor will be perturbed and the input transistor can leave saturation ($V_{drain} = V_{dd} - V_{laser}$). However, as seen in Figure 4.39, this was not a problem.

Figure 4.39 shows that the current flowing through the laser leg is just equal to the laser threshold current when the differential input is at its minimum value, and that it increases to over 30 mA when ΔV_{in} is at its maximum value. When $\Delta V_{in} = 0$, the current in the laser leg is at the laser threshold plus half of I_{ss}. Similarly, when $\Delta V_{in} = -1$ V, the current through the reference leg is equal to I_{ss} (the simulated value of I_{ss} is actually closer to 11 mA than 10 mA). When $\Delta V_{in} = +1$ V, the current in the reference leg drops to zero. Figure 4.40 shows that the output power is at a minimum when the differential input voltage is minimum and at a maximum when ΔV_{in} is maximum. The differential circuit inherently has a discriminating or thresholding function. This can be seen in Figure 4.40 by the fact that once the differential input exceeds a certain value (either negative or positive) the output power does not change. This type of behavior is highly desirable for digital optoelectronic systems.

4.7 Alternate Techniques — Hybrid Transmitter

While a monolithically integrated transmitter is clearly desirable, the difficulty of fabricating such a device has led many researchers to adopt a hybrid approach. There are many different methods of hybrid optoelectronic integration, including flip-chip bonding [4.51] – [4.54], silicon waferboard [4.55], epitaxial lift-off [4.56], and conventional wire bonding [4.57] – [4.58]. Because of the difficulty in coupling light between the laser and an optical fiber, flip-chip bonding is especially attractive because of the passive, self-aligning characteristics of the solder-bumping process.

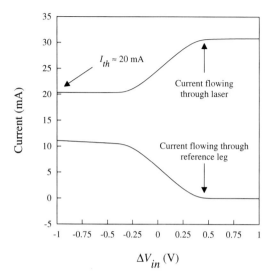

Figure 4.39: Differential circuit drive currents.

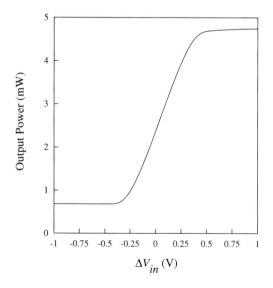

Figure 4.40: Transmitter power characteristic.

Unlike electronic packaging in which a circuit/die is completely designed and fabricated (essentially) independently of the packaging, the unique problem of optical coupling makes optoelectronic integration extremely difficult.

In this section, the design and simulation of a hybrid transmitter will be illustrated using roughly the same laser driver topology as the monolithic transmitter. The key difference in the approach of this section will be the use of a standard simulator that does not have a laser diode model. As demonstrated in Section 4.4, it is not always necessary to have models for optoelectronic devices in order to simulate optoelectronic circuits. One commonly used approach is to model the laser diode as a resistor in series with a voltage source, both in parallel with a capacitor (Figure 4.41). The capacitor C_{laser} is used to model the junction capacitance of the laser diode, the resistor R_{laser} represents the series resistance of the laser, and the voltage source V_{laser} is used to model the electrical voltage drop across the laser junction. (It should be noted that all three of these values are, in the traditional approach, made constant).

This approach treats only the electrical properties of the laser and completely ignores the optical aspects. It is useful, however, in situations in which the laser driver and the laser diode are developed by two separate groups. The circuit designer is usually provided with preliminary information about the laser's electrical and optical properties. In this scenario, the circuit designer is often forced to build a laser driver that has the capability to provide the specified threshold current in the bias circuit, over a range of threshold currents caused by process variations. He must then use the system specification for optical output power and the slope efficiency to calculate the amount of current required in the high and low states, with these specifications, again, subject to process variations. With this information, the circuit designer must design in the capability to provide an appropriate amount of modulation current (I_{ss} in the previous example). Another situation in which this approach might be used is one in which the laser diode is not being designed, but is, instead, an off-the-shelf component. In this case, the electrical properties must be

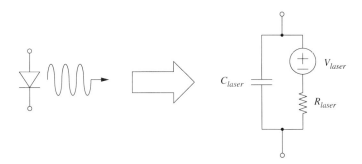

Figure 4.41: Laser diode electrical equivalent circuit.

obtained from a commercial data sheet, and the laser driver circuit designed appropriatcly.

While often the only choice when optical models are not available, this approach fails to account for any of the optical properties of the laser diode including the laser's turn-on time, the optical relaxation oscillations, and the effects of physical and material parameters on the laser output. It is not possible to assess the effect on the output power of modifying the cavity dimensions, nor is it possible to *predict* the laser's efficiency or threshold current.

As stated previously, the driver designed in this section will be very similar to the driver in the monolithic design. In fact, its topology will be nearly identical to that of Figure 4.37, but with the equivalent circuit of Figure 4.41 in place of the laser diode model. This driver, however, will be composed of MESFETs for illustrative purposes. In addition, while the modulation current I_{ss} was made constant in the previous case, in this design it will be made adjustable in the same manner as the bias current. This will allow maximum flexibility in case the laser's actual characteristics deviate from their specified values; such deviation often occurs during the development cycle, as will be illustrated in the next section.

4.7.1 Device characteristics

The MESFET device characteristics to be used in the driver design, as extracted from the Raytheon (Statz) model [4.59], are summarized in Table 4.2. These parameters were extracted from a 0.6-μm MESFET process and are listed for use in a general SPICE-like simulator. As is standard practice, several parameters in Table 4.2 are normalized to a unit (1 μm) gate width. This allows for easy scaling of transistor widths through the AREA argument of the MESFET model.

The laser characteristics are preliminary, as is often the case in research and development, and are depicted in Table 4.3. The task, then, is to design a laser driver circuit that can operate over all potential ranges of laser and MESFET parameters. As stated previously, to accommodate these ranges, the gate voltage on both the bias and modulation transistors will be made adjustable.

4.7.2 System specifications

As seen in the previous section, the device specifications were very loose; for example, the laser threshold current had a 300% variation from its lowest to its highest value and the MESFET threshold voltage had a 15% possible variation. This makes design of the driver circuit very difficult; however, by designing in several variable voltage sources, the system will be able to operate over the proper range. Once the device specifications become more concrete, these variable gate voltage sources can be replaced by gate-to-source-tied transistors, in the same fashion as the modulation transistor (which sinks I_{ss}) in the monolithic transmitter design.

Parameter	Symbol	Value
Gate Length	L	0.6 μm
Threshold Voltage	VTO	-1.46 V ± 15%
Transconductance Parameter	BETA	83.86 μA/V^2
Channel-Length Modulation	LAMBDA	0.080735 V^{-1}
Saturation Voltage Parameter	ALPHA	4.108351 V^{-1}
Drain Ohmic Resistance	RD	537.0317 Ω
Gate Ohmic Resistance	RG	248.2472 Ω
Source Ohmic Resistance	RS	529.2483 Ω
Zero-Bias Gate-Source Junction Capacitance	CGSO	0.85 fF
Zero-Bias Gate-Drain Junction Capacitance	CGDO	0.25 fF
Drain-Source Junction Capacitance	CDS	0.264 fF
Gate Junction Saturation Current	IS	2 fA

Table 4.2: The 0.6-μm MESFET parameters [4.60].

Quantity	Symbol	Best Case	Worst Case
Threshold Current	I_{th}	5 mA	15 mA
Differential Quantum Efficiency	η	1 mW/mA	0.25 mW/mA
Series Resistance	R_s	≤ 1 Ω	≤ 2 Ω
Junction Capacitance	C_j	30 pF	50 pF
Turn-on Voltage	V_{on}	1.6 V - 15%	1.6 V + 15%

Table 4.3: Projected laser characteristics [4.61].

Before proceeding any further, it will be helpful to present a specification on the transmitter's output power. The mean optical power is defined as the arithmetic mean of the high and low powers, in units of dBm:

$$P_{mean}(\text{dBm}) = \frac{1}{2}\left[P_{high}(\text{dBm}) + P_{low}(\text{dBm})\right] \tag{4.36}$$

For this design, a nominal mean optical power of 500 µW (-3 dBm) is desired with minimum and maximum acceptable values of 250 µW (-6 dBm) and 1 mW (0 dBm), respectively. Furthermore, it will be required that the extinction (or contrast) ratio between the high and low optical powers be 6 W/W (or 7.8 dB). Thus, in watts, the optical power requirements can be stated as

$$P_{high}(\text{W}) = \sqrt{6}\, P_{mean}(\text{W})$$

$$P_{low}(\text{W}) = \frac{1}{\sqrt{6}}\, P_{mean}(\text{W}) \tag{4.37}$$

From the specification on the maximum and the minimum mean optical powers, it can be deduced that $612\ \mu\text{W} < P_{high} < 2.45\ \text{mW}$, while the optical power in the low state would have one-sixth the range of P_{high} due to the extinction ratio of six. For this particular example, an arbitrary selection of 1.53 mW in the high state and, therefore, 255 µW in the low state, is chosen, resulting in values that fall right in the middle of the allowable range.

Next, the electrical requirements must be detailed. The most pertinent quantity is the specification of differential input signals. For this example, the input will consist of an 800 mV swing on each signal. It will be assumed that the input is AC coupled; thus, it can be assumed that both the positive and negative input voltages will range from - 400 mV to + 400 mV. The differential input will thus range from - 800 mV to + 800 mV.

4.7.3 Laser driver circuit design

To review the performance of the differential amplifier (Figure 4.37), a modulation current I_{ss} is switched through the laser diode whenever the differential input voltage swings positive. The magnitude of this current was controlled by the 110-µm-wide, gate-to-source-tied transistor at the bottom of the circuit. In the monolithic design, tight device specifications were available; thus, it was possible to precisely size the *modulation transistor* at 110 µm and then tie the gate to the source to create a constant current source I_{ss}. In this case, I_{ss} was set to about 10 mA by the choice of modulation transistor size. Since the differential quantum efficiency η of the laser in the present hybrid transmitter can potentially fluctuate by over 400%, and since a strict range of output powers is specified, it is necessary to make the gate voltage of the modulation transistor an externally adjustable quantity $V_{modulation}$ (or V_{mod}, for short).

The main consideration in sizing the modulation transistor is that it must be able to provide sufficient modulation current to produce 1.53 mW of optical output in the high state and 255 μW of power in the low state, under all possible sets of conditions and under the full range of possible device parameter values. According to the laser processing variations of Table 4.3, this translates to a range of 1.275 mA to 5.1 mA for I_{ss} (which will henceforth be referred to as $I_{modulation}$ or I_{mod}). Since increasing the MESFET threshold voltage decreases the transistor current, the upper bound of the threshold current variation will be designated as the worst case and the modulation transistor will be sized for this situation. Similarly, a large laser voltage will reduce the drain voltage of the positive input transistor, thereby reducing the amount of current that this FET will sink. Consequently, this will be viewed as the worst-case laser voltage. Since the modulation current I_{mod} is controlled by the modulation voltage V_{mod}, in addition to satisfying the current requirements under all conditions, it would also be desirable for the I_{mod} - V_{mod} characteristic of the modulation transistor to be easily tunable.

For this design example, a 25-μm width was chosen for the modulation transistor. The I_{mod} - V_{mod} characteristic under nominal processing conditions is depicted in Figure 4.42. In this simulation, both the laser voltage and the MESFET threshold voltage are set to their nominal values of 1.6 V and -1.46 V, respectively. As seen in this figure, the range of modulation currents attainable is about 0.5 mA – 8 mA over a modulation voltage range of - 1 V to 0.8 V. This completely

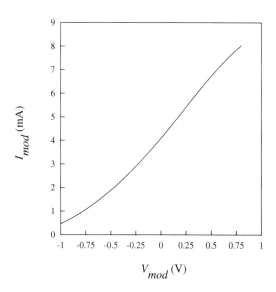

Figure 4.42: Modulation transistor nominal characteristic.

encompasses the range of required modulation currents of 1.275 mA to 5.1 mA that was stated earlier. A similar simulation under worst-case conditions shows that even in this situation, the modulation transistor is tunable, in a fairly linear manner, over the desired range. This greatly facilitates testing and operation of the transmitter. In successive design iterations, the laser and FET parameters will most likely become more stable. When this happens, it will be possible to use gate-to-source-tied MESFETs for the modulation and bias transistors (as in the design of the monolithic transmitter), rather than the adjustable-gate FETs used in the current design. It should be noted that in the worst case, a value of laser resistance $R_{laser} = 2\ \Omega$ was used.

Similar considerations are involved in the design of the bias transistor. As detailed in the design of the monolithic transmitter, it is necessary to sink a DC current through the laser to keep it biased slightly above threshold. As with the modulation transistor, potential variations in the laser and the MESFET characteristics necessitate a tunable, rather than fixed, gate voltage on the bias transistor. Given the range of potential operation of the laser diode, as well as the specifications on the output power levels, the bias transistor must be capable of sinking currents in the range of 5.255 mA – 16.02 mA. The variations of the laser voltage and MESFET threshold voltage are treated in a manner similar to that used for the modulation transistor. In Figure 4.43, the relationship between the bias current and the bias voltage is depicted. As with the modulation transistor, the bias transistor should be able to sink currents over the entire range dictated by system and device specifications as well as have a somewhat linear I-V characteristic for ease of

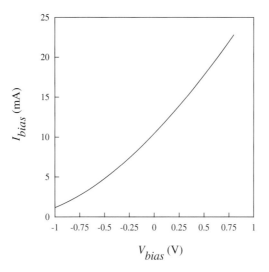

Figure 4.43: Bias transistor nominal characteristic.

tunability. Figure 4.43 shows that under nominal conditions, the bias transistor is capable of sinking currents ranging from 0.5 mA to more than 20 mA over a voltage tuning range of - 1 V to + 0.8 V. In addition, the I_{bias} - V_{bias} characteristic is, again, very smooth. If the slope of this characteristic is excessively steep, then very small changes in V_{bias} will create large swings in I_{bias}, making testing very difficult. To obtain this characteristic, a 55-µm-wide bias transistor was used. Simulations under worst-case considerations provided similar results.

The next design issue is the implementation of the two differential input signal transistors. As is well known from circuit theory, the slope of the transition in differential operation (Figure 4.39) can be increased by increasing the amount of current through the (input) signal transistors. This is equivalent to increasing their widths. Increasing this slope is beneficial since it will allow the same amount of output swing for a smaller differential input voltage. This also has the benefit of increasing the noise margins by making the circuit more tolerant to voltage fluctuations while in either the high or low states. Increasing the size of the signal transistors, however, also increases their capacitance; thus, there is a trade-off between the DC and transient performances. Figure 4.44 shows a simulation with 50-µm signal transistors using: a laser voltage, threshold current and slope efficiency of 1.6 V, 5 mA, and 1 mW/mA, respectively; a MESFET threshold voltage of -1.46 V; a bias voltage of - 0.452 V; and a modulation voltage of - 0.681 V. In Figure 4.44, the driver sinks a bias current of about 5.25 mA through the laser leg, then superimposes a 1.3 mA modulation current. With the assumed 1 mW/mA efficiency, this would result in an optical swing of 250 µW to 1.55 mW.

In Figure 4.45, the current swings through the laser for three different signal transistor sizes are compared. As stated earlier, the wider the transistor, the steeper the transition; thus, the steepest curve would correspond to the 100-µm signal FET while the curve with the most gradual slope would correspond to the 50-µm signal transistor. This figure shows that complete switching from high to low can be achieved for a ΔV_{in} of roughly ±300 mV, ±400 mV, and ±500 mV, for the 100-µm, 75-µm and 50-µm signal transistors, respectively.   However, the system specifications state that an 800 mV differential input will be available for switching; thus, the choice of signal transistor widths will be set by the amount of noise margin desired. The 100-µm choice, for example, would provide a safety range of about 800 mV - 300 mV = 500 mV, while a 50-µm signal FET would provide a buffer of only 300 mV. On the other hand, large signal FETs will have large capacitance, degrading the high-speed response. Given the wide range of possible device parameters, a choice of 75-µm signal transistors is made.

4.7.4 Supply voltages

The modulation and bias transistor simulations were performed with 0 V on the input (signal) transistors and V_{dd} and V_{ss} values of 5 V and 0 V, respectively. The

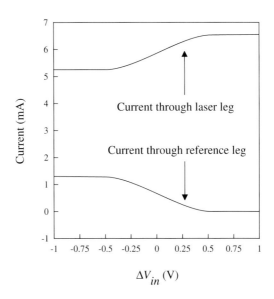

Figure 4.44: Laser and reference currents for 50-μm signal transistors.

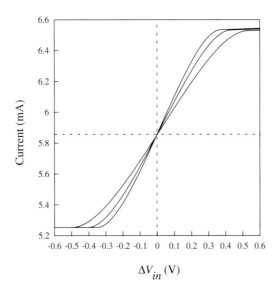

Figure 4.45: Comparison of laser currents for 50-μm, 75-μm, and 100-μm signal transistors.

signal transistor simulations were also performed with a 5-V power supply. It is obviously desirable to operate the transmitter with a V_{ss} value of 0 V because it allows single-power supply operation. However, since V_{bias} and V_{mod} are already required to be tunable, eliminating V_{ss} is only an incremental improvement. With successive iterations in device fabrication, the device parameters will become tighter and the tunable voltage sources can be eliminated by tying the bias and modulation transistor gates to their sources and resizing. Once completed, this will allow true single-supply operation. The final laser driver circuit is depicted in Figure 4.46.

4.7.5 Transient simulation

With the transistor sizes and supply voltages set, a transient simulation of the driver circuit of Figure 4.46 is depicted in Figure 4.47. For this simulation, the worst-case laser threshold current of 15 mA was assumed. The differential input was basically a pulse train of square waves. From this figure, the hybrid transmitter is seen to operate at roughly 1 Gb/s, under the conditions of a 1-Ω series resistance, a 1.6-V junction voltage, and a 30-pF junction capacitance. It is quite clear from Figure 4.46 that the simulation of Figure 4.47 captures only the electrical properties of the laser; to link the transient simulation results into optical power, it is necessary to scale the laser current by the laser's differential quantum efficiency.

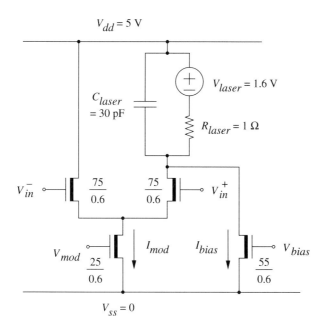

Figure 4.46: Hybrid transmitter simulated without laser diode model.

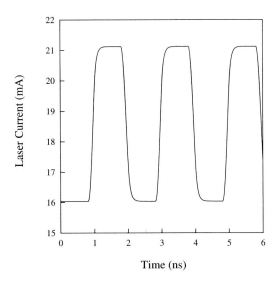

Figure 4.47: Transient characteristic of hybrid transmitter driver circuit.

4.8 Summary

In this chapter, several design examples were presented to complement the description of optoelectronic device models in Chapter 3. The first part of this chapter involved the computer-aided design of MSM-HEMT photoreceivers. Two designs were presented, one involving depletion-only transistors and one involving enhancement-depletion HEMTs. An often used alternate method for simulating photoreceivers was also depicted. This practice, while relatively inaccurate, is useful when photodetector models are not available. A method for simulating power was also illustrated. Since this technique essentially measures the average power drop (I·V product) across voltage sources, and does not rely on any transistor characteristics, it is applicable to any device technology. The next section involved the design of monolithic transmitters. Several relevant issues involved in transmitter design were discussed, including photomonitoring and high-speed modulation. A typical design cycle was presented including parameter extraction and the sizing of transistors for a differential laser driver.   Finally, various aspects of hybrid transmitter design were presented. A laser driver circuit similar to the one used in the monolithic transmitter was designed without using a model for the laser diode. Instead, the simulation modeled only the electrical properties of the laser, thus losing all ability to predict the optical properties. It was emphasized that while this approach is undesirable, it is very popular due to the general lack of availability of laser diode models.

4.9 References

[4.1]   G. F. Williams, "Lightwave receivers," in *Topics in Lightwave Transmission Systems*. San Diego, CA: Academic Press, 1991.

[4.2]   J. L. Gimlett, "Low-noise 8 GHz PIN/FET optical receiver," *Electronics Letters*, vol. 23, no. 6, pp. 281-283, 1987.

[4.3]   A. S. Sedra and K. C. Smith, *Microelectronic Circuits*. New York: Holt, Rinehart and Winston, 1987.

[4.4]   A. A. Ketterson, J.-W. Seo, M. H. Tong, K. L. Nummila, J. J. Morikuni, K.-Y. Cheng, S. M. Kang, and I. Adesida, "A MODFET-based optoelectronic integrated circuit receiver for optical interconnects," *IEEE Transactions on Electron Devices*, vol. 40, no. 8, pp. 1406-1416, 1993.

[4.5]   M. Abraham, "Design of Butterworth-type transimpedance and bootstrap-transimpedance preamplifiers for fiber-optic receivers," *IEEE Transactions on Circuits and Systems*, vol. CAS-29, no. 6, pp. 375-382, 1982.

[4.6]   P. R. Gray and R. G. Meyer, *Analysis and Design of Analog Integrated Circuits*. New York: John Wiley & Sons, Inc., 1984.

[4.7]   M. S. Ghausi, *Electronic Circuits*. New York: Van Nostrand Reinhold Company, 1971.

[4.8]   D. C. W. Lo, Y. K. Chung, and S. R. Forrest, "A high performance monolithic $In_{0.53}Ga_{0.47}As$ voltage-tunable transimpedance amplifier," *IEEE Photonics Technology Letters*, vol. 2, no. 9, pp. 675-677, 1990.

[4.9]   G. F. Williams and H. P. LeBlanc, "Active feedback lightwave receivers," *IEEE Journal of Lightwave Technology*, vol. LT-4, no. 10, pp. 1502-1508, 1986.

[4.10]   J. J. Brown, D. C. W. Lo, J. T. Gardner, Y. K. Chung, C. D. Lee, and S. R. Forrest, "$In_{0.53}Ga_{0.47}As$ junction field-effect transistors as tunable feedback resistors for integrated receiver preamplifiers," *IEEE Electron Device Letters*, vol. 10, no. 12, pp. 588-590, 1989.

[4.11]   G.-K. Chang, H. Schumacher, R. F. Leheny, P. Gerskovich, and S. Nelson, "On-chip characterization of fully integrated photoreceivers using high-yield ion-implanted GaAs IC technology," *GaAs IC Symposium Technical Digest*, pp. 57-60, 1987.

[4.12]   J. S. Wang, C. G. Shih, S. H. Chang, J. R. Middleton, P. J. Apostolakis, and M. Feng, "11 GHz bandwidth optical integrated receivers using GaAs MESFET and MSM technology," *IEEE Photonics Technology Letters*, vol. 5, no. 3, pp. 316-318, 1993.

[4.13] J. J. Morikuni, M. H. Tong, K. Nummila, J.-W. Seo, A. A. Ketterson, S. M. Kang, and I. Adesida, "A monolithic integrated optoelectronic photoreceiver using an MSM detector," *IEEE International Solid-State Circuits Conference (ISSCC) Digest*, pp. 178-179, 1993.

[4.14] W. S. Lee, D. A. H. Spear, P. J. G. Dawe, and S. W. Bland, "Monolithic integration of an InP/InGaAs four-channel transimpedance optical receiver array," *Electronics Letters*, vol. 26, no. 22, pp. 1833-1834, 1990.

[4.15] G.-K. Chang, W. P. Hong, J. L. Gimlett, R. Bhat, C. K Nguyen, G. Sasaki, and J. C. Young, "A 3 GHz transimpedance OEIC receiver for 1.3-1.55 μm fiber-optic systems," *IEEE Photonics Technology Letters*, vol. 2, no. 3, pp. 197-199, 1990.

[4.16] A. L. Guiterrez-Aitken, P. Bhattacharya, Y. C. Chen, D. Pavlidis, and T. Brock, "High-performance PIN-MODFET transimpedance photoreceiver," *IEEE Photonics Technology Letters*, vol. 5, no. 8, pp. 913-915, 1993.

[4.17] B. J. Van Zeghbroeck, C. Harder, J.-M. Halbout, H. Jäckel, H. Meier, W. Patrick, P. Vettiger, and P. Wolf, "5.2 GHz monolithic GaAs optoelectronic receiver," *IEEE International Electron Devices Meeting (IEDM) Technical Digest*, pp. 229-232, 1987.

[4.18] A. H. Gnauck, C. A. Burrus, and D. T. Ekholm, "A transimpedance APD optical receiver operating at 10 Gb/s," *IEEE Photonics Technology Letters*, vol. 4, no. 5, pp. 468-470, 1992.

[4.19] V. Hurm, J. Rosenzweig, M. Ludwig, W. Benz, M. Berroth, A. Huelsmann, G. Kaufel, K. Koehler, B. Raynor, and J. Schneider, "8.2 GHz bandwidth monolithic integrated optoelectronic receiver using MSM photodiode and 0.5 μm recessed-gate AlGaAs/GaAs HEMTs," *Electronics Letters*, vol. 27, no. 9, pp. 734-735, 1991.

[4.20] Y. Akatsu, Y. Miyagawa, Y. Miyamoto, Y. Kobayashi, and A. Akahori, "A 10 Gb/s high sensitivity, monolithically integrated p-i-n-HEMT optical receiver," *IEEE Photonics Technology Letters*, vol. 5, no. 2, pp. 163-165, 1993.

[4.21] N. Scheinberg, R. J. Bayruns, and T. M. Laverick, "Monolithic GaAs transimpedance amplifiers for fiber-optic receivers," *IEEE Journal of Solid-State Circuits*, vol. 26, no. 12, pp. 1834-1839, 1991.

[4.22] J. P. Uyemura, *Fundamentals of MOS Digital Integrated Circuits*. Reading, MA: Addison-Wesley Publishing Company, 1988.

[4.23] R. E. Saad and R. F. Souza, "Method to find the transimpedance gain of optical receivers using measured S-parameters," *SPIE High-Frequency Analog Fiber Optic Systems*, vol. 1371, pp. 142-148, 1990.

[4.24] J. W. Nilsson, *Electric Circuits*. Reading, MA: Addison-Wesley Publishing Co., 1983.

[4.25] D. A. Hodges and H. G. Jackson, *Analysis and Design of Digital Integrated Circuits*. New York: McGraw-Hill Publishing Co., 1983.

[4.26] S. M. Kang, "Accurate simulation of power dissipation in VLSI circuits," *IEEE Journal of Solid-State Circuits*, vol. SC-21, no. 5, pp. 889-891, 1986.

[4.27] A. A. Ketterson, M. Tong, J.-W. Seo, K. Nummila, J. J. Morikuni, S.-M. Kang, and I. Adesida, "A high-performance AlGaAs/InGaAs/GaAs pseudomorphic MODFET-based monolithic optoelectronic receiver," *IEEE Photonics Technology Letters*, vol. 4, no. 1, pp. 73-76, 1992.

[4.28] M. Nakamura, N. Suzuki, and T. Ozeki, "The superiority of optoelectronic integration for high-speed laser diode modulation," *IEEE Journal of Quantum Electronics*, vol. QE-22, no. 6, pp. 822-826, 1986.

[4.29] T. Hayashi, K. Katsura, and H. Tsunetsugu, "New hybrid integrated laser diode-drivers using microsolder bump bonding: SPICE simulation of high speed modulation characteristics," *IEEE Journal of Lightwave Technology*, vol. 12, no. 11, pp. 1963-1970, 1994.

[4.30] A. Alping, R. Tell, and S. T. Eng, "Photodetection properties of semiconductor laser diode detectors," *IEEE Journal of Lightwave Technology*, vol. LT-4, no. 11, pp. 1662-1668, 1986.

[4.31] T. Fukuzawa, M. Nakamura, M. Hirao, T. Kuroda, and J. Umeda, "Monolithic integration of a GaAlAs injection laser with a Schottky-gate field effect transistor," *Applied Physics Letters*, vol. 36, no. 3, pp. 181-183, 1980.

[4.32] J. K. Carney, M. J. Helix, and R. M. Kolbas, "Gigabit optoelectronic transmitters," *IEEE GaAs IC Symposium Technical Digest*, pp. 48-51, 1983.

[4.33] K. Kasahara, J. Hayashi, and H. Nomura, "Gigabit per second operation by monolithically integrated InGaAsP/InP LD-FET," *Electronics Letters*, vol. 20, no. 15, pp. 618-619, 1984.

[4.34] N. Bar-Chaim, K. Y. Lau, I. Ury, and A. Yariv, "Monolithic optoelectronic integration of a GaAlAs laser, a field-effect transistor, and a photodiode," *Applied Physics Letters*, vol. 44, no. 10, pp. 941-943, 1984.

[4.35] H. Matsueda and M. Nakamura, "Monolithic integration of a laser diode, photo monitor, and electric circuits on a semi-insulating GaAs substrate," *Applied Optics*, vol. 23, no. 6, pp. 779-781, 1984.

[4.36] C. S. Hong, D. Kasemset, M. E. Kim, and R. A. Milano, "Integrated quantum-well-laser transmitter compatible with ion-implanted GaAs integrated circuits," *Electronics Letters*, vol. 20, no. 18, pp. 733-735, 1984.

[4.37] T. Sanada, S. Yamakoshi, H. Hamaguchi, O. Wada, T. Fujii, T. Horimatsu, and T. Sakurai, "Monolithic integration of a low threshold current quantum well laser and a driver circuit on a GaAs substrate," *Applied Physics Letters*, vol. 46, no. 3, pp. 226-228, 1985.

[4.38] F. Brillouet, A. Clei, A. Kampfer, S. Biblemont, R. Azoulay, and N. Duhamel, "Laser-MESFET optoelectronic integration on GaAs: A simple technological process," *Electronics Letters*, vol. 22, no. 23, pp. 1258-1260, 1986.

[4.39] H. Nakano, S. Yamashita, T. P. Tanaka, M. Hirao, and M. Maeda, "Monolithic integration of laser diodes, photomonitors, and laser driving and monitoring circuits on a semi-insulating GaAs," *IEEE Journal of Lightwave Technology*, vol. LT-4, no. 6, pp. 574-582, 1986.

[4.40] O. Wada, H. Nobuhara, T. Sanada, M. Kuno, M. Makiuchi, T. Fujii, and T. Sakurai, "Optoelectronic integrated four-channel transmitter array incorporating AlGaAs/GaAs quantum-well lasers," *IEEE Journal of Lightwave Technology*, vol. 7, no. 1, pp. 186-197, 1989.

[4.41] Y. H. Lo, P. Grabbe, M. A. Iqbal, R. Bhat, J. L. Gimlett, J. C. Young, P. S. D. Lin, A. S. Gozdz, N. A. Koza, and T. P. Lee, "Multigigabit/s 1.5 μm λ/4-shifted DFB-OEIC transmitter and its use in transmission experiments," *IEEE Photonics Technology Letters*, vol. 2, no. 9, pp. 673-674, 1990.

[4.42] K.-Y. Liou, S. Chandrasekhar, A. G. Dentai, E. C. Burrows, G. J. Qua, C. H. Joyner, and C. A. Burrus, "A 5 Gb/s monolithically integrated lightwave transmitter with 1.5 μm multiple quantum well laser and HBT driver circuit," *IEEE Photonics Technology Letters*, vol. 3, no. 19, pp. 928-930, 1991.

[4.43] D. B. Slater Jr., P. M. Enquist, J. A. Hutchby, F. E. Reed, A. S. Morris, R. M. Kolbas, R. J. Trew, A. S. Lujan, and J. W. Swart, "Monolithically integrated SQW laser and HBT laser driver via selective OMVPE regrowth," *IEEE Photonics Technology Letters*, vol. 5, no. 7, pp. 791-794, 1993.

[4.44] Y. J. Yang, T. G. Dziura, T. Bardin, S. C. Wang, R. Fernandez, and A. S. H. Liao, "Fabrication on GaAs optoelectronic integration of a surface-emitting laser and a MESFET," *OFC/IOOC '93 Technical Digest*, pp. 164-165, 1993.

[4.45] S. J. Pearton and N. J. Shah, "Heterostructure field-effect transistors," in *High-Speed Semiconductor Devices.* New York: John Wiley & Sons, Inc., 1990.

[4.46] R. G. Smith, "Optical power handling capacity of low loss optical fibers as determined by stimulated Raman and Brillouin scattering," *Applied Optics*, vol. 11, no. 11, pp. 2489-2494, 1972.

[4.47] D. E. Quinn, "Optical fibers," in *Fiber Optics Handbook For Engineers and Scientists.* New York: McGraw-Hill Publishing Co., 1990.

[4.48] M. R. Matthews, B. M. MacDonald, and K. R. Preston, "Optical components – the new challenge in packaging," *IEEE Transactions on Components, Hybrids and Manufacturing Technology*, vol. 13, no. 4, pp. 798-806, 1990.

[4.49] M. S. Cohen, M. F. Cina, E. Bassous, M. M. Oprysko, and J. L. Speidell, "Passive laser-fiber alignment by index method," *IEEE Photonics Technology Letters*, vol. 3, no. 11, pp. 985-987, 1991.

[4.50] C. A. Armiento, "Optoelectronic integration of transmitter and receiver arrays on silicon waferboard," *Packaging, Interconnects, Optoelectronics for the Design of Parallel Computer Workshop Report*, pp. 66-72, 1992.

[4.51] M. J. Wale and C. Edge, "Self-aligned flip-chip assembly of photonic devices with electrical and optical connections," *IEEE Transactions on Components, Hybrids and Manufacturing Technology*, vol. 13, no. 4, pp. 780-786, 1990.

[4.52] P. Schmid and H. Melchior, "Coplanar flip-chip mounting technique for picosecond devices," *Review of Scientific Instruments*, vol. 55, no. 11, pp. 1854-1858, 1984.

[4.53] M. J. Goodwin, A. J. Moseley, M. Q. Kearley, R. C. Morris, G. J. G. Kirby, J. Thompson, R. C. Goodfellow, and I. Bennion, "Optoelectronic component arrays for optical interconnection of circuits and subsystems," *IEEE Journal of Lightwave Technology*, vol. 9, no. 12, pp. 1639-1645, 1991.

[4.54] M. Wale and M. Goodwin, "Flip-chip bonding optimizes opto-ICs," *IEEE Circuits and Devices*, vol. 8, no. 6, pp. 25-31, 1992.

[4.55] C. A. Armiento, "Silicon waferboard: Challenges and opportunities for low-cost optoelectronic subsystems," *LEOS '92 Conference Proceedings*, pp. 574-575, 1992.

[4.56] A. Ersen, I. Schnitzer, E. Yablonovitch, and T. Gmitter, "Direct bonding of GaAs films on silicon circuits by epitaxial liftoff," *Solid-State Electronics*, vol. 36, no. 12, pp. 1731-1739, 1993.

[4.57] J. F. Ewen, K. P. Jackson, R. J. S. Bates, and E. B. Flint, "GaAs fiber-optic modules for optical data processing networks," *IEEE Journal of Lightwave Technology*, vol. 9, no. 12, pp. 1755-1763, 1991.

[4.58] T. Iwama, T. Horimatsu, Y. Oikawa, K. Yamaguchi, M. Sasaki, T. Touge, M. Makiuchi, H. Hamaguchi, and O. Wada, "4 × 4 OEIC switch module using GaAs substrate," *IEEE Journal of Lightwave Technology*, vol. 6, no. 6, pp. 772-778, 1988.

[4.59] H. Statz, P. Newman, I. W. Smith, R. A. Pucel, and H. A. Haus, "GaAs FET device and circuit simulation in SPICE," *IEEE Transactions on Electron Devices*, vol. ED-34, no. 2, pp. 160-169, 1987.

[4.60] M. Feng, "Device model parameters for the iPOINT chipset design," Internal Memorandum, University of Illinois at Urbana-Champaign, February 10, 1993.

[4.61] J. J. Coleman, "Laser requirements," Internal Memorandum, University of Illinois at Urbana-Champaign, 1993.

5

ALTERNATE CIRCUIT SIMULATION METHODS

In Chapters 3 and 4, the iSMILE circuit simulator was presented as a method for implementing new circuit models. The main merit of this approach was that the model could be defined independently of the simulator internals. The user needed to specify only the internal and external parameters, the node topology, and the constitutive device equations; the simulator would then automatically compile and append the new circuit model into its library of available devices. Often, however, close access to the simulator internals is required in the implementation of new models. For example, intricate device models may require access to variables that are not accessible using standard techniques. In addition, situations often arise in which it is desirable to use numerical integration schemes other than the backward Euler algorithm coded in iSMILE. iSMILE is also rather limited in the types of analyses it can perform. While iSMILE can perform AC, DC, and transient analyses, conventional circuit simulators, such as SPICE, are capable of performing not only these, but other analyses as well, such as noise, Fourier, distortion, and transfer function analyses. For these reasons, circuit model developers sometimes choose to hard-code their new model into a conventional simulator such as SPICE. While the effort can be painstaking, once the new model is embedded, it gives the user access to the full-blown simulation engine of the host simulator.

In this chapter, alternate methods of optoelectronic circuit simulation will be discussed, centering mainly on those that use SPICE as the simulation core. Two approaches will be presented; the first involves the traditional approach, namely, the direct embedding of the device model into the SPICE source code, which is then recompiled. To illustrate this method, a new model for the MSM photodetector will be discussed. The second approach investigates a novel method of simulating laser diodes which allows for the model to be implemented in an input deck subcircuit rather than embedded in the SPICE source code.

5.1 MSM Photodetector

Unlike the MSM model of Bianchi et al. presented in Chapter 3, which was implemented into the iSMILE simulator, the MSM model presented in this section of Xiang et al. [5.1] was implemented into the standard SPICE simulator. Recent versions of SPICE actually contain a documented method through which new device models can be added without directly modifying the SPICE source code. Although this method is much more involved than using iSMILE, it does, however, allow the user to develop new models essentially independently of the simulator internals. The approach of Xiang in [5.1], however, was to embed the model directly into the simulator source code. While the model topology seems simple enough to implement as a netlist in the SPICE input file, upon closer inspection, it will become apparent that the device equations are sufficiently complex to necessitate an embedded model.

As an example, in [5.1] Xiang tailors the MSM model to the design and fabrication of MSM photodetectors for long-wavelength operation. In contrast to short-wavelength detectors (~ 0.8 μm), which can use GaAs-based materials as the active medium, long-wavelength (1.31 μm $\sim$ 1.55 μm) detectors require an active material such as InGaAs. The use of $In_xGa_{1-x}As$ is advantageous since the mole fraction x can be varied to allow detection of wavelengths inaccessible to GaAs-based systems. However, the Schottky barrier height of InGaAs (~ 0.2 V) is much lower than that of GaAs (~ 0.7 V). Since the metal fingers of an MSM detector make direct contact with the active region, this poses a serious drawback because the dark, or leakage, current increases with decreasing barrier height. In addition, a low barrier height can lead to high photocurrent gains [5.2]. To circumvent this problem, heterostructure technology is used to increase the Schottky barrier height through the addition of "barrier-enhancement" layers.

To illustrate this principle, consider Figure 5.1 in which a 1-μm-thick $In_{0.53}Ga_{0.47}As$ layer is used as the active region to detect 1.31-μm light. Because of the low Schottky barrier height of $In_{0.53}Ga_{0.47}As$ (~ 0.2 V), a thin 700 Å $In_{0.52}Al_{0.48}As$ layer is grown on top of the InGaAs layer which drastically reduces the dark current; the Schottky barrier height of this material is approximately 0.7 V [5.1]. A layer of graded InAlGaAs is sandwiched between the InGaAs and the InAlAs; the resulting smoothing of the heterojunction energy band structure reduces charge storage effects at the InAlAs/InGaAs interface [5.3], [5.4]. As discussed in

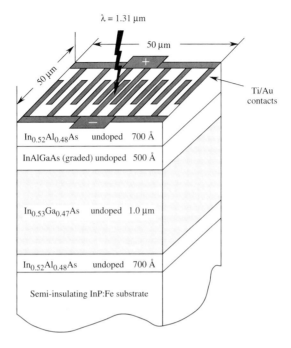

$\lambda = 1.31\ \mu m$

50 μm

50 μm

Ti/Au contacts

$In_{0.52}Al_{0.48}As$ undoped   700 Å

InAlGaAs (graded) undoped   500 Å

$In_{0.53}Ga_{0.47}As$ undoped   1.0 μm

$In_{0.52}Al_{0.48}As$ undoped   700 Å

Semi-insulating InP:Fe substrate

Figure 5.1: Long-wavelength MSM photodetector cross-section.

Chapter 3, while thickening the active layer increases responsivity, it decreases bandwidth. The speed reduction is due mainly to the decrease in electric field with distance; since the carrier velocity is proportional to the electric field, long tails appear in the impulse response. Thus, the 1-μm active layer was chosen as a reasonable compromise. In this particular example, a 700 Å $In_{0.52}\,Al_{0.48}As$ buffer layer is grown between the 1-μm active layer and the semi-insulating Fe-doped InP substrate in order to reduce the propagation of defects from the substrate to the absorption layer. A quarter wavelength-thick film of Si_3N_4 is used to passivate the device; this decreases reflectivity at the air-semiconductor interface, leading to better absorption [5.1].

5.1.1 Dark current

Traditional MSM models, such as Bianchi's in Chapter 3, consider Schottky emission at the metal-semiconductor interface as the only source of dark current; however, it is well known that other current transport mechanisms exist [5.5]. A distinguishing aspect of Xiang's dark current model is that it includes three different sources: Schottky emission, Frenkel-Poole emission, and tunneling current. A considerable difference is that Xiang covers current mechanisms that cause leakage both through metal-semiconductor junctions as well as through the insulating layers that form from both surface oxidation and passivation on the surface of the MSM.

As discussed in Chapter 3, thermionic Schottky emission is proportional to the number of electrons with enough energy to surmount a potential barrier. This form of leakage current can occur both at metal-semiconductor interfaces, as was seen in Bianchi's model, as well as at metal-insulator interfaces. To review, thermionic emission can be expressed as

$$I \propto \exp\left[\frac{-q}{kT}(\phi - \Delta\phi)\right] \qquad (5.1)$$

where q is the electron charge, ϕ is the potential barrier height, and $\Delta\phi$ is the amount of Schottky barrier lowering due to image forces [5.1]. If the electric field is represented as E and the dynamic dielectric constant as ε, the amount of Schottky barrier lowering can be expressed as

$$\Delta\phi = \sqrt{\frac{qE}{4\pi\varepsilon}} \qquad (5.2)$$

The metal-insulator Schottky barrier is much higher than that of the metal-semiconductor interface; thus, while Schottky emission makes a significant contribution to the metal-semiconductor dark current, its effect on metal-insulator leakage is negligible [5.1].

In [5.1], the two dominant insulator leakage mechanisms are identified as Frenkel-Poole emission and interface trap-assisted tunnel emission [5.5], [5.6]. Frenkel-Poole current is caused by field-enhanced thermal excitation of trapped electrons into the conduction band. The actual number of these electrons depends on the potential of the trap. Frenkel-Poole emission experiences a potential barrier-lowering effect similar to that of Schottky emission; however, the two mechanisms are different in that under Frenkel-Poole emission, the positive ion does not move with respect to the image charge [5.1], [5.5]. Because of the similarities between Frenkel-Poole and Schottky thermionic emission, Frenkel-Poole emission can also be modeled by (5.1), provided that the amount of barrier lowering is represented by

$$\Delta\phi = \sqrt{\frac{qE}{\pi\varepsilon}} \qquad (5.3)$$

Using (5.1), (5.2), and (5.3), Xiang combines the dark currents due to Schottky and Frenkel-Poole emission into a single expression:

$$I_{emission} = C \cdot \exp\left[-\frac{q}{kT}\left(\phi - \sqrt{\frac{qE}{\alpha\pi\varepsilon\varepsilon_o}}\right)\right] \qquad (5.4)$$

In this equation, E is the electric field calculated as the voltage difference between electrodes divided by the electrode spacing, C is a fitting parameter, ϕ is the effective potential barrier that Xiang uses to account for both the Schottky barrier and the trap potential well, and ε is the combined relative dynamic permittivity for both the semiconductor and the insulator. The factor α is used to quantify the degree of barrier lowering and is given to be 4.0 for thermionic Schottky emission and 1.0 for Frenkel-Poole emission [5.1], [5.6], [5.8]; the quantities α and ε are often extracted together as one parameter.

The third leakage mechanism in Xiang's model is tunnel emission. Tunneling occurs when a particle encounters a finite potential barrier, such as a metal-semiconductor or metal-insulator interface. When a particle approaches the barrier, its wave function decays exponentially inside the barrier, as is well known from quantum mechanics. Tunneling can occur not only at interfaces, but also from trap levels to the conduction band in the insulator. As documented in [5.1], tunneling can also be assisted by interface traps which serve as "stepping stones" for carriers. Xiang's model considers all of these effects empirically as

$$I_{tunnel} = A \cdot \exp\left(-\frac{B}{V}\right) \qquad (5.5)$$

where V is the bias voltage and A and B are fitting parameters.

While mechanisms such as Schottky emission are controlled by the electrodes at the interface, Frenkel-Poole and tunnel emission current are controlled by the properties of the insulator [5.7]. In unpassivated devices, a thin layer of oxide (< 100 Å) forms on the surface of the InAlAs layer; this produces a Frenkel-Poole emission current whose magnitude is comparable to that of the tunneling current. In [5.1], Xiang states that the tunneling current is controlled by the electron and hole occupation factors of the interface traps and the time constants for electron and hole tunneling [5.8]. Since, through the Fermi Golden Rule, the time constants are dependent on the electric field, they decrease exponentially at low fields then become constant at high fields. The tunneling current, on the other hand, increases at low fields and saturates at high fields; increasing the field past this saturation point alters only the Frenkel-Poole emission current. Thus, at low bias voltages, tunnel current dominates oxide leakage, while at high voltages Frenkel-Poole emission dominates. Tunneling is often not a significant factor in passivated devices since the passivation layer is relatively thick; thus, Frenkel-Poole emission tends to be the dominant insulator leakage mechanism in passivated devices.

The parameters A, B, C, and $\alpha\varepsilon$ are extracted from experimental data as shown in Figure 5.2, which depicts dark currents for different sized MSMs, both passivated and unpassivated, of dimensions (finger width $\times$ finger spacing) 1 μm $\times$ 1 μm, 2 μm $\times$ 2 μm, and 3 μm $\times$ 3 μm. The parameters extracted for the unpassivated devices are depicted in Table 5.1. The unpassivated devices of Figure 5.2a are found to be dominated by tunneling current below 4 V; above 4 V, Frenkel-Poole emission

Dimension	C (kA)	ϕ (V)	$\alpha\varepsilon$	A (nA)	B (V)
1 μm × 1 μm	26.4	0.938	1.70	1.87	0.98
2 μm × 2 μm	24.5	0.921	1.18	1.72	1.01
3 μm × 3 μm	23.7	0.902	1.18	1.66	0.97

Table 5.1: Extracted dark current for unpassivated MSM detectors.

becomes dominant. Since Frenkel-Poole emission is proportional to the electric field, it is inversely proportional to the electrode spacing. This is because the electric field in the insulator decreases with increasing electrode spacing.

The dark currents for devices passified with a 1650-Å quarter-wavelength anti-reflective Si_3N_4 layer are depicted in Figure 5.2b. These dark currents are seen to be higher than their equivalent unpassivated counterparts. The dominant mechanism at work here is Frenkel-Poole emission current, which is enhanced by the thick layer of insulation between the electrodes formed by passivation. As stated previously, tunneling is negligible in this passivated device. Table 5.2 depicts the extracted parameters for the passivated MSMs.

The parameters in Tables 5.1 and 5.2 are determined by the material and geometrical properties of the MSM, as clearly evidenced by $\alpha\varepsilon$ and ϕ. The values of A indicate lower tunneling currents with larger finger spacing (smaller fields), as expected, while the values of B are constant since the device structure and fabrication process were identical for all of the unpassivated devices. The parameter C is clearly seen to vary with finger spacing [5.1]. For both the passivated and unpassivated detectors, the results of the MSM model (5.4), (5.5) match the measured, experimental results quite well.

Although the passivated MSMs suffer from higher dark currents as seen in Figure 5.2, unpassivated MSMs are much more susceptible to surface oxidation and contamination over time. When the surface structure eventually degrades, leakage currents will rise and the lifetime of the device will ultimately be shortened.

Dimension	C (kA)	ϕ (V)	$\alpha\varepsilon$
1 μm × 1 μm	67.3	0.854	1.84
2 μm × 2 μm	53.7	0.847	1.46
3 μm × 3 μm	39.5	0.865	0.94

Table 5.2: Extracted dark current for passivated MSM detectors.

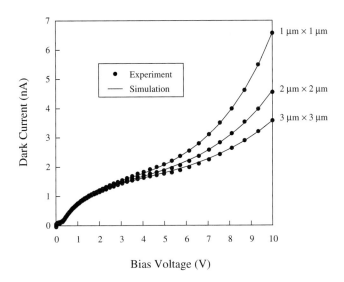

Figure 5.2a: Dark current of unpassivated InGaAs MSM.

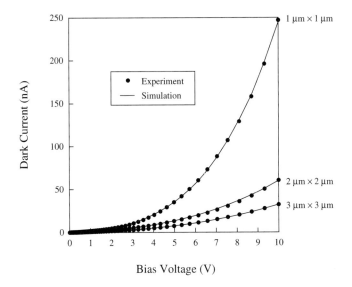

Figure 5.2b: Dark current of passivated InGaAs MSM.

5.1.2 DC characteristic

By illuminating the MSM with constant light, the DC I-V characteristic of Figure 5.3 is obtained. As expected, under low bias voltages (~ 0.6 V in this case), the metal-semiconductor junctions are not fully depleted, making diffusion the primary transport mechanism for photogenerated electron-hole pairs. As depletion is approached with increasing bias, the photocurrent rises exponentially. Once the flat-band voltage is reached, drift overtakes diffusion as the main transport mechanism. The flat-band voltage, which is defined as the voltage at which the active region under the electrodes is fully depleted, has been derived in [5.2] for one-dimensional MSMs and can be used as a good approximation for the geometries under consideration in this chapter:

$$V_{FB} = \frac{qN_{dope}}{2\varepsilon_s} L^2 \tag{5.6}$$

In (5.6), N_{dope} is the doping concentration in the InGaAs, ε_s is the semiconductor permittivity, and L is the electrode spacing. The doping concentration in the MSMs used here is between 1×10^{15} cm^{-3} and 2×10^{15} cm^{-3}, from which a flat-band voltage of 0.65 V is determined from (5.6).

Figure 5.3 also exhibits the photocurrent gain that often occurs after the flat-band voltage is exceeded. In the model of [5.1], Xiang attributes this to the charging of interface traps at the Schottky contacts and charge localization at the InAlAs/InGaAs heterostructure notch. He uses (5.5) to model the DC gain mechanism through the tunneling current, which is enhanced in the presence of excess carriers.

The formulation used in [5.1] to model the DC photocurrent is

$$I_{ph} = \frac{A_e}{A_t} \cdot (1-R)(1-e^{-\alpha d}) \cdot \frac{q\eta}{h\nu} \cdot P \tag{5.7}$$

where A_e is the MSM area exposed to light (the nonshadowed area), A_t is the total area, R is the reflectivity, α is the absorption coefficient, d is the absorption layer thickness, η is the internal quantum efficiency, $h\nu$ is the photon energy, and P is the incident optical power. All parameters in (5.7), except optical power, are assumed to be constant.

When the MSM is biased below the flat-band voltage, Xiang uses (5.8) to model the barrier height.

$$V_{barrier} = \frac{(V_{FB}-V)^2}{4V_{FB}} \tag{5.8}$$

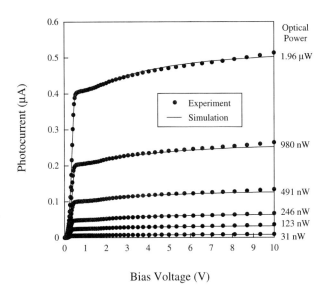

Figure 5.3a: DC photoresponse for unpassivated MSM.

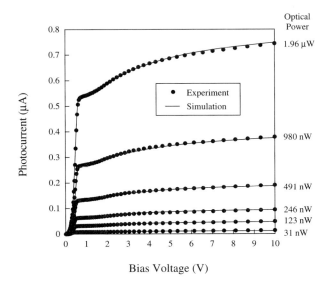

Figure 5.3b: DC photoresponse for passivated MSM.

By substituting (5.8) into the expression for thermionic emission, the DC photocurrent below the flat-band voltage is modeled in [5.1] by (5.9). The second term in this equation is the tunneling current used to model the photocurrent gain and I_{ph} is the photocurrent as given by (5.7).

$$I_{DC} = I_0\left\{\exp\left[\frac{q}{nkT}\cdot\frac{V(2V_{FB}-V)}{4V_{FB}}\right]-1\right\}+I_{ph}A\exp\left(-\frac{B}{V}\right) \qquad V < V_{FB} \qquad (5.9)$$

Above the flat-band voltage, the photocurrent is modeled as

$$I_{DC} = I_{ph}\left[1+A\exp\left(-\frac{B}{V}\right)\right] \qquad\qquad V \geq V_{FB} \qquad\qquad (5.10)$$

In these equations, n is a fitting parameter to control the rate of the photocurrent increase below flat-band, V_{FB} is the flat-band voltage, and I_0 is chosen to ensure continuity between (5.9) and (5.10) at $V = V_{FB}$. The extracted values for the MSMs in Figure 5.3 are given in Table 5.3.

By analyzing the slope of the DC photocurrent for $V > V_{FB}$ in Figure 5.3, it is clear that through the gain, the photocurrent, and hence the tunneling current, is linearly proportional to the optical input power (Figure 5.4). This indicates that the

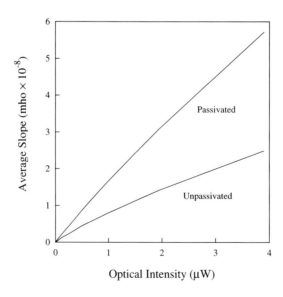

Figure 5.4: Average slope of DC photoresponse past flatband.

MSM	Responsivity (A/W)	V_{FB} (V)	n	A (A)	B (V)
Passivated	0.264	0.68	0.85	0.534	3.19
Unpassivated	0.197	0.445	0.81	0.421	2.58

Table 5.3: MSM DC model parameters.

interface trap charge density and charge localization at the heterojunction are also linearly dependent on the light intensity. The passivated MSM in Figure 5.4 exhibits a steeper slope than the unpassivated MSM mainly because of the reduction in reflectivity caused by the Si_3N_4. This linear relationship is reflected in the MSM model as evidenced by the DC photoresponses shown in Figure 5.3.

5.1.3 Extraction of parasitics

There are two limiting factors in the MSM frequency response: transit time effects and parasitics. Sources of parasitics in the MSM are elements such as the interdigitated electrodes and the bond/probe pads. While transit time effects will be discussed in the next section, S parameters can be used to characterize parasitic behavior in the MSM. Xiang's equivalent-circuit model for the MSM is depicted in Figure 5.5. In this model, R_{in} and C_{in} are the intrinsic resistance and capacitance, respectively, of the active area, while R_p, L_p, and C_p are used to model the external parasitics. The parameter P is the incident optical power and $I(P)$ models the photo and dark currents.

The MSM parasitics are extracted by placing the MSM in the dark, biasing it above the flat-band voltage, terminating it with a 50-Ω impedance matching resistance, and measuring S parameters. Both the measured and simulated S_{11} for three MSMs of sizes 1 μm × 1 μm, 2 μm × 2 μm, and 3 μm × 3 μm are depicted in Figure 5.6 over the range 1 GHz – 35 GHz. The measured S_{11} is used to numerically

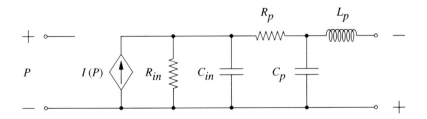

Figure 5.5: Equivalent circuit model for the MSM.

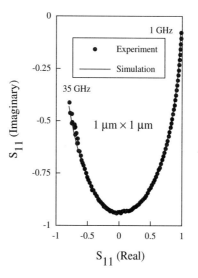

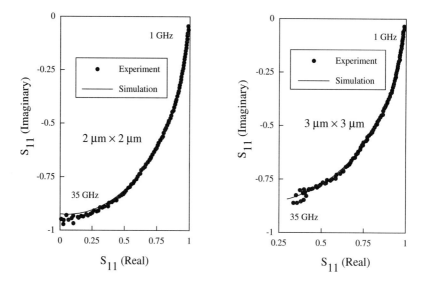

Figure 5.6: Measured and simulated S_{11} used to extract parasitics.

Parameter	1 µm × 1 µm	2 µm × 2 µm	3 µm × 3 µm
R_{in} (kΩ)	3.88	5.27	6.06
C_{in} (fF)	26.7	15.4	18.7
L_p (nH)	0.114	0.077	0.085
C_p (fF)	96.3	53.3	33.4
R_p (Ω)	52.0	64.6	55.9

Table 5.4: Extracted MSM parasitics.

optimize the model parameters R_p, L_p, and C_p until a fit is obtained. In the model of [5.1], Xiang was able to achieve an average error between simulation and measurement of less than 1.5%. The optimized parasitics are depicted in Table 5.4.

By measuring S parameters, the intrinsic capacitance can be extracted at various voltages and at different levels of optical illumination. The intrinsic capacitance consists of depletion-layer capacitance and capacitance due to charge storage. While the depletion capacitance is a function of applied voltage, the charge storage capacitance generally depends on the incident power and subsequently photogenerated carriers. In Figure 5.7, the intrinsic capacitance is plotted for a 1 µm × 1 µm MSM under different levels of illumination. Since the capacitance does not vary significantly with input power, Xiang concludes that the capacitance is dominated by the depletion capacitance.

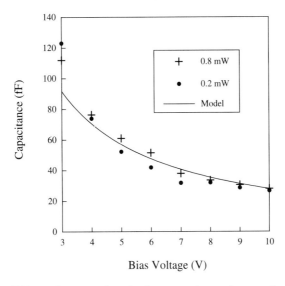

Figure 5.7: Voltage and optical power dependence of capacitance.

Because of the placement and biasing of the metal fingers, two depletion regions exist in the MSM: a forward-biased depletion region and a reverse-biased one. From standard semiconductor theory, it is clear that the capacitance of the forward-biased junction is much larger than that of the reverse-biased junction. Thus, since the two capacitances appear in series, in [5.1] it is assumed that the capacitance across the reverse-biased junction dominates the voltage-dependent capacitance when the bias voltage is greater than flat-band. The capacitance is modeled in [5.1] as

$$C = \frac{C_o}{\left(1 - \dfrac{V_r}{V_{bi}}\right)^m} \tag{5.11}$$

where C_o is the zero-bias capacitance, V_{bi} is the built-in voltage, m is the junction capacitance grading coefficient, and V_r is the voltage across the reverse-biased region:

$$V_r = V_{bi} - V \tag{5.12}$$

The voltage V is the total positive voltage applied across the terminals. Based on Figure 5.7, C_o is found to be 307 fF and m to be 0.90. Similar results were obtained in [5.9].

While there is always a small amount of capacitance contributed by the measurement setup, the capacitance C_p is dominated by the positioning of the metal fingers. Although the standard conformal mapping technique is traditionally used to approximate the capacitance of the interdigitated electrodes [5.10], this technique is overly optimistic for the devices that Xiang considers. The theory of Lim and Moore in [5.10] assumes that the thickness of the electrodes is zero; this yields a value that is lower than the actual measured capacitance.

5.1.4 Time-domain response

The *types* of equations used to implement the dark current, photocurrent, and capacitance in this model are similar, in nature, to those of Bianchi's MSM model in Chapter 3. Bianchi's model was implemented using the simulator-independent tool, iSMILE; however, in the introduction to this chapter, it was stated that there are situations in which access to the simulator internals is necessary for model development. In this section, the transient behavior of Xiang's model is analyzed. To more accurately model the time-domain response of the MSM, in [5.1] Xiang represents the photocurrent not by sets of RC branches as in Bianchi's model, but rather by a convolution of the incident optical power with the MSM impulse response. Evaluation of the time-domain convolution integral is a very computationally intense operation; thus, for maximum accuracy and computational efficiency, Xiang has embedded his model directly into the SPICE source code.

To numerically generate the impulse response, the drift-diffusion equations are used to model carrier transport. Neglecting heterojunction and interface trap effects, these are expressed as

$$\varepsilon \nabla \cdot E = q\,(p - n + N_d - N_a) \tag{5.13}$$

$$\frac{\partial p}{\partial t} = -\frac{p}{\tau_p} + g_p - \frac{1}{q}\nabla \cdot J_p \tag{5.14}$$

$$\frac{\partial n}{\partial t} = -\frac{n}{\tau_n} + g_n + \frac{1}{q}\nabla \cdot J_n \tag{5.15}$$

$$J_p = qpv_p - qD_p\nabla p \tag{5.16}$$

$$J_n = qnv_n + qD_n\nabla n \tag{5.17}$$

$$v_p = \frac{\mu_{po}E}{1 + \dfrac{\mu_{po}E}{v_{sp}}} \tag{5.18}$$

$$v_n = \begin{cases} \mu_{no}E & E < E_o \\[2ex] \dfrac{\mu_{no}E}{\sqrt{1 + \left(\dfrac{E - E_o}{E_c}\right)^2}} & E \geq E_o \end{cases} \tag{5.19}$$

Equation (5.13) is Gauss's law, where p and n are the hole and electron concentrations, and N_d and N_a are the donor and acceptor concentrations. Equations (5.14) and (5.15) are the hole and electron continuity equations, (5.16) and (5.17) are the current density equations, and (5.18) and (5.19) are empirical formulas for carrier velocity in the semiconductor [5.11].

In [5.1], it is assumed that the active region is depleted and the diffusion coefficient is constant with respect to the electric field. This assumption is justified by the fact that the diffusion current constitutes only a minor portion of the total current. Xiang does not use Einstein's relation since it only holds for systems in equilibrium. Another liberty taken by Xiang is the treatment of the MSM structure as one dimensional, an approach valid as long as the active layer thickness is less than the finger width. The total current is given by the conduction current plus the displacement current

$$J(t) = J_c(x,t) + \varepsilon \frac{\partial}{\partial t} E(x,t) = \frac{1}{L} \int_0^L J_c(x,t)\, dx \tag{5.20}$$

where L is the length of the device and J_c is the conduction current given by the sum of the drift and diffusion currents. In this model, carriers that reach the metal fingers are assumed to be collected instantaneously; the Schottky barrier is not considered in the transient analysis. The transient behavior of the device is determined by discretizing and solving the partial differential continuity equations, (5.14) and (5.15), numerically. A further assumption is that the photogenerated carriers are distributed uniformly throughout the device.

As an example, a carrier concentration of 10^{15} cm^{-3} is photogenerated at time zero, then removed immediately afterward; the reaction of the MSM to this stimulus is interpreted as the impulse response. The transient impulse photoresponse for a 1 µm × 1 µm MSM under various DC bias voltages is depicted in Figure 5.8. As in Bianchi's model, the slow-hole tail is evident in this figure. The changing electric field has a large effect on the hole velocity; a plot of this effect is depicted in Figure 5.9. This figure matches the experimental data observed in [5.12]; this reference also reported an increase, then saturation, of the peak amplitude of the impulse response with bias voltage. Although Xiang's model does exhibit this characteristic, as shown in Figure 5.9, the magnitude of this effect as predicted by the model is much smaller than experimentally observed values. In [5.1], it is postulated that this is due to variation of the intrinsic capacitance with bias voltage, voltage-dependent space-charge effects, and the photocurrent gain which is easily observed in the DC characteristic. It is found, however, that the last two of these three mechanisms are small when compared to the RC variation caused by intrinsic capacitance variations. Space-charge effects can cause appreciable nonlinearity in the impulse response when the excess carrier concentration is much higher than the doping concentration. Under strong illumination, Xiang's one-dimensional model may thus be unsuitable, and a two-dimensional model may become necessary [5.13] – [5.15].

Using the results of Figure 5.8, the impulse response of a given MSM can be determined numerically through the device simulation techniques illustrated above. However, from this figure, it is apparent that the impulse response can be approximated by the sum of the intrinsic electron and hole current responses $h_e(t)$ and $h_h(t)$ in (5.21) and (5.22).

$$h_e(t) = \frac{1 - \dfrac{t}{\tau_e}}{\tau_e} \left[u(t) - u(t - \tau_e) \right] \tag{5.21}$$

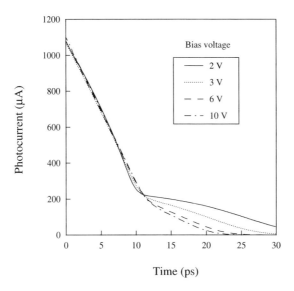

Figure 5.8: Simulation of intrinsic photocurrent impulse response.

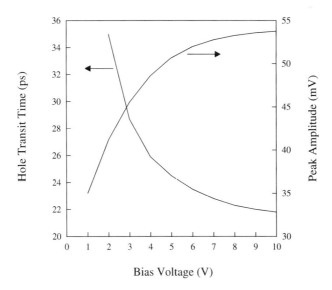

Figure 5.9: Variation of hole transit time and impulse response height with bias.

$$h_h(t) = \frac{1 - \dfrac{t}{\tau_h}}{\tau_h} [u(t) - u(t - \tau_h)] \tag{5.22}$$

In these equations, τ_e and τ_h are the electron and hole transit times, respectively, and $u(t)$ is the unit step function. It is clear in these equations that the electron and hole impulse responses are linearly decreasing functions of time. For circuit simulation, it is often easier and more efficient to use the approximations of (5.21) and (5.22) to model the impulse response rather than to numerically generate the impulse response through device simulation (5.13) – (5.20). Alternatively, device simulation can be used to determine more accurate values for τ_e and τ_h than could ordinarily be obtained from standard references; these values could then be used in the simple linear approximations of (5.21) and (5.22). For illustrative purposes, the approximations (5.21) and (5.22) will be used to model the impulse response in this section.

Once the impulse response is determined, by whatever technique the user deems most appropriate, the transient response of the MSM is determined in Xiang's model by convolving it with an arbitrary optical input. The difficulty of implementing the convolution integral at either the input deck level or in a model-independent simulator such as iSMILE is one of the motivating factors for embedding the model directly into the SPICE source code. Evaluation of the convolution integral is carried out numerically through the trapezoidal rule:

$$x_n = x_{n-1} + \frac{\Delta t}{2}(\dot{x}_n + \dot{x}_{n-1}) \tag{5.23}$$

In this equation, x_n is the intrinsic photocurrent from the MSM and is given by

$$x_n = i = \int_{t-\tau}^{t} p(t_o) h(t - t_o) dt_o \tag{5.24}$$

Equation (5.24) is a generalized form to be applied to both the electron and hole impulse responses (5.21) and (5.22), as indicated by the transit time τ. In this equation, $p(t)$ is the optical input power. The time derivative of the photocurrent, which is needed by the trapezoidal rule, is calculated as

$$\frac{dx}{dt} = \frac{di}{dt} = -\frac{1}{\tau^2}[\varphi(t) - \varphi(t - \tau)] + \frac{p(t)}{\tau} \tag{5.25}$$

$$\varphi(t) = \int_{0}^{t} p(t_o) dt_o \tag{5.26}$$

As previously stated, these equations were embedded directly into the SPICE code. This not only allows access to simulator internals, such as time, but direct embedding also reduces simulation time. It is noted that the time step Δt should be less than the transit time τ in order to obtain accurate simulation results.

The circuit elements were extracted from the measurements shown previously, and used to simulate the impulse photoresponse for the three devices (Figure 5.10). As expected, an increase in tail length and decrease in peak amplitude is observed when spacing is increased. The ripple in the tail of the experimental data is attributed to reflections caused by impedance mismatches in the measurement setup.

5.1.5 Frequency-domain response

The AC response of the MSM can be calculated by taking the Laplace transform of the impulse response. When the simple linear approximation for the impulse response is used (5.21), (5.22), the AC response for each carrier becomes

$$H(\omega) = \frac{1 - \cos(\omega\tau)}{\omega^2\tau^2} + j\frac{\dfrac{\sin(\omega\tau)}{\omega\tau} - 1}{\omega\tau} \tag{5.27}$$

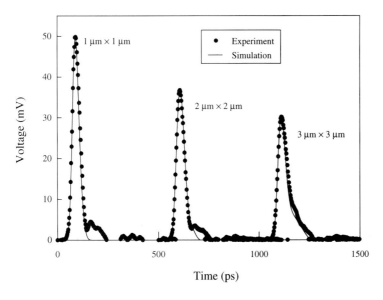

Figure 5.10: Impulse response for three MSMs,
each subject to a 30-ps, 220-fJ input pulse.
All circuit element values extracted from S parameter measurements.

When the impulse response is determined numerically from device simulation, the AC response can be determined through numerical transform techniques. From Figure 5.11 it is clear that the magnitude decreases rapidly when $\omega\tau$ exceeds unity [5.1]. In this figure, the 3-dB point occurs at $\omega\tau \approx 3.48$, which corresponds to a phase shift of 0.35π. Figure 5.12 shows the frequency response using (5.27) with given transit times. The simulations are depicted for three MSMs with 50-Ω loads; the data show a decrease of the bandwidth with increasing electrode spacing, a trend which agrees with experiment. This fact indicates that the device is transit-time limited.

5.2 Laser Diode

The approach taken by Xiang in the previous section was to embed the MSM model directly into and recompile the SPICE source code. In this section, a novel technique for laser diode simulation developed by Javro et al. [5.16], [5.17] will be discussed. A common problem in the simulation of laser diodes is that the rate equations have multiple solutions. Indeed, in Chapter 3, it was stated that one of the solutions to the rate equations can result in a negative output power. While this solution is numerically stable, it results in a nonphysical situation. In the model of Gao et al. the problem was solved by a heuristic algorithm which sampled intermediate Newton-Raphson solution steps and steered the solver away from the negative power solutions. In [5.16] and [5.17], Javro introduces a new laser diode model which is essentially the original rate equation representation, but numerically transformed, or mapped, into a separate variable space. The rate equations in this new space have only one solution, thus eliminating the "negative power" problem. With only one solution, there is no need to go through the techniques used by Gao to constantly monitor the intermediate numerical solutions. Thus, Javro does not need to embed his model into the SPICE source code, as Xiang did. Because of the simplicity of this new model, Javro does not even need to use a model development tool like iSMILE, as Gao did; rather, he is able to implement his model as a SPICE subcircuit. This section will discuss both the transformation of the rate equations as well as Javro's SPICE subcircuit implementation.

5.2.1 Linear rate equation model

While the quantum-mechanical rate equations used by Gao in Chapter 3 are very rigorous, a more common and often used model is the linearized rate equation representation used in [5.18] – [5.20] and depicted in (5.28) – (5.30). The parameters used in the rate equations are summarized in Table 5.5.

$$\frac{dN}{dt} = \frac{I}{qV_o} - \frac{N}{\tau_n} - g_o(N-N_o)(1-\varepsilon S)S + \frac{N_e}{\tau_n} \tag{5.28}$$

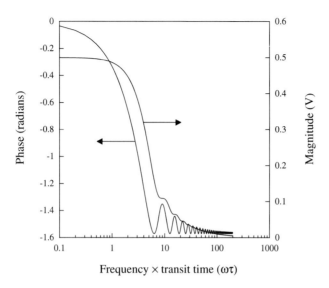

Figure 5.11: Magnitude and phase of AC response of one carrier.

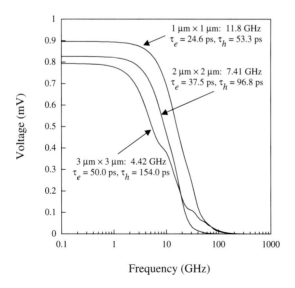

Figure 5.12: Simulated AC response of three MSMs under a 10-V bias. The 3-dB frequencies are indicated next to the device sizes.

Symbol	Quantity	Typical Value
S	Average photon density in cavity	cm^{-3}
N	Average carrier density in cavity	cm^{-3}
I	Injected current	mA
Γ	Optical confinement factor	0.44
g_0	Linearized gain constant	3×10^{-6} cm^3/s
N_0	Material transparency carrier density	1.2×10^{18} cm^{-3}
ε	Gain saturation coefficient	3.4×10^{-17} cm^3
β	Spontaneous emission coupling coefficient	4×10^{-4}
τ_n	Carrier lifetime	3 ns
τ_p	Photon lifetime	1 ps
V_0	Volume of active region	9×10^{-11} cm^3
q	Electronic charge	1.6×10^{-19} C
N_e	Equilibrium carrier density	5.41×10^{10} cm^{-3}
P	Power exiting one facet	mW
λ_0	Free-space laser wavelength	1.502 μm
η	Quantum efficiency	0.1
h	Planck's constant	6.624×10^{-34} J·s
c	Speed of light in vacuum	3×10^{10} cm/s

Table 5.5: Parameters used in the linearized rate equation model. Typical values are listed for constants (after [5.16], [5.19]). For variables (N, S, I, P), only units are listed.

$$\frac{dS}{dt} = \beta\Gamma\frac{N}{\tau_n} + \Gamma g_0 (N-N_0)(1-\varepsilon S)S - \frac{S}{\tau_p} \tag{5.29}$$

$$\frac{S}{P} = \frac{\Gamma\tau_p\lambda_0}{V_0\eta hc} \equiv \vartheta \tag{5.30}$$

The parameters in (5.28) – (5.30) describe the optical characteristics of the quantum well. In addition, an equation relating carrier density to voltage must be implemented. In [5.17], it is explained that when the laser is lasing, the carrier concentrations are degenerate and should, thus, be described by their full Fermi-integral representations (5.31) – (5.33).

$$N = N_c F_{1/2}\left(\frac{F_c}{kT}\right) \tag{5.31}$$

$$F_j(\eta) = \frac{1}{j!}\int_0^\infty \frac{x^j}{1 + \exp(x - \eta)}dx \tag{5.32}$$

$$N_c = \frac{1}{2}\left(\frac{8m^*kT}{\pi h^2}\right)^{3/2} \tag{5.33}$$

In these equations, N is the carrier density, N_c is the conduction band effective density of states, $F_{1/2}$ is the Fermi integral of order 1/2, F_c is the quasi-Fermi level in the conduction band, m^* is the electron effective mass, k is Boltzmann's constant, h is Planck's constant, and T is the absolute temperature; similar formulations can be made for the valence band. However, these equations are very difficult to solve, and Javro shows in [5.17] that a nondegenerate representation is usually accurate enough in most situations:

$$N = N_e \exp\left(\frac{qV}{nkT}\right) \tag{5.34}$$

5.2.2 Transformation of rate equations

As alluded to previously, while there is only one DC solution to the rate equations (5.28) – (5.29) that makes physical sense, there are actually three numerically stable solutions. Figure 5.13 graphically depicts the three solutions to the rate equations, subject to the values in Table 5.5, with $\vartheta = 3.686 \times 10^{17}$ s/cm^3/J. The familiar L-I curve can be seen in this figure, as well as a negative-power solution and a high-power solution. The negative-power solution can be easily observed *analytically* by solving the rate equations, without the gain saturation term (i.e., $\varepsilon \to 0$), in steady state. By eliminating the electron density N, the rate equations become second order in the photon density; two solutions for S thus arise from the $\pm$ sign in the quadratic formula. Addition of the gain saturation term makes the equations third order in S, resulting in the high-power solution.

From Figure 5.13, it is apparent that the negative and positive solution regimes are very close when the injected current is at its threshold value ($\sim$ 10 mA in this case). When solving the rate equations numerically with Newton-Raphson iteration, it is possible for the estimate of the next value to overshoot, or overestimate, the actual value. This can be seen from the basic formulation of the Newton-Raphson algorithm in Figure 5.14. In this figure, the previous solution point (x_0, y_0) and the derivative at this point are used to estimate the next solution, denoted as y_{N-R} (for Newton-Raphson). Clearly, the actual value of y (denoted as y_{actual}) differs from that

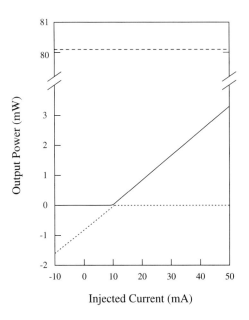

Figure 5.13: Three solution regimes of rate equations.

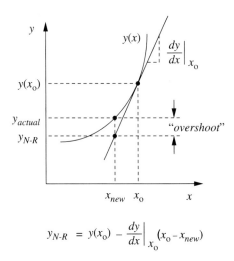

$$y_{N-R} = y(x_o) - \frac{dy}{dx}\bigg|_{x_o}(x_o - x_{new})$$

Figure 5.14: Overshoot caused by Newton-Raphson algorithm.

predicted by the Newton-Raphson algorithm — this is labeled as "overshoot" in Figure 5.14. As is well documented, this problem can be overcome by taking smaller and smaller steps between DC iterations; that is, by making $\Delta x \equiv x_{new} - x_0$ as small as possible. Of course, this increases the simulation time, as well, but it is often a necessary trade-off.

In Figure 5.13, if the overshoot is significant at or about the threshold current (~ 10 mA), the simulation engine can be fooled into crossing over into the negative solution regime. Again, this regime is *numerically* stable but is *physically* impossible. Thus, many implementations of the rate equation model in circuit simulators (such as Gao's) must use some sort of monitoring mechanism to guide the solver into the correct regime. In [5.16], Javro maps the rate equations into a separate variable space in which the rate equations have only one solution. The first step in this method is to remove the negative solution regime through the mapping

$$P = (m + \delta)^2 \tag{5.35}$$

where m is a variable function of time $m(t)$, and δ is a constant. The purpose of δ will become apparent shortly. By replacing S with P through (5.30), the rate equations using (5.35) are expressed as (5.36) and (5.37).

$$\frac{dN}{dt} = \frac{I}{qV_0} - g_0 (N - N_0) [1 - \varepsilon \vartheta (m + \delta)^2] \vartheta (m + \delta)^2 - \frac{N}{\tau_n} + \frac{N_e}{\tau_n} \tag{5.36}$$

$$2\frac{dm}{dt} = \Gamma g_0 (N - N_0) [1 - \varepsilon \vartheta (m + \delta)^2] (m + \delta) + \frac{\Gamma \beta N}{\vartheta \tau_n (m + \delta)} - \frac{(m + \delta)}{\tau_p} \tag{5.37}$$

Examination of (5.37) reveals that the factor $(m + \delta)$ appears in the denominator; thus, if δ were not included in (5.35), division by zero would occur in the solution of (5.37) when $m = 0$. This is a situation which often arises since all variables are usually initialized to zero at the beginning of each simulation. The value of δ is not important, its presence being required only to prevent division by zero; in [5.16], δ is set to 5×10^{-8}. Since $P = (m + \delta)^2$, even if $m < 0$, there is no way for the power to become negative.

The second step in Javro's approach is to eliminate the high-power solution caused by the addition of the gain saturation term. In [5.16] he shows that in the high-power solution regime, the carrier concentration becomes negative (Figure 5.15). Thus, the high-power regime can be eliminated by transforming the carrier population into a function that cannot become negative. As stated previously, Javro assumes a simple Boltzmann distribution for the carrier-voltage relationship (5.38); this representation, clearly, cannot become negative.

$$N = N_e \exp\left(\frac{qV}{nkT}\right) \tag{5.38}$$

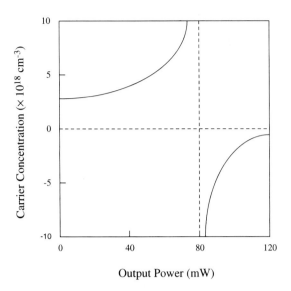

Figure 5.15: Negative carrier concentration,
third (high-power) solution regime.

The rate equation mapping is completed by substituting (5.38) for each occurrence
of N in (5.36) and (5.37); this results in (5.39) and (5.40).

$$\frac{d}{dt}\left[N_e\exp\left(\frac{qV}{nkT}\right)\right] = \frac{I}{qV_o} - \frac{N_e}{\tau_n}\exp\left(\frac{qV}{nkT}\right) + \frac{N_e}{\tau_n} \tag{5.39}$$

$$-g_o\left[N_e\exp\left(\frac{qV}{nkT}\right) - N_o\right][1 - \varepsilon\vartheta(m+\delta)^2]\,\vartheta(m+\delta)^2$$

$$2\frac{dm}{dt} = \Gamma g_o\left[N_e\exp\left(\frac{qV}{nkT}\right) - N_o\right][1 - \varepsilon\vartheta(m+\delta)^2]\,(m+\delta) \tag{5.40}$$

$$+\frac{\Gamma\beta N_e}{\vartheta\tau_n(m+\delta)}\exp\left(\frac{qV}{nkT}\right) - \frac{(m+\delta)}{\tau_p}$$

By rearranging terms in (5.39), the recombination current can be modeled as a diode
current:

$$I_{recombination} = \frac{qV_oN_e}{\tau_n}\left[\exp\left(\frac{qV}{nkT}\right) - 1\right] \tag{5.41}$$

The modified rate equations (5.39) – (5.40) can then be recast into the form of (5.42) and (5.43).

$$I = I_{D1} + I_{D2} + [2g_o \vartheta \tau_n I_{D1} - qV_o g_o \vartheta (N_o - N_e)] \cdot \\ [1 - \varepsilon \vartheta (m + \delta)^2] (m + \delta)^2 \tag{5.42}$$

$$2\tau_p \frac{dm}{dt} + m = \left[\left(\frac{2\Gamma \beta \tau_p}{qV_o \vartheta} \right) I_{D1} + \left(\frac{\Gamma \beta \tau_p}{\vartheta \tau_n} \right) N_e \right] (m + \delta)^{-1} - \delta \tag{5.43}$$

$$+ \left[\left(\frac{2\Gamma g_o \tau_n \tau_p}{qV_o} \right) I_{D1} - \Gamma g_o \tau_p (N_o - N_e) \right] [1 - \varepsilon \vartheta (m + \delta)^2] (m + \delta)$$

In these equations, the currents through diodes D_1 and D_2 are

$$I_{D1} = \frac{1}{2} I_{recombination} = \frac{1}{2} \cdot \frac{qV_o N_e}{\tau_n} \left[\exp\left(\frac{qV}{nkT}\right) - 1 \right] \tag{5.44}$$

$$I_{D2} = \frac{1}{2} \cdot \frac{qV_o N_e}{\tau_n} \left[\exp\left(\frac{qV}{nkT}\right) - 1 \right] + \tag{5.45}$$
$$\frac{1}{2} \cdot \frac{qV_o N_e}{\tau_n} \left\{ \left[\frac{q}{nkT} (2\tau_n) \right] \exp\left(\frac{qV}{nkT}\right) \cdot \frac{dV}{dt} \right\}$$

Equation (5.44) is one half of the recombination current while (5.45) consists of the other half of the recombination current plus the storage capacitance that results from the time derivative of the electron density. The first part of (5.45) is obviously identical to (5.44). The second half of (5.45) is the current flow through the diffusion capacitance of diode D_2; as a review, the diffusion capacitance of a diode is given by

$$C_d = \frac{dQ_d}{dV} = \frac{q}{nkT} \tau_d I_s \exp\left(\frac{qV}{nkT}\right) \tag{5.46}$$

while the current through this diffusion capacitance is given by

$$I_d = C_d \frac{dV}{dt} = \left[\frac{q}{nkT} \tau_d I_s \exp\left(\frac{qV}{nkT}\right) \right] \cdot \frac{dV}{dt} \tag{5.47}$$

Thus, (5.45) can be modeled as a diode with an associated diffusion capacitance described by a transit time $2\tau_n$. The reason for splitting up the recombination current ($I_{recombination}$) into these two terms (I_{D1} and I_{D2}) will become apparent in the next section.

5.2.3 Equivalent subcircuit implementation

Once transformed into a variable space that has only one solution, the rate equations can be solved very simply through standard circuit techniques. There is no need to access simulator internals to monitor intermediate stages in the Newton-Raphson iteration, as Gao did. Thus, in [5.17], Javro shows that it is possible to implement the transformed laser model directly into a SPICE subcircuit. That is, it is possible to compose the laser diode model at the input deck level, rather than embedding it into the SPICE source code as Xiang did in Section 5.1. The subcircuit used by Javro in [5.17] is depicted in Figure 5.16. Here, the two diodes represented by (5.44) and (5.45) are clearly depicted, as are four nonlinear dependent sources B_1, B_2, B_3, and B_{ph}. The current-voltage relationships described by these sources are given by

$$I_{B1} = [(g_0 \vartheta \tau_n) 2 I_{D1} - q V_0 g_0 \vartheta (N_0 - N_e)] [1 - \varepsilon \vartheta (m + \delta)^2] (m + \delta)^2 \quad (5.48)$$

$$I_{B2} = \left[\left(\frac{\Gamma g_0 \tau_n \tau_p}{q V_0} \right) 2 I_{D1} - \Gamma g_0 \tau_p (N_0 - N_e) \right] [1 - \varepsilon \vartheta (m + \delta)^2] (m + \delta) \quad (5.49)$$

$$I_{B3} = \left[\left(\frac{\Gamma \beta \tau_p}{q V_0 \vartheta} \right) 2 I_{D1} + \left(\frac{\Gamma \beta \tau_p N_e}{\vartheta \tau_n} \right) \right] (m + \delta)^{-1} - \delta \quad (5.50)$$

$$V_{ph} = (m + \delta)^2 \quad (5.51)$$

Thus, the reason for splitting the spontaneous emission current into two components was to be able to replace the numerous occurrences of the recombination current (5.41) by twice the current through diode D_1.

The values of the other components are given by

$$C_{ph} = 2 \tau_p \quad (5.52)$$

$$R_{ph} = 1 \quad (5.53)$$

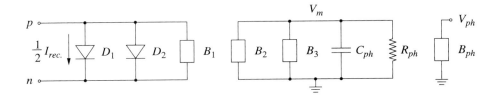

Figure 5.16: Javro's laser diode equivalent circuit.

If each of the terms in (5.44), (5.45), and (5.48) – (5.53) is interpreted as an element through which current flows and across which there is a potential difference, the combination of these expressions through Kirchhoff's laws will result in the rate equations (5.39) and (5.40).

Since all of the elements described by (5.44), (5.45), (5.48) – (5.53) are standard elements available in the SPICE input deck, the topology shown in Figure 5.16 is easily implementable as a SPICE input deck subcircuit. However, the element equations given by (5.44), (5.45), and (5.48) – (5.53) contain analytical forms of the rate equation parameters. Standard SPICE subcircuit syntax requires the expressions to be numerical, rather than analytical. Thus, in [5.16], Javro constructs a simple preprocessor that takes, as input, a list of rate equation parameters, evaluates (5.44), (5.45), and (5.48) – (5.53) using these parameters and produces the SPICE subcircuit with the right hand sides of (5.44), (5.45), and (5.48) – (5.53) expressed numerically, rather than analytically. An input file containing the values of Table 5.5 is passed to Javro's preprocessor, resulting in the subcircuit depicted in Figure 5.17. Figures 5.18 and 5.19 illustrate DC L-I and transient pulse response characteristics that result from using this subcircuit.

5.3 Summary

In this chapter two alternate methods for circuit-level optoelectronic simulation were investigated. The first example depicted the traditional approach of embedding the device model directly into the SPICE source code and recompiling. This was essentially the only method available for new model implementation before the

```
.subckt  Laser    p    n    pf
Vd1   nd1 n    0
D1    p    nd1 dspon1_Laser
D2    p    n    dspon2_Laser
B1    p    n    I=(6651.369*I(Vd1)-19.17989)*V(pf)*
                 (1-12.5637*V(pf))
B2    0    m    I=(549.3134*I(Vd1)-1.584)*(V(m)+5e-8)*
                 (1-12.5637*V(pf))
B3    0    m    I=(6.606921e-05*I(Vd1)+8.589149e-15)/
                 (V(m)+5e-8)  -  5e-8
Cph  0    m    2e-12
Rph  0    m    1
Bout pf  0    V=(V(m)+5e-8)*(V(m)+5e-8)
.ends

.model   dspon1_Laser D (Is=1.300023e-10 N=2)
.model   dspon2_Laser D (Is=1.300023e-10 N=2 TT=6e-09)
```

Figure 5.17: SPICE subcircuit generated by Javro's preprocessor.

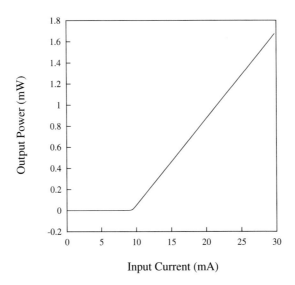

Figure 5.18: DC L-I characteristic using SPICE subcircuit.

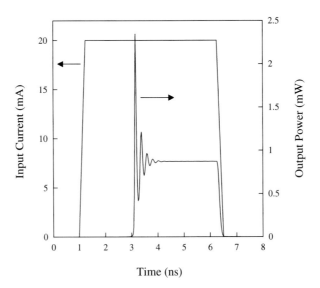

Figure 5.19: Transient pulse response using SPICE subcircuit.

advent of model-independent simulators, such as iSMILE. In the case of the MSM, however, use of a model-independent simulator was precluded by Xiang's use of numerical convolution to calculate the detector's transient response. In cases where access to the simulator internals is required, the model developer often has no recourse other than to modify the simulator source code. Xiang's new MSM *model* was emphasized in this chapter, the actual details of embedding models in SPICE being left to standard references such as [5.22] and [5.23]. By including mechanisms such as Frenkel-Poole and trap-assisted tunnel emission currents, Xiang included alternative dark current formulations to augment the traditional Schottky thermionic emission model. By incorporating S parameter extraction and impulse response convolution, Xiang was able to fit his MSM model to measured laboratory data over the DC, AC, and transient simulation regimes.

In the second half of this chapter, a novel approach to modeling laser diodes was discussed. The method of Javro et al. involved a mapping of the linearized rate equations into a solution space that had only one numerical solution. This allowed the model to avoid problems associated with the multiple numerical solution regimes of the standard model. Javro showed in [5.16] that the standard model actually contains three solution regimes: the desired regime, a negative-power solution, and a high-power region. By prudent variable mapping, Javro was able to eliminate the negative- and high-power solutions. The presence of the negative-power regime is one of the main reasons that previous equivalent-circuit implementations of laser diode models could not be entered at the input-deck level. Since Javro's model contains only one solution regime, it is easily implemented in a SPICE subcircuit, a fact which allows the model developer to avoid the complex task of model embedding. Since SPICE subcircuits require numerical values, rather than analytical expressions, for the circuit elements, Javro's preprocessor was also depicted in this chapter. This tool simply converts a set of rate-equation input parameters into the appropriate laser diode SPICE subcircuit.

While there are undoubtedly alternate methods of implementing new optoelectronic equivalent-circuit models (including the use of commercial tools), the methods detailed thus far are representative of the most common techniques in use. In Chapters 3–5, both conventional and novel approaches to the problem were presented.

5.4 References

[5.1]   A. Xiang, W. Wohlmuth, P. Fay, S. M. Kang, and I. Adesida, "Modeling of InGaAs MSM photodetector for circuit-level simulation," *IEEE Journal of Lightwave Technology*, vol. 14, no. 5, pp. 716-723, 1996.

[5.2]   S. M. Sze, D. J. Coleman, and A. Loya, "Current transport in metal-semiconductor-metal (MSM) structures," *Solid State Electronics*, vol. 14, pp. 1209-1218, 1978.

[5.3]   D. Kuhl, K. Hieronymi, E. H. Böttcher, T. Wolf, D. Bimberg, J. Kuhl, and M. Klingenstein, "Influence of space charges on the impulse response of InGaAs metal-semiconductor-metal photodetectors," *IEEE Journal of Lightwave Technology*, vol. 10, no. 6, pp. 753-759, 1992.

[5.4]   J. H. Burroughes and M. Hargis, "1.3 μm InGaAs MSM photodetector with abrupt InGaAs/AlInAs interface," *IEEE Photonics Technology Letters*, vol. 3, no. 6, pp. 532-534, 1991.

[5.5]   S. M. Sze, *Physics of Semiconductor Devices*. New York: John Wiley & Sons, Inc., 1981.

[5.6]   J. Frenkel, "On pre-breakdown phenomena in insulators and electronic semiconductors," *Physical Review*, vol. 54, p. 647, 1938.

[5.7]   S. M. Sze, "Current transport and maximum dielectric strength of silicon nitride films," *Journal of Applied Physics*, vol. 38, no. 7, pp. 2951-2956, 1967.

[5.8]   T. E. Chang, C. Huang, and T. Wang, "Mechanisms of interface trap-induced drain leakage current in off-state n-MOSFETs," *IEEE Transactions on Electron Devices*, vol. 42, no. 4, pp. 738-743, 1995.

[5.9]   M. Tong and A. Ketterson, Private Communication, University of Illinois at Urbana-Champaign, 1990.

[5.10] Y. C. Lim and R. A. Moore, "Properties of alternately charged coplanar parallel strips by conformal mappings," *IEEE Transactions on Electron Devices*, vol. ED-15, no. 3, pp. 173-180, 1968.

[5.11] C. S. Chang and H. P. Fetterman, "Electron drift velocity versus electric field in GaAs," *Solid State Electronics*, vol. 29, pp. 1295-1296, 1986.

[5.12] E. H. Böttcher, D. Kuhl, F. Hieronymi, E. Dröge, T. Wolf, and D. Bimberg, "Ultrafast semi-insulating InP:Fe-InGaAs:Fe-InP:Fe MSM photodetectors: Modeling and performance," *IEEE Journal of Quantum Electronics*, vol. 28, no. 10, pp. 2343-2357, 1992.

[5.13] W. L. Engl, *Process and Device Modeling*. Elsevier, Amsterdam: 1986.

[5.14] G. Lucovsky, R. F. Schwarz, and R. B. Emmons, "Transit-time considerations in p-i-n diodes," *Journal of Applied Physics*, vol. 35, no. 3, pp. 622-628, 1964.

[5.15] A. F. Salem, A. W. Smith, and K. F. Brennan, "Theoretical studies on the effect of an AlGaAs double heterostructure on metal-semiconductor-metal photodetector performance," *IEEE Transactions on Electron Devices*," vol. 52, no. 7, pp. 1112-1119, 1994.

[5.16] S. A. Javro and S. M. Kang, "Transforming Tucker's linearized laser rate equations to a form that has a single solution regime," *IEEE Journal of Lightwave Technology*, vol. 13, no. 9, pp. 1899-1904, 1995.

[5.17] S. Javro, "Modeling the laser," Internal Memorandum, University of Illinois at Urbana-Champaign, 1994.

[5.18] R. S. Tucker and D. J. Pope, "Large signal circuit model for simulation of injection laser modulation dynamics," *IEE Proceedings,* vol. 128, no. 5, pt. I, pp. 180-184, 1981.

[5.19] H. J. A. daSilva, R. S. Fyath, and J. J. O'Reilly, "Sensitivity degradation with laser wavelength chirp for direct detection optical receivers," *IEE Proceedings*, vol. 136, no. 4, pt. J, pp. 209-218, 1989.

[5.20] K. Hansen and A. Schlachetzki, "Transferred-electron device as a large-signal laser driver," *IEEE Journal of Quantum Electronics*, vol. 27, no. 3, pp. 423-427, 1991.

[5.21] J. Vlach and K. Singhal, *Computer Methods for Circuit Analysis and Design.* New York: Van Nostrand Reinhold, 1983.

[5.22] G. Massobrio and P. Antognetti, *Semiconductor Device Modeling with SPICE.* New York: McGraw-Hill Publishing Co., 1993.

[5.23] T. I. Quarles, "The SPICE3 Implementation Guide," Electronics Research Laboratory Report, no. ERL-M44, University of California at Berkeley, 1989.

6

DEVICE AND MIXED-MODE SIMULATION

Contributed by B. Onat and M. S. Ünlü
Department of Electrical, Computer and Systems Engineering and
Center for Photonics Research
Boston University

*I*t was emphasized, in previous chapters, that the majority of existing optoelectronic simulation work has been centered on the very detailed, device-level simulation of individual structures. It has been further emphasized that the focus of this book is on circuit- and higher-level methods of simulation, as a means for designing *integrated* optoelectronic circuits and systems. However, situations often arise in which the increased detail afforded by device simulation is necessary. In these cases, it is desirable to use a *general-purpose* device simulator, capable of simulating a variety of structures, rather than a specialized tool designed to model only one particular device. Furthermore, it is also desirable to retain a degree of connection between the device simulator and the circuit simulator so that integrated systems can still be designed. That is, a mixed-mode (device/circuit) simulation environment is ideal.

In this chapter, contributed by B. Onat and M. S. Ünlü of Boston University, exactly such a tool is described. While the goal of this book is to concentrate on optoelectronic simulation methodologies *apart* from traditional device simulation techniques, this chapter was included because of its mixed-mode nature, its applicability to general optoelectronic devices, and its implementation in a standard, widely used CAD tool, SPICE.

6.1 Introduction

Improvements in performance often come at the expense of novel devices with increased complexity; however, actual device and integrated circuit fabrication is expensive and time consuming. For such structures, detailed analysis of complex phenomena, not possible through experiments or simple analytic models, can be achieved through device-level simulations. Such approaches also provide valuable insight into the physical mechanisms which ultimately determine the operational performance of the device. Most existing device simulators are based on the drift-diffusion (DD) description of carrier transport and employ specialized techniques to obtain the one-, two-, or three-dimensional transient and steady-state solutions.

The simulation of optoelectronic devices, however, presents some specific challenges that are not well addressed by existing device simulation tools. One important requirement is that optical effects, such as absorption and photo-generation, be explicitly representable at any local point within the device. The simulation tool should enable simple modeling of optoelectronic devices which may have localized optical absorption in specific regions and should be capable of simulating fast transients. The tool should also allow simple integration of circuit-level and device-level simulation tasks. Finally, simple extraction of device and circuit model parameters is desirable to aid in the development of compact simulation models for novel optoelectronic devices.

There exists a growing consensus that process, device, and circuit simulation tools, which are currently used as separate entities, need to be integrated into a combined simulation environment for the efficient design and optimization of semiconductor devices and integrated circuits. Such integration will shorten the development cycle for new technologies by providing accurate predictions of various physical phenomena on the device and circuit level, and by allowing the investigation of detailed device behavior within the circuit environment. The majority of the proposed tool integration approaches seek either to create specific mixed-mode simulation environments, or to combine specialized device and circuit simulation tools through user-friendly interfacing environments.

In this chapter, a novel approach is presented for incorporating the transient solution of the one-dimensional semiconductor drift-diffusion equations *within a general circuit simulation tool* [6.8], eliminating the need for interfacing various device and circuit simulation environments. This approach allows the simple representation of localized carrier transport models using equivalent circuit elements such as voltage-controlled current sources and capacitors. The availability of local photo-generation models at every grid point enables accurate transient simulation of various optoelectronic devices without any restrictions on light absorption and carrier generation characteristics. The approach is conceptually simple and easy to implement in many existing circuit simulation environments with only a moderate amount of additional effort. Besides the inherent advantage of providing mixed-mode device and circuit simulation capability embedded in a circuit simulator, this approach can also be used for extracting the device- and circuit-level model parameters.

In Section 6.2, background information about conventional mixed-mode simulation techniques and earlier work on the circuit representation of semiconductor devices is given. Sections 6.3 and 6.4 present the discretization of the time-dependent semiconductor equations, a detailed derivation of model parameters, and the implementation of the drift-diffusion model using equivalent-circuit elements. The application of the new approach is examined in Section 6.5 with several device and mixed-mode simulation examples.

6.2 Background

While some circuit-level models have been developed for optoelectronic devices, a greater degree of accuracy is often desired than that which can be accomplished with such models. This is especially true under transient, large-signal conditions. On the other hand, various well-established circuit models exist for the simulation of many electronic devices. Therefore, to best model optical devices operating within an electronic circuit environment, mixed-mode device/circuit models are necessary. For example, in a photoreceiver, an electronic circuit is integrated with a photodetector. When the photodetector is illuminated by an optical pulse, the photocurrent response is detected and processed by the circuit's electronics. In addition to processing the signal, the electronic circuit is also responsible for setting the electrical bias point of the optoelectronic device. When large signals are involved, the bias points of both the electrical and the optical devices can vary with time. Thus, a mixed-mode device/circuit simulator has the responsibility of accurately modeling the device response, changing bias conditions, and the effect of these bias point changes on the device characteristics. This situation is depicted in Figure 6.1.

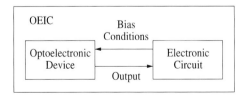

Figure 6.1: Block diagram showing device-circuit interaction.

6.2.1 Conventional mixed-mode simulation

Many different mixed-mode simulation techniques have been demonstrated [6.1], [6.13], [6.16]. In these conventional approaches, two separate simulation tools are used, both installed on one platform and running simultaneously (Figure 6.2). Each simulator is dedicated to either the circuit simulation or the device simulation portion of the mixed-mode integrated circuit. These independent simulators are interfaced via a third simulation tool dedicated to controlling the mixed-mode simulation and ensuring its self-consistency.

For a given set of bias conditions, each of the circuit and device simulation tools solves for its variables locally. Once a steady-state solution is obtained, the common voltage and current variables are compared by the tool interface. Local variables are iteratively solved for until the common variables arrive at a consistent solution for both simulators. Thus, long simulation times can often be a serious setback for conventional mixed-mode simulators. Since large-signal input values can alter device bias points, convergence under these conditions is nontrivial and requires extensive communication between the independent simulators and the tool interface before self consistency is achieved.

6.2.2 Circuit representation of devices

A solution to the mixed-mode device/circuit simulation problem would be to integrate the device simulation directly into a circuit simulator. Circuit simulators have been in existence for many years [6.5] and have evolved into quite powerful tools. Circuit representation of semiconductor devices, however, is not new, and has been implemented in the past. Sah et al. represented semiconductor devices in terms of lumped elements in a method known as the *Circuit Technique for Semiconductor*

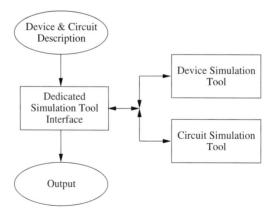

Figure 6.2: Conventional mixed-mode simulation environment.

Analysis (CTSA) [6.2], [6.14]. This method featured analytical as well as numerical solutions to the five main semiconductor equations: the electron and hole continuity equations, the electron and hole current equations, and Poisson's equation. Also incorporated was a sixth equation, the trapping-charge kinetic equation [6.14]. Boundary conditions were applied only at points representing external contacts. The spatial variation of impurity doping profiles and the dependence of mobility on dopant impurity concentration were included, as well. The recombination terms in the continuity equations were modeled by multiple-energy level Shockley-Read-Hall recombination centers.

In the CTSA model, all physical phenomena are represented by conventional circuit elements via an exact transmission-line circuit model (Figure 6.3) [6.2]. The circuit shown in this figure represents the steady-state behavior of a one-dimensional device. The two series branches in this configuration represent the electron and hole conduction currents, while the center node in each section models the electrostatic potential for the grid point. The resistors and capacitors in this figure are not only nonlinear but have values that vary with both time and space, as well. The presence of the nonlinear bridging capacitors at every grid node significantly complicates simulation. The small-signal circuit representation of the CTSA model is even more complex. In this case, an additional voltage-controlled current source, proportional to the generation-recombination rate, is connected in parallel with each resistor shown in Figure 6.3. Implementation of this complex structure, along with the bridging capacitors between each node, into a circuit simulator is, indeed, challenging.

The model described in this chapter is a simple circuit representation of semiconductor optoelectronic devices, in which the bridging capacitors and the nonlinear circuit elements between adjacent grid nodes are eliminated. Thus, the complexity of the circuit decreases, allowing the device model to be incorporated *effectively* into a circuit-simulation environment.

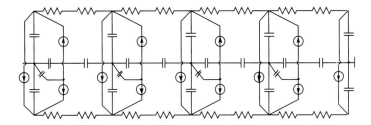

Figure 6.3: Classical representation of semiconductor equations.

6.3 Discretization of Semiconductor Equations

Carrier transport in semiconductors can be described by three coupled partial differential equations, namely, Poisson's equation, the electron and hole current equations, and the electron and hole continuity equations. Assuming that the carrier flow and the potential variation within a semiconductor structure are confined to one dimension x, the normalized carrier transport equations are given by (6.1) – (6.3).

$$\frac{\partial^2 V}{\partial x^2} - \frac{1}{r}(n - p + N_A - N_D) = 0 \tag{6.1}$$

$$\frac{\partial n}{\partial t} = \frac{1}{q} \cdot \frac{\partial J_n}{\partial x} - R_n + G_n \tag{6.2}$$

$$\frac{\partial p}{\partial t} = -\frac{1}{q} \cdot \frac{\partial J_p}{\partial x} - R_p + G_p \tag{6.3}$$

The normalized electron and hole current densities J_n and J_p are defined as

$$J_n = \mu_n \cdot \frac{\partial n}{\partial x} - n\mu_n \cdot \frac{\partial (V + V_n)}{\partial x} \tag{6.4}$$

$$J_p = -\mu_p \cdot \frac{\partial p}{\partial x} - p\mu_p \cdot \frac{\partial (V - V_p)}{\partial x} \tag{6.5}$$

In these equations, the normalized variables V, n, p, N_D, and N_A represent the local electrostatic potential and the electron, hole, donor, and acceptor concentrations, respectively, and q is the electron charge. The parameter r represents the normalized dielectric constant and R and G represent carrier recombination and generation rates, respectively. (A discussion about the normalization is contained in the appendix). Two potentials V_n and V_p are also introduced as band parameters to account for the heterostructure bandgap variations, and are defined in (6.6) and (6.7).

$$qV_n = \chi - \chi_r + kT\ln\left(\frac{N_C}{N_{Cr}}\right) \tag{6.6}$$

$$qV_p = -\left[\chi - \chi_r + E_G - E_{Gr} - kT\ln\left(\frac{N_V}{N_{Vr}}\right)\right] \tag{6.7}$$

In these equations, N_C and N_V are the effective density of states in the conduction and valence bands, respectively, χ is the electron affinity, E_G is the bandgap, and the subscript r denotes the reference material, for example GaAs or AlGaAs.

To obtain the solution of the semiconductor equations (6.1) – (6.3) by using the finite difference method, the equations need to be discretized on a nonuniform one-dimensional grid (Figure 6.4). Usually, the discretization of the time-dependent semiconductor equations is carried out in both the time domain and space domain [6.17]. In the following, the dependent variables will be discretized with respect to the space (position) variable x, but they will be left continuous in time, resulting in a semi-discrete form. The goal of this formulation is to facilitate a simple implementation of the device model within the circuit simulation environment. Three new local variables are defined as $x_1 = V(j)$, $x_2(j) = V(j) - \psi_n(j)$, and $x_3(j) = -V(j) + \psi_p(j)$, where the normalized variables V, ψ_n, and ψ_p represent the local electrostatic potential, and the electron and hole quasi-Fermi potentials, respectively. The Scharfetter-Gummel approximation has been used to express the current densities J_n and J_p in terms of the carrier concentrations [6.9]:

$$J_n\left(j+\frac{1}{2}\right) = \frac{\mu_{n(j+\frac{1}{2})}\Delta_n(j)}{h_{j+1}}\left[\frac{n(j+1)\exp(\Delta_n(j)) - n(j)}{\exp(\Delta_n(j)) - 1}\right] \tag{6.8}$$

$$J_p\left(j+\frac{1}{2}\right) = -\frac{\mu_{p(j+\frac{1}{2})}\Delta_p(j)}{h_{j+1}}\left[\frac{p(j+1)\exp(\Delta_p(j)) - p(j)}{\exp(\Delta_p(j)) - 1}\right] \tag{6.9}$$

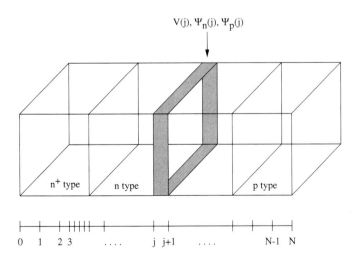

Figure 6.4: Partitioning of a p-n-n^+ structure on a nonuniform, one-dimensional grid.

In (6.8) and (6.9), the following quantities are defined:

$$\Delta_n(j) = V(j) + V_n(j) - V(j+1) - V_n(j+1) \tag{6.10}$$

$$\Delta_p(j) = V(j+1) - V_p(j+1) - V(j) + V_p(j) \tag{6.11}$$

The grid spacing at the jth grid point is defined as h_j. The carrier concentrations n and p are related to the local potential variables as

$$\begin{aligned} n(j) &= \exp[V(j) + V_n(j) - \psi_n(j)] \\ &= \exp[x_2(j) + V_n(j)] \end{aligned} \tag{6.12}$$

$$\begin{aligned} p(j) &= \exp[-V(j) + V_p(j) + \psi_p(j)] \\ &= \exp[x_3(j) + V_p(j)] \end{aligned} \tag{6.13}$$

Substituting (6.8) – (6.13) into the device equations (6.1) – (6.3), a set of time-dependent space-discretized equations are obtained on a nonuniform one-dimensional grid (6.14) – (6.16).

$$\begin{aligned} x_1(j) &= \frac{h_j h_{j+1}}{r_{j+\frac{1}{2}} h_j + r_{j-\frac{1}{2}} h_{j+1}} \cdot \frac{h_j + h_{j+1}}{2} \\ &\quad \cdot [\exp(x_3(j) + V_p(j)) - \exp(x_2(j) + V_n(j)) + C_j] \\ &\quad + \frac{r_{j+\frac{1}{2}} h_j x_1(j+1) + r_{j-\frac{1}{2}} h_{j+1} x_1(j-1)}{r_{j+\frac{1}{2}} h_j + r_{j-\frac{1}{2}} h_{j+1}} \end{aligned} \tag{6.14}$$

Here, (6.14) corresponds to the discretized Poisson equation at the jth grid node, and (6.15) and (6.16) correspond to the electron and hole continuity equations, respectively. Introduction of the band gap parameters V_n and V_p allows simple representation of the band gap variation effects in heterostructures [6.9]. Notice that a photon-induced carrier generation term G_j is available at every grid point x_j; this term can be modeled by an independent current source.

The extension of this approach to two dimensions is straightforward. Assuming a rectangular space-grid structure, the Poisson equation and the electron and hole continuity equations can easily be discretized at each grid point. Each resulting discretized equation contains the electrostatic potential and Fermi potential variables of the four (instead of two) neighboring grid points. Thus, the resulting set of time-continuous and space-discretized equations will again be characterized by next-neighbor relations in the space domain.

$$\frac{\partial x_2(j)}{\partial t} = \frac{2\mu_{n(j+\frac{1}{2})}}{(h_j + h_{j+1})h_{j+1}} [x_1(j) - x_1(j+1) + V_n(j) - V_n(j+1)]$$

$$\cdot \left\{ \frac{\exp[x_2(j+1) - x_2(j) + x_1(j) - x_1(j+1)] - 1}{\exp[x_1(j) - x_1(j+1) + V_n(j) - V_n(j+1)] - 1} \right\}$$

$$+ \frac{2\mu_{n(j-\frac{1}{2})}}{(h_j + h_{j+1})h_j} [x_1(j) - x_1(j-1) + V_n(j) - V_n(j-1)] \qquad (6.15)$$

$$\cdot \left\{ \frac{\exp[x_2(j-1) - x_2(j) + x_1(j) - x_1(j-1)] - 1}{\exp[x_1(j) - x_1(j-1) + V_n(j) - V_n(j-1)] - 1} \right\}$$

$$- \frac{R_j - G_j}{\exp[x_2(j) + V_n(j)]}$$

$$\frac{\partial x_3(j)}{\partial t} = \frac{2\mu_{p(j+\frac{1}{2})}}{(h_j + h_{j+1})h_{j+1}} [x_1(j+1) - x_1(j) + V_p(j) - V_p(j+1)]$$

$$\cdot \left\{ \frac{\exp[x_3(j+1) - x_3(j) + x_1(j+1) - x_1(j)] - 1}{\exp[x_1(j+1) - x_1(j) + V_p(j) - V_p(j+1)] - 1} \right\}$$

$$+ \frac{2\mu_{p(j-\frac{1}{2})}}{(h_j + h_{j+1})h_j} [x_1(j-1) - x_1(j) + V_p(j) - V_p(j-1)] \qquad (6.16)$$

$$\cdot \left\{ \frac{\exp[x_3(j-1) - x_3(j) + x_1(j-1) - x_1(j)] - 1}{\exp[x_1(j-1) - x_1(j) + V_p(j) - V_p(j-1)] - 1} \right\}$$

$$- \frac{R_j - G_j}{\exp[x_3(j) + V_p(j)]}$$

6.3.1 Recombination models

The carrier generation/recombination rate at any grid point is assumed to be a function of two basic mechanisms, that is, trap-related phonon transitions and photon transitions. The four partial processes involved in phonon transitions are

(1) Electron capture from the conduction band by an unoccupied trap
(2) Hole capture, that is, release of an electron from an occupied trap to the valence band
(3) Electron emission from an occupied trap to the conduction band
(4) Hole emission, that is, electron capture from the valence band by an unoccupied trap

The well-known Shockley-Read-Hall model accounts for the total generation/recombination rate due to these phonon transitions [6.12]. These mechanisms can be represented as

$$R_j = \frac{n(j)\,p(j) - n_i^2}{\tau_p\,[n(j) + n_1] + \tau_n\,[p(j) + p_1]} \tag{6.17}$$

where $n(j)$ and $p(j)$ represent the electron and hole concentrations at the jth grid point, and τ_n and τ_p represent the carrier lifetimes. The recombination term R_j can also be expressed in terms of the electrostatic potential and the Fermi potentials at the jth grid point [6.17]

$$R_j = \frac{\exp[x_2(j) + V_n(j)]\,\exp[x_3(j) + V_p(j)] - 1}{\tau_p\{\exp[x_2(j) + V_n(j)] + n_1\} + \tau_n\{\exp[x_3(j) + V_p(j)] + p_1\}} \tag{6.18}$$

In the model presented in this chapter, (6.18) has been utilized to describe recombination in semiconductor devices. Both of the carrier lifetimes were assumed to be 1 ns.

The photon-induced transitions, on the other hand, are assumed to be exclusively in the form of optical *generation* of carriers, the phenomenon in which an electron gains energy from incident photons and moves from the valence band to the conduction band. This process is represented by the photon-induced carrier generation term G_j at each grid point.

6.3.2 Mobility models

In (6.14) – (6.16), the electron and hole mobilities have been modeled considering variations with static variables, such as doping and alloy composition, as well as dynamic dependence on the local electric field. The field dependence is taken into account by a suitable local mobility model based on carrier velocity saturation. The model presented in this chapter utilizes approximations of these effects, derived for the GaAs materials system [6.9], which are suitable for numerical simulation. To approximate the hole mobility for an arbitrary doping and composition in GaAs, an empirical formula, $\mu_{p\text{-}GaAs}$ is used (6.19). In this equation, E_{Qp} is the effective field for holes, readily calculated by the model, $\mu_{p\text{-}L}$ is the low-field and intrinsic mobility for the material, and T is the temperature. In (6.19), the field and temperature-dependent portions are, due to lack of more accurate information, assumed to be the same for those of p-type silicon.

$$\mu_{p-GaAs}(T, N_D + N_A, E_{Qp}) = \frac{\mu_{p-L}}{[1 + 3.17 \times 10^{-17}(N_D + N_A)]^{0.266}}$$

$$\cdot \frac{1}{\left[1 + \frac{|E_{Qp}|}{1.95 \times 10^4}\right]} \cdot \left(\frac{300}{T}\right)^{2.7} \quad \frac{cm^2}{V-s} \tag{6.19}$$

The absence of a well-structured experimental characterization of the electron mobility, as a function of composition, carrier compensation, doping, and applied field, makes the development of an accurate model difficult. An electron mobility model developed in [6.9] has explicitly considered only electrons which occupy the Γ and X valleys. The electron mobility in GaAs is taken as

$$\mu_{n-GaAs}(T, N_D + N_A, E_{Qp}) = \frac{\mu_{LD} + v_{sat}|E_{Qn}|^3/E_{C-GaAs}^4}{\left(1 + \left[\frac{|E_{Qn}|}{E_{C-GaAs}}\right]^4\right)} \tag{6.20}$$

where the low-field doping concentration-dependent mobility is

$$\mu_{LD} = \frac{\mu_{n-L}}{[1 + 5.51 \times 10^{-17}(N_D + N_A)]^{0.233}} \cdot \left(\frac{300}{T}\right)^{2.3} \quad \frac{cm^2}{V-s} \tag{6.21}$$

the saturation velocity is

$$v_{sat} = (1.28 - 0.0015T) \times 10^7 \quad \frac{cm}{s} \tag{6.22}$$

and the critical field is

$$E_{C-GaAs} = \left(5.4 - \frac{T}{215}\right) \quad \frac{kV}{cm} \tag{6.23}$$

In this model, GaAs is used as the reference system. Thus, all InGaAs and AlGaAs materials simulated are assumed to have the same electric field dependency as GaAs. For small mole fractions of Al and In, this is a valid approximation. The intrinsic/low-field mobilities μ_{LD} and μ_{p-L} are calculated for each grid to accurately characterize the material used.

In Figure 6.5, the dynamic behavior of the electron mobility due to the variation in the local electric field is illustrated for a photodiode under large-signal optical excitation. The mobility variations of three different grid points in a GaAs p-i-n photodiode are shown. Figure 6.5 indicates the importance of incorporating

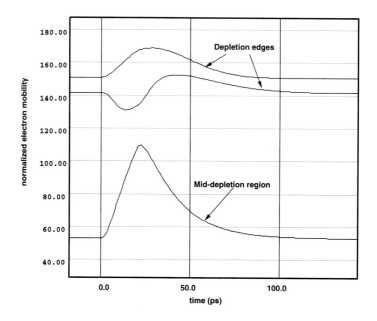

Figure 6.5: Transient behavior of the electron mobility at three different grid points due to variation of the electric field in a photodiode.

the field dependence, especially for investigating the transient effects of semiconductor devices.

6.4 Model Implementation

The semi-discrete carrier transport equations (6.14) – (6.16) describe the transient behavior of the electrostatic potential and the Fermi potentials at any grid point. These equations have the following general form:

$$x_1(j) = f_1[x_1(j+1), x_1(j-1), x_2(j), x_3(j), V_n(j), V_p(j)] \qquad (6.24)$$

$$\frac{\partial x_2(j)}{\partial t} = f_2[x_1(j), x_1(j+1), x_1(j-1),$$
$$x_2(j), x_2(j+1), x_2(j-1), x_3(j), \qquad (6.25)$$
$$G(j), V_p(j), V_n(j), V_n(j-1), V_n(j+1)]$$

$$\frac{\partial x_3(j)}{\partial t} = f_3[x_1(j), x_1(j+1), x_1(j-1),$$

$$x_3(j), x_3(j+1), x_3(j-1), x_2(j), \qquad (6.26)$$

$$G(j), V_n(j), V_p(j), V_p(j-1), V_p(j+1)]$$

It is seen that the electrostatic and Fermi potentials at any grid point are described by three coupled nonlinear differential equations in the time domain which involve the corresponding potential variables as well as the neighboring grid node potentials. With a simple variable assignment, (6.24) – (6.26) can be represented by an equivalent circuit, which can then be solved using a conventional transient circuit simulation program, such as SPICE. [This is analogous to the mapping of optical variables in Chapter 3 to node voltages and, again, exploits the fact that a circuit simulator is essentially an arbitrary differential equation solver]. For a simple equivalent-circuit interpretation of these equations, first, let the variables x_1, x_2, and x_3 represent three node voltages. Also, let the right-hand sides of (6.24) – (6.26), that is, the functions f_1, f_2, and f_3, represent the currents of three nonlinear voltage-controlled current sources. Then, (6.25) and (6.26) can be interpreted as the state equations of linear capacitors with terminal currents of f_2 and f_3, respectively, while (6.24) describes the terminal voltage of a unit resistor with a terminal current of f_1. The equivalent-circuit diagrams corresponding to (6.14) – (6.16) are given in Figure 6.6. It is seen that each space-grid point is represented by three circuit nodes, where the node potentials correspond to the electrostatic potential and the two Fermi potentials associated with the grid point.

The conversion of the semiconductor drift-diffusion equations into an equivalent-circuit representation was first proposed by Sah [6.14], as described previously, and later pursued by others [6.2], [6.4]. In this classical transmission-line representation of the one-dimensional semiconductor equations, the equivalent circuit consists of a ladder-like structure of connected nonlinear capacitors and current sources. The electrical coupling of adjacent grid nodes is accomplished by capacitive current components in the equivalent transmission-line circuit, as seen in Figure 6.3. The presence of bridging, nonlinear capacitors between nodes, however, results in a strongly connected network topology, which may ultimately complicate the numerical solution. In the modeling approach presented here, in contrast, the local circuit nodes representing grid points are linked through node voltages and dependent sources only. This new formulation eliminates the bridging capacitors and the nonlinear circuit elements used in previous circuit element representations. The modularity of the new modeling approach can also be used to adjust the grid point density of the simulated structure [6.7]. The boundary conditions are established assuming purely ohmic contacts at both external boundaries. The electrostatic potential at the boundary is determined by the externally applied voltage, which, in turn, depends on the external load circuit. It is assumed that no

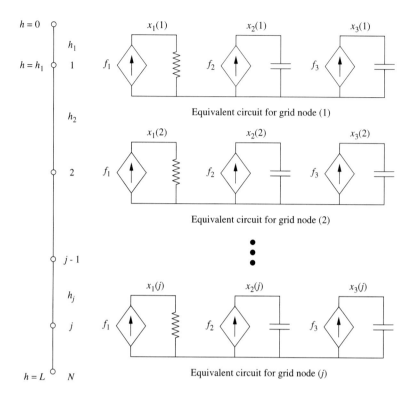

Figure 6.6: The one-dimensional finite difference grid and the equivalent-circuit representation of the variables associated with each grid point. All resistors and capacitors are taken as unity.

voltage drop occurs at the boundary. The conditions for the carrier concentrations at the ohmic boundaries are established by increasing the recombination rate. This is accomplished by gradually reducing the electron and hole lifetimes in the Shockley-Read-Hall recombination term by three orders of magnitude for the grid points in the immediate neighborhood of the boundaries.

The calculation of currents is also carried out using the equivalent-circuit representation. The total conduction current component across the device is found as the sum of the electron and hole currents J_n and J_p.

$$J_c = J_n(x_j) + J_p(x_j) \tag{6.27}$$

The displacement current component is calculated as

$$J_d = -\frac{1}{A}C_{junction} \cdot \frac{d}{dt}V_{space-charge} \tag{6.28}$$

where $C_{junction}$ represents the junction capacitance, and the space-charge voltage is found by integrating the local carrier densities. The displacement current component for a given grid node can be computed directly by taking the time derivative of the local electric field

$$J_d = -\varepsilon \cdot \frac{dE}{dt} \tag{6.29}$$

where ε is the dielectric constant. In the presented model, the electric field is obtained by taking the difference in electrostatic potential at the neighboring nodes and time differentiation is implemented by a simple circuit involving a voltage source and a capacitor. At every time step, therefore, all the currents, as well as the total current, are available.

Alternatively, the displacement current can also be evaluated by integrating the time derivative of carrier concentrations in the depletion region:

$$J_d = \frac{q}{\varepsilon A} C_{junction} \int_0^{x_m} dy \int_0^y \left[\frac{\partial p(x,t)}{\partial t} - \frac{\partial n(x,t)}{\partial t} \right] dx \tag{6.30}$$

This formulation is particularly useful when the transient variation of electrostatic potential is difficult to evaluate. In the present model, the drift current is calculated directly via (6.29).

Given the device structure, grid spacing, and the relevant model parameters, the equivalent-circuit representation of the semiconductor drift-diffusion equations can thus be readily generated. A conventional circuit simulator is then used to solve these equations in the time domain and to obtain the variation of the electrostatic potential and the Fermi potentials at each grid point. The approach presented here also makes full use of the numerical stability and efficiency of well-established circuit-simulation routines.

The exponential terms which appear in the space-discretized Poisson equation (6.14) and the electron and hole continuity equations (6.15), (6.16) essentially possess the same characteristics as the compact diode model or the BJT Ebers-Moll model equations, which are handled quite well numerically in most conventional circuit simulators. The Bernoulli equation appearing in (6.15) and (6.16), which has the general form

$$f(x) = \frac{x}{\exp(x) - 1} \tag{6.31}$$

presents a challenge in numerical evaluation for very small values of x. This situation is encountered in the case of neutral regions because the numerator and the denominator both approach zero. However, since the function does not possess a singularity at $x = 0$, the problem can easily be circumvented by assigning the function its known limit, $f(0) = 1$, for values of x smaller than a certain lower boundary [6.7]. The availability of the absolute value function for nonlinear controlled sources in SPICE provides an alternative option in which (6.31) is modified to

$$f(x) = \frac{|x| + \delta}{|\exp(x) - 1| + \delta} \tag{6.32}$$

where δ is a small, positive number. The value of δ (typically $10^{-6} \sim 10^{-8}$) can be arbitrarily set to control the precision of the equations. A more detailed numerical error analysis is not presented here mainly because the error analysis results derived for conventional circuit simulation tools can be applied directly to this case.

The simple equivalent-circuit representation allows detailed modeling of the internal device structure by locally specifying the physical device parameters. This model allows the investigation of the transient behavior of optoelectronic devices using the local photo-generation term G_j [6.14], [6.20], [6.21].

It was previously mentioned that mixed device/circuit simulation capability using the same simulation tool is very desirable for accurate and realistic assessment of novel devices and integrated circuits. Since the external voltage and current conditions of the simulated device are available at any time point t, and since the discretized semiconductor equations are being solved using circuit simulation routines, the device simulation approach presented here lends itself directly to mixed-mode device/circuit simulation. This can be accomplished by using compact, circuit-level models for the electrical components in tandem with the device-level semiconductor equations, *all in the same simulation input file*. Figure 6.7 shows the major components of the combined device/circuit simulation environment.

To facilitate initial DC convergence during the simulation of some semiconductor structures, a simple ramping scheme is used. During preliminary studies [6.7], the entire structure is initially defined as having the same uniform doping density, with zero potential drop across its external terminals. DC convergence for this uniform structure can easily be achieved, since the local electrostatic potential and the electron and hole densities will be constant and equal at all grid nodes. Next, the local doping densities and the external boundary conditions are gradually changed (ramped) in the time domain, until they reach the actual levels intended for the semiconductor structure. At the point, the system is in steady-state (= DC), and the actual excitation can be applied as either an electrical or optical pulse waveform.

For implementation in SPICE [6.20], [6.21] DC convergence for a *p-n* junction can usually be achieved without any preconditioning or ramping for moderate doping densities and applied biases. Although there is no formula for the maximum allowable values, and they are usually interdependent, typical numbers for *p-n* junctions are on the order of 10^{17} cm^{-3} and 50 kT/q for doping densities and applied

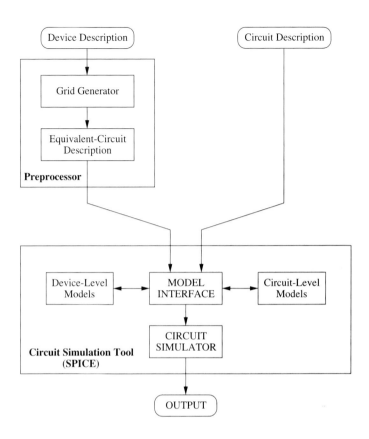

Figure 6.7: The mixed-mode device/circuit simulation environment.
No additional control programs or tools are required.

biases, respectively. In the event that DC convergence cannot be achieved, ramping
of one or more device parameters can also be utilized.

It was also found that abrupt changes in the electrical or optical excitation
waveforms occasionally caused numerical convergence problems during simulation.
To avoid such problems, a "precursor" pulse several orders of magnitude smaller
than the actual excitation pulse may be applied first. This small excitation ensures
that the carrier density levels rise sufficiently above the steady-state levels so that the
subsequent excitation pulse does not cause numerical problems.   Since the
magnitude of the precursor pulse is very small, it does not influence the transient
response of the device in any significant way. It was also noted that within SPICE, if
the rising edge of the optical pulse has a staircase structure of arbitrarily small time
step size, convergence is assured. In this scheme, pulses with very small rise time
(subpicosecond) can be accurately represented.

In order to achieve the circuit representation of a given device, seven numerical values of the device characteristics must be given. Once the number of grid points into which the device is partitioned is specified, the following must be specified:

(1) Grid spacing
(2) Dielectric constant
(3) Intrinsic/Low-field electron mobility
(4) Intrinsic/Low-field hole mobility
(5) Conduction-band discontinuity
(6) Valence-band discontinuity
(7) Intrinsic concentration

These values, for the simulations described in the next section, were obtained from [6.10].

For a 50-node device, inserting all of the 350 variables manually would be neither time efficient nor reliable. Thus, a C program has been developed that reads the device description from a data file, computes all of the coefficients of the device models, and creates a SPICE-ready input file. This is the preprocessor depicted in Figure 6.7, which changes the device description to a circuit representation. This preprocessor has been optimized to create as few nodes in the circuit file as possible and to set boundary and initial conditions. Therefore, the computation load is minimized and the convergence of SPICE is facilitated. The interface allows the device bias voltage, generated grid numbers, transient analysis specifications, and tolerances to be imported into the circuit file.

6.5 Simulation Examples

The physical phenomena incorporated in the device simulation approach will be demonstrated in this section through several examples. The examples presented in this section were implemented in SPICE [6.20], [6.21], version 3F2.

6.5.1 Carrier transport models

The first example investigates the drift, diffusion, and recombination of excess carriers generated by a localized light pulse in an n-type semiconductor sample. The semiconductor has a length of $L = 2$ μm and a doping concentration of $N_D = 10^{16}$ cm^{-3}. A voltage of 50 kT/q is applied across the sample, and a light pulse focused on the midpoint of the sample is used to generate excess carriers (Figure 6.8). Note that this arrangement corresponds to the well-known Haynes-Shockley experiment for the measurement of carrier drift mobility in semiconductors. Figure 6.9 shows the excess hole concentration within the sample as a function of position and time. It is observed that the generated excess carriers diffuse away from the midpoint and recombine, while the concentration peak moves (drifts) toward the negatively biased end of the sample under applied electric field. This example

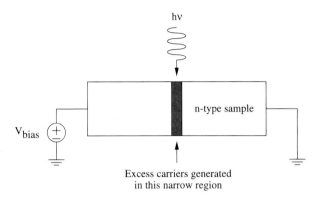

Figure 6.8: Uniform n-type semiconductor sample subjected to a localized optical pulse excitation, leading to excess carrier generation in a narrow region.

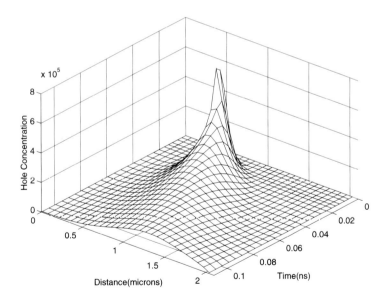

Figure 6.9: Simulated variation of hole concentration as a function of position and time.

qualitatively demonstrates that the equivalent-circuit models introduced previously accurately represent the transient drift, diffusion, and recombination behavior of carriers in a semiconductor sample.

6.5.2 GaAs p-i-n photodiode

This example investigates the drift and diffusion of excess carriers generated by a localized light pulse in a GaAs p-i-n diode structure (Figure 6.10). The total thickness of the device is $L = 1.5$ μm and the low-doped, normally depleted region is 1 μm thick. A total of 50 variably spaced grid points were taken, with a finer grid spacing at the metallurgical junctions. A small voltage of 10 kT/q was applied across the sample to reverse bias the diode, and a light pulse with 0.1 ps rise and fall times and a 50 ps duration was focused on the depletion region of the diode.

To demonstrate the concept, uniform optical generation is assumed to take place in a 0.5-μm-thick region located in the middle of the depletion region. Assuming a 40% quantum efficiency (typical for a 0.5-μm GaAs absorber), this optical excitation corresponds to 5×10^{21} photons/cm^{2}·s (i.e., 1 kW/cm^{2}). Although this power density seems to be extremely large, the corresponding total power is on the order of 10 mW for a 30 × 30 μm device, and for the given pulse duration, the total energy per pulse is only 50 fJ.

Figures 6.11 and 6.12 show the variation of the electron and hole concentrations, respectively, within the sample as a function of position and time. As can be seen from the carrier concentration plots, the optical generation is large enough to create electron-hole densities comparable to those in the neutral regions, demonstrating the large-signal capability of this model. A comparison of the transient variation of the electron and hole concentrations reveals that the electrons, due to higher mobility, traverse the depletion region much faster than the holes.

In the model, the field dependences of the material parameters were taken into account as dynamic variables, as mentioned previously. Transient consideration of

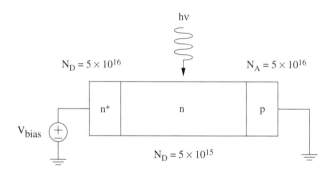

Figure 6.10: The simulated GaAs p-i-n photodiode structure.

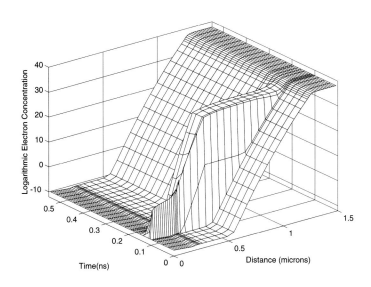

Figure 6.11: Variation of electron concentration in GaAs p-i-n photodiode.

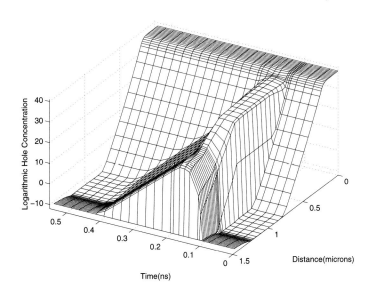

Figure 6.12: Variation of hole concentration in GaAs p-i-n photodiode.

these dependences is extremely important to such properties as mobility. Figures 6.13 and 6.14 show the time and space dependence of the electrostatic potential and electric field, respectively, under the same optical excitation for the photodiode. For the given excitation level, the variation of electric field during the pulse is very prominent. Therefore, the corresponding variation in carrier dynamics, especially in the electron mobility, is very significant (see, for example, Figure 6.5).

The terminal current for the photodiode is calculated at every time step using dedicated circuits inside the equivalent-circuit model. The conduction and displacement current components for the photocurrent response of the GaAs p-i-n photodiode are shown in Figure 6.15. For this simulation, a 20-ps, 1-μW square optical pulse was absorbed at the n-side of the depletion region. The delay between the optical pulse and the conduction current is due to the long transit time for holes. In addition, a significant fraction of the holes generated in the proximity of the depletion edge diffuses into the neutral n-region as shown in Figure 6.12. For an absorption profile close to the n-region, therefore, the response speed is further limited by the slow diffusion process.

For two different absorption profiles, the photocurrent response of the GaAs p-i-n is shown in Figure 6.16. The photocurrent with the faster response time corresponds to absorption closer to the p-contact (~ 0.15 μm from the depletion edge) resulting in a shorter distance for holes to traverse than electrons. In this case, the rise and fall times of the current are 8 ps and 9 ps, respectively. Therefore, a photodetector with such an absorption profile will have a bandwidth approaching 50 GHz. In contrast, for absorption at the opposite end of the depletion region (~ 0.15 μm from the n-region) the photocurrent response has a fall time of 40 ps due to the longer transit time of holes. This corresponds to a bandwidth of 9 GHz.

6.5.3 Resonant cavity enhanced photodetectors

As described earlier, the availability of photogeneration terms at every grid point in the model enables the study of novel photodetector structures with arbitrary light absorption profiles. In this section, the resonant cavity enhancement (RCE) effect and its implications on high-speed devices are presented. The superiority of the RCE detection scheme is confirmed by comparing its performance to that of a conventional p-i-n photodiode via the simulation tool described in this chapter.

A larger quantum efficiency η, that is, the probability of detecting incident photons, is an important property for high-performance photodetectors. The quantum efficiency of conventional detector structures is governed by the absorption coefficient of the semiconductor material. Thus, for high quantum efficiencies, thick active regions are required. However, thick active regions result in long transit times for the photogenerated carriers and correspondingly reduced device speeds. It is thus desirable to enhance the quantum efficiency without increasing the active layer thickness, in order to optimize the gain-bandwidth product.

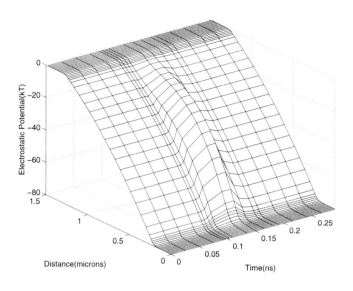

Figure 6.13: Variation of electrostatic potential for GaAs p-i-n photodiode.

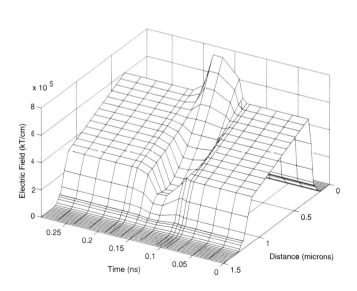

Figure 6.14: Variation of electric field in GaAs p-i-n photodiode.

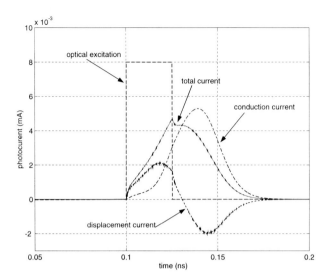

Figure 6.15: Detail of the photocurrent displaying the transient behavior of the displacement current and conduction current components, as well as the optical generation term.

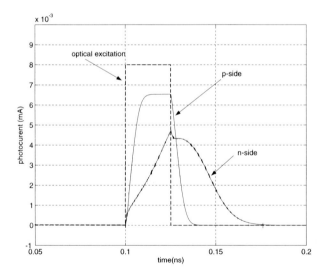

Figure 6.16: Current response of GaAs p-i-n for two different absorption profiles. The faster response corresponds to absorption closer to the p-type contact region and the slower response is for absorption closer to the n-type contact.

In RCE photodetectors (Figure 6.17), the detection/absorption region is integrated into an asymmetric Fabry-Pérot cavity formed by two mirrors, one on the top and one on the bottom. These mirrors usually take the form of distributed Bragg reflectors (DBRs), which consist of periodically alternating layers of bandgap materials with correspondingly alternating indices of refraction. In [6.24], the quantum efficiency of the device of Figure 6.17 is defined as

$$\eta = \left\{ \frac{(1 + R_2 e^{-\alpha d})}{1 - 2\sqrt{R_1 R_2} e^{-\alpha d} \cos(2\beta L + \psi_1 + \psi_2) + R_1 R_2 e^{-2\alpha d}} \right\} \cdot (1 - R_1)(1 - e^{-\alpha d}) \tag{6.33}$$

where R_1 and R_2 are the effective reflectivities of the top and bottom mirrors, respectively, α is the absorption coefficient of the active layer, d is the thickness of the active region, $L = L_1 + L_2 + d$, ψ_1 and ψ_2 denote phase shifts due to light penetrating into the DBR mirror regions [6.24], and β is defined as

$$\beta = \frac{2n\pi}{\lambda_o} \tag{6.34}$$

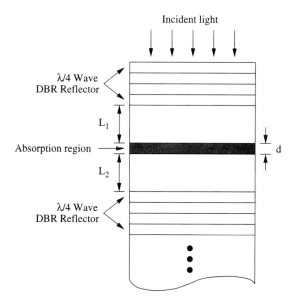

Figure 6.17: Schematic diagram of an RCE photodetector.

In (6.34), n is the refractive index and λ_o is the free-space wavelength. Notice from (6.33) and (6.34) that the quantum efficiency is periodically enhanced at resonant wavelengths determined by

$$2\beta L + \psi_1 + \psi_2 = 2m\pi \quad m = 1, 2, 3\ldots \tag{6.35}$$

This effect is demonstrated in Figure 6.18 for various top mirror reflectivities, for a bottom mirror reflectivity of $R_2 = 0.9$, and an absorption parameter of $\alpha d = 0.1$. The spacing of the cavity modes (i.e., resonant wavelengths) is defined as the free spectral range FSR [6.23]. The flat line in Figure 6.18 indicates the value given by $R_1 = 1 - e^{\alpha d}$, the maximum η value attainable for a conventional photodetector.

On the right-hand side of (6.33), the term inside the braces represents the cavity enhancement effect [6.24]. This term becomes unity when $R_2 = 0$, giving the quantum efficiency for a conventional detector without the RCE structure:

$$\eta = (1 - R_1)(1 - e^{-\alpha d}) \tag{6.36}$$

The peak quantum efficiency η_p achieved at resonant wavelengths can be derived by imposing the resonance condition (6.35) with $m = 1$ to (6.33), resulting in (6.37).

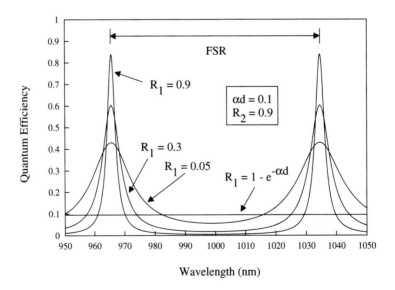

Figure 6.18: Wavelength dependence of quantum efficiency. Bottom mirror reflectivity is set at $R_2 = 0.9$, top mirror reflectivity is varied. Absorption parameter is fixed at $\alpha d = 0.1$.

$$\eta_p = \left\{ \frac{1 + R_2 e^{-\alpha d}}{(1 - \sqrt{R_1 R_2} e^{-\alpha d})^2} \right\} \cdot (1 - R_1)(1 - e^{-\alpha d}) \qquad (6.37)$$

At wavelengths that satisfy the resonance condition (6.35), incoming light interferes constructively with the reflected light from the bottom mirror. The resulting resonant cavity effect enhances the internal optical field amplitudes at the resonant wavelengths [6.6], [6.18]. Because of the internal optical field reflections at the mirror interfaces, the RCE effect can also be viewed as multiple-pass detection over a thin absorbing region. For a high Q (quality factor) cavity, (high mirror reflectivities and a thin absorption region), a drastic increase in η can be obtained.

For a comparison of conventional and RCE detectors, two heterojunction p-i-n photodiode structures have been selected. These structures follow the design rules suggested in [6.22] for highest performance. The intrinsic region is assumed to be lightly n-doped and its width is selected to be 0.72 μm, corresponding to the maximum bandwidth-efficiency product condition for a 10×10 μm^2 conventional p-i-n photodiode. This depletion width is slightly larger than the optimum value for the RCE detector. Therefore, this selection favors the conventional p-i-n structure.

The conventional device that was modeled (#1) has a 0.64-μm-thick, normally depleted n$^-$-GaAs absorbing region and $Al_{0.06}Ga_{0.94}As$ contact layers. A schematic diagram of this device is shown in Figure 6.19. The AlGaAs contact layers allow for high-speed operation by removing the limitation imposed by the diffusion of current out of absorbing but undepleted regions, an inherently slow process. Within a wavelength range (825 nm $< \lambda \le$ 870 nm) determined by the absorption edge of the AlGaAs and GaAs layers, photogeneration will occur only in GaAs regions ($\alpha = 10^4$ cm^{-1}). The carriers that may be generated in the GaAs substrate and cap layers are blocked by the AlGaAs barriers; that is, they do not contribute to the photocurrent. Thus, the device speed is limited solely by the transit time of photogenerated carriers across the 0.72-μm depletion region. A small Al mole fraction was chosen to avoid large band discontinuities. The heterojunctions were graded and positioned inside the depletion region to prevent charge trapping. The total band discontinuity is 75 meV, one-third of which is assumed to be at the valence band. For this conventional detector structure, a nearly ideal antireflection coating is assumed (surface reflectivity ≈ 0.05) resulting in a quantum efficiency of 0.45 within the 825–875 nm wavelength range.

The RCE photodiode structure (#2) has GaAs contacts (n and p) and depletion regions as shown in Figure 6.19. A 0.08-μm-thick $In_{0.07}Ga_{0.93}As$ absorption region is placed in the depletion region, extending the spectrum of wavelength sensitivity to 920 nm. Within a wavelength range of 870–920 nm, only this InGaAs region absorbs the incident light ($\alpha = 10^4$ cm^{-1}), and the remainder of the detector is transparent.

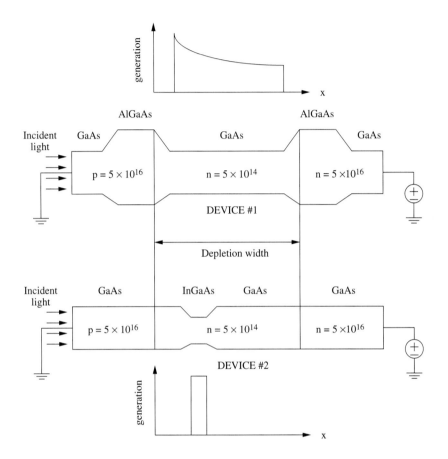

Figure 6.19: The schematic flat-band diagram
(applied and built-in fields are not shown) of the modeled p-i-n photodiodes
and a qualitative representation of the photogeneration terms.

Device #1 is the conventional AlGaAs/GaAs p-i-n heterojunction photodiode.
The photogeneration occurs in the depleted GaAs region with
an exponentially decaying profile.

Device #2 is the GaAs RCE photodiode with an InGaAs absorbing region.
The generation is localized in the InGaAs region.
The band discontinuities are exaggerated to clearly identify the graded regions.

The position of the absorbing layer is optimized so that the carriers have to traverse the depletion region a distance proportional to their velocities. Therefore, holes (which have a slower mobility) pass through a short depletion region, and electrons through a long depletion region. This ensures the arrival of photogenerated carriers at both contacts simultaneously, eliminating "tails" in the photocurrent response which limits the bandwidth of conventional detectors. For this optimization, the device simulation is first run with no optical excitation and the electron and hole mobilities are calculated in the depletion region for the electrical bias to be used in transient analysis. These mobilities are then used to evaluate the ratio of electron and hole velocities and, thus, the position of the absorption region for the fastest photodetector response. As a result, the distance of the absorption layer from the p-region is determined to be approximately half of that from the n-region.

A Fabry-Pérot resonant cavity is formed by an ideal bottom mirror ($R_2 \approx 1.0$) and a high-reflectivity ($R_1 = 0.7$) top mirror. The quantum efficiency of this RCE detector is estimated at $\eta = 0.9$ at wavelengths around 900 nm. The electrical contacts are assumed to be between the multilayer mirrors to prevent high resistances. Since the current does not flow through the mirror regions, the large band discontinuities required for mirror formation do not affect the electrical performance.

Figures 6.20 and 6.21 illustrate the time evolution of the hole $p(x,t)$ and the electron $n(x,t)$ concentrations, respectively, for the InGaAs/GaAs heterojunction photodiode structure (Device #2, Figure 6.19). Steady-state carrier distributions, as can be observed at $t = 0$, clearly show the depletion region and the location of the heterojunctions. The wavelength of the optical excitation was chosen so that photogeneration would occur in the InGaAs region only. The optical excitation is a square pulse of about 9×10^{14} photons/device for a 10×10 μm^2 device (≈ 180 W/cm^2) with negligible rise and fall times (0.01 ps) and a full-width at half-maximum (FWHM) of about 10 ps. It is assumed that 90% of the incident photons are absorbed in the thin InGaAs region. Such a high quantum efficiency is quite possible with the RCE structure [6.18], [6.6], [6,19].

The transient variations of the electron and hole concentrations (Figures 6.20 and 6.21) reveal the advantages of the design with a thin absorbing layer embedded in the depletion region. Both electrons and holes are swept under the electric field in the depletion region immediately after photogeneration. Carrier diffusion into the neutral regions is negligible and the carrier concentrations return to their steady-state values shortly after termination of the optical excitation. Figure 6.22 shows the short-circuit photocurrent of the same photodiode (Device #2). The time variation of the displacement and conduction current components are illustrated in comparison with the generation term. Since there are no assumptions and approximations in the time variation of the semiconductor equations in the presented model, the calculated current is accurate, even under large pulse excitations.

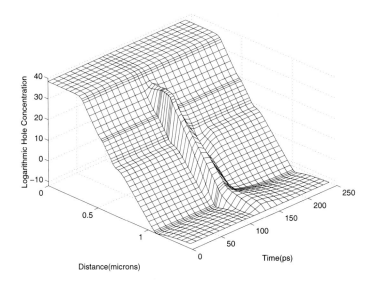

Figure 6.20: Variation of hole concentration in Device #2 as a function of time and position when light is absorbed across the depletion region. The steady-state distribution can be observed at $t = 0$.

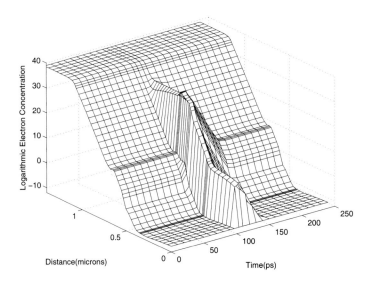

Figure 6.21: Variation of electron concentration in Device #2 as a function of time and position when light is absorbed across the depletion region.

The ultimate photodetector response time can be determined by observing the transient current under optical pulses of varying duration. Within the convergence capabilities of the circuit simulator, it is possible to calculate the output current for optical pulses less than 1 ps FWHM. Since the estimated transit times for conventional and RCE devices designs are about 5 to 10 ps, the transient response for the two devices (Devices #1 and #2) are compared for a 5-ps optical pulse in Figure 6.23. To ensure small-signal excitation, the pulse amplitude was chosen to be 1% of that used in the previous example (Figures 6.20–6.22). A larger (magnitude) and sharper (less spread) current is observed for the RCE detector.

Closer inspection of the individual current components reveals the expected features and superiority of the RCE detector:

(1)   The fall time is comparable to the rise time for the RCE detector whereas it is significantly longer in the conventional structure due to the long transit time of holes.

(2)   The RCE detector not only has a larger photocurrent under identical optical excitation (because of enhanced quantum efficiency), but the response is also faster than its conventional counterpart. Therefore, the bandwidth-efficiency product is doubly improved.

(3)   Due to the simultaneous arrival of both electrons and holes at the contacts, the time spread of the conduction current for the RCE design is much smaller than that of the conventional structure, in which the carriers reach the contacts at different times, as determined by the location of the photogeneration.

For a comparison of the RCE and conventional detectors in the frequency domain, a Fast Fourier Transform (FFT) was used. The simulated temporal responses depicted in Figure 6.23 were converted into the frequency domain by FFT and are depicted in Figure 6.24. Fourier analysis of the optical excitation term gives a flat spectrum (within 1 dB) up to approximately 100 GHz verifying the accuracy of the frequency-domain representation of the current pulses. Furthermore, to eliminate the influence of the finite width of the optical excitation pulse, it was deconvolved from the simulated current responses. From Figure 6.24, the transit-time limited 3-dB bandwidth of the conventional p-i-n detector is calculated to be 52 GHz, compared to 70 GHz for the RCE p-i-n. This corresponds to a 35% improvement in bandwidth. In addition, there is a drastic improvement in the transit-time limited bandwidth-efficiency (BWE) product. For the conventional p-i-n, the BWE is 23 GHz, which is slightly smaller than the theoretical limit of 27 GHz [6.22] based on the transit time of holes. The BWE of the RCE device is 63 GHz which represents a nearly threefold improvement. If smaller device areas for the RCE detector are considered [6.22], then the BWE can be extended to 100 GHz.

For a direct comparison of the bandwidths for the RCE and conventional structures, the peak value of the normalized short-circuit current (detector response) is plotted in Figure 6.25 as a function of the inverse FWHM. When the pulse

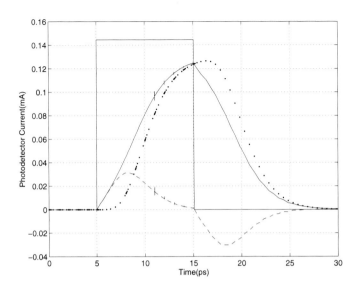

Figure 6.22: The short-circuit current output evaluated for the RCE structure. The photogeneration term and displacement (dashed line) and conduction (dotted line) current components are illustrated.

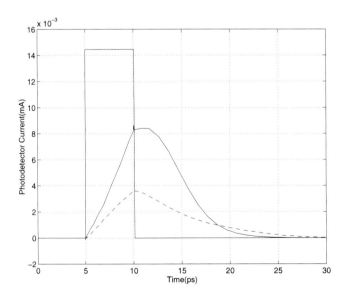

Figure 6.23: Transient short-circuit photocurrent under a 5-ps FWHM optical pulse for conventional p-i-n (Device #1, dashed line) and RCE p-i-n (solid line).

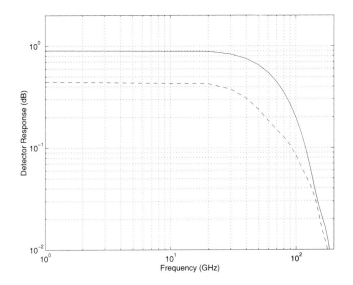

Figure 6.24: Frequency responses of conventional (dashed) and RCE (solid) p-i-n detectors obtained by FFT of transient responses in Figure 6.23. The optical excitation term was deconvolved to extract the impulse responses.

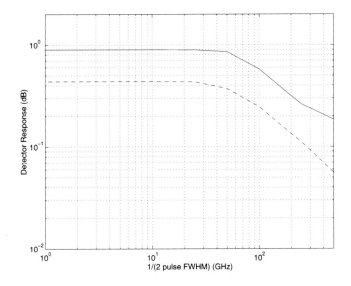

Figure 6.25: Detector response (peak value of the normalized detector current) versus the inverse of the optical pulse FWHM for the conventional (dashed) and RCE (solid) structures.

duration is large, the output current reaches its maximum value as determined by the internal quantum efficiency, that is, $\eta = 0.45$ for the conventional device and $\eta = 0.90$ for the RCE structure. As the optical pulse becomes smaller, the electrical current is unable to reach its steady-state value. Although the ratio of electron-hole pairs to the number of incident photons remains the same, the *peak* current decreases since the current is spread over time. The peak short-circuit current drops to half of its maximum at FWHMs of 4.5 ps and 3.5 ps for the conventional and RCE structures, respectively. Therefore, the bandwidth of the RCE p-i-n is approximately 30% larger than that of the conventional diode (Figure 6.25), confirming previous findings through Fourier analysis. Note that both of these comparisons are made for a single optical pulse of varying duration. If repetitive pulses or sinusoidal excitation are considered, the bandwidth difference will be even larger due to the large fall time of the conventional p-i-n.

To this end, the fall times of the RCE and conventional devices are compared in Figure 6.26. The optical input pulse was intentionally chosen long enough for both devices to reach their steady-state current levels, and was terminated abruptly (fall time = 0.01 ps). Despite the higher steady-state value of the RCE detector, its photocurrent decays faster and drops to lower values than that of the conventional

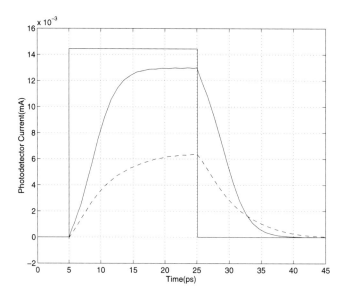

Figure 6.26: Transient short-circuit photocurrent for conventional (dashed)
and RCE (solid) detectors.
The optical pulse has a fall time of 0.01 ps and is long enough to allow
both devices to reach their steady-state current levels.

device. The fall time τ_f is defined as the time required for the photocurrent to drop from 90% to 10% of its steady-state value. With this definition, the fall time for the RCE p-i-n is 7.1 ps, which is 65% faster than that of the conventional structure (11.7 ps).

6.5.4 AlGaAs/GaAs heterojunction p-i-n

In this section, an AlGaAs/GaAs heterojunction p-i-n externally biased by the simple RC circuit shown in Figure 6.27 is simulated. While this simulation demonstrates both mixed-mode and large-signal capability, the latter is emphasized in this example.

A small voltage of $V_{ext} = 0.5$ V is applied to the circuit which reverse biases the photodiode. The junction capacitance of the device is represented by the capacitor C shown in Figure 6.27. The total thickness of the device is $L = 1.6$ μm, with a low-doped intrinsic region 0.8 μm thick. The highly doped GaAs cap layers, representing realistic ohmic contact regions, are 0.1 μm each. Utilizing variable grid spacing, 50 grid nodes were taken. The model provides the means to observe the device variables as well as the circuit (I,V) variables during the transient simulation.

Figure 6.28 shows the typical I-V characteristics of a photodiode under stepped illumination. The DC load line is imposed by the simple bias circuit, where V_{ext} and $V_{ext} \div R$ are the points where the load line intersects the horizontal and vertical axes, respectively. The bias condition on the photodetector is determined by the intercept of the load line with the diode I-V characteristics corresponding to steady-state optical excitation. Under small-signal conditions, the variation in the optical signal does not have a significant effect on the bias conditions; thus, the terminal voltages can be assumed constant. In contrast, for large optical excitations, the I-V characteristics of the photodiode will vary significantly, as shown in Figure 6.28, resulting in changing operating conditions. Thus, bias conditions can no longer be assumed constant and must be considered as dynamic variables. In the extreme case,

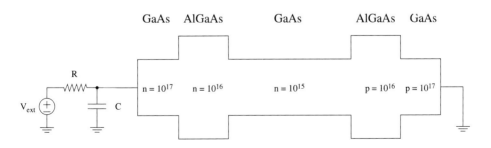

Figure 6.27: Photoreceiver circuit using a GaAs/AlGaAs p-i-n heterojunction structure biased by a simple RC circuit.

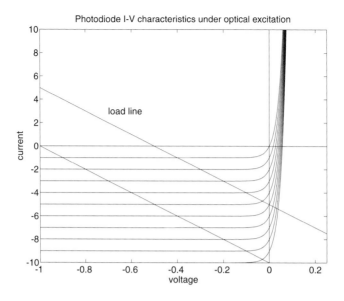

Figure 6.28: Photodiode I-V characteristics under increasing illumination.

the optical signal can be large enough to drive the photodetector into saturation [6.11]. Saturation refers to the state of operation in which the input power is so large that the electrical output current and voltage no longer follow the input linearly. Saturation not only distorts the signal waveform but also increases the transit time (as will be shown later) which also decreases the bandwidth of the device.

For a demonstration of the concept, a detector structure with uniform optical generation in the depletion region is assumed. The optical pulse has 0.01-ps rise and fall times and a 20-ps duration, corresponding to 5×10^{20} photons/cm^2·s (i.e., 100 W/cm^2). The total power for a 30×30 μm^2 device is on the order of 1 mW. Figures 6.29 and 6.30 show the simulated electron and hole concentrations, respectively. The AlGaAs/GaAs heterojunctions can be easily distinguished by the dip in the carrier concentrations caused by the band discontinuity between the larger and smaller bandgap materials. Examination of the dynamic behavior of the carriers reveals that the electron concentration recovers to its steady-state condition right after the pulse has been terminated. The hole concentration, however, regains its steady-state value more slowly, due to its lower mobility. The transient variation of the local electrostatic potential is depicted in Figure 6.31. The "valley" perceived on the top of the potential profile represents the decrease in detector bias voltage caused by the voltage drop across the resistor due to the photocurrent. Since the electric field in the device also decreases during optical generation, the drift component of the carrier transport is reduced. This example is clearly a large-signal case, since the bias conditions are significantly altered by the optical generation.

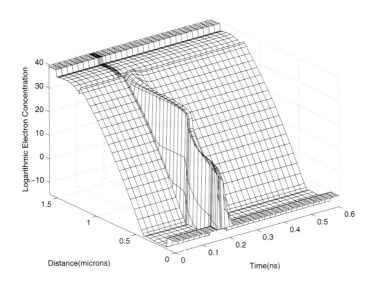

Figure 6.29: Variation of electron concentration in AlGaAs/GaAs p-i-n.

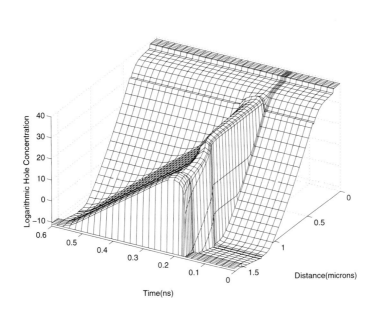

Figure 6.30: Variation of hole concentration in AlGaAs/GaAs p-i-n.

To demonstrate the dependence of the carrier response times to the electric field even more distinctly, another large signal simulation, in which the photodetector is saturated by a 200-ps pulse, is depicted in Figure 6.32. This figure shows the impact of such a pulse excitation (100 W/cm^2) on the electrostatic potential. The transient bias drop is very significant (25 kT/q); in fact, the drop in the detector bias is even greater than the external applied voltage. This results in switching from reverse to forward biasing via the optical pulse. It should be noted that once the optical pulse is terminated, the electrostatic potential slowly recovers to its steady-state value. Figure 6.33 depicts the hole generation. The photogenerated carrier concentration is almost equivalent to the doping concentration of the p-GaAs cap layer. Even though the optical pulse is terminated at 0.35 ns, the hole concentration still remains stationary in the absorption region for another 0.1 ns. This is a result of the vanishing drift component of the carriers. Once the electrostatic potential (hence the electric field) regains its significance (≈ 0.5 ns), the hole concentration decreases. Even the electrons, which have large mobilities at low fields (Figure 6.34), are kept stationary in the absorption region due to the nearly complete quenching of internal electric field during the pulse.

These examples clearly indicate the importance of large-signal analysis, proving that large-signal photogeneration can drastically affect device speed. Figure 6.35 displays the broadening photocurrent responses for increasing optical generation rates. The initial photogeneration (the generation rate for which the smallest photocurrent in Figure 6.35 is obtained) is increased by factors of 5, 10, and 20. When five times the generation rate of the initial pulse is applied, the photocurrent response does not increase linearly and pulse broadening is observed. Therefore, it is clear that the generation rate is sufficient to saturate the photodiode. The ringing seen in the photocurrent responses for increasing photogeneration is due to numerical instability in the calculation of the displacement current components. This instability may be avoided by reducing the time step size during the photogeneration process. For example, similar instabilities were observed in the generation of Figure 6.15 and can be eliminated by reducing the time step. However, since observation of large-signal effects was the primary goal of these examples, such time step reduction was not performed.

6.5.5 Integrated photoreceiver

As previously discussed, the simulation of optoelectronic devices in a circuit simulator allows for mixed-mode device/circuit simulation. This capability is also demonstrated with the following example of a simple photoreceiver circuit consisting of a GaAs p-i-n photodiode and a single-transistor BJT amplifier circuit (Figure 6.36). The relevant parameters in the SPICE model for the BJT are $\beta_f = 500$, $I_s = 10^{-14}$ A, $C_{je} = 30$ fF, $C_{js} = 20$ fF, $r_c = 10$ Ω, $r_b = 50$ Ω, and $t_r = 0.01$ ns. The circuit is adjusted to reverse bias the photodiode and saturate the BJT. In this

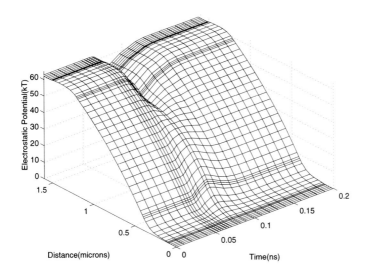

Figure 6.31: Variation of electrostatic potential in AlGaAs/GaAs p-i-n
for an optical pulse of 20-ps duration.

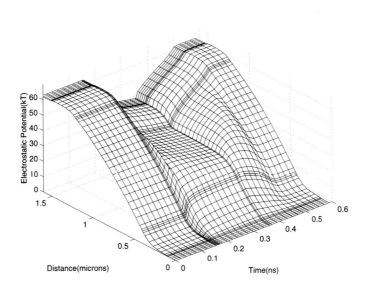

Figure 6.32: Variation of electrostatic potential in AlGaAs/GaAs p-i-n
for an optical pulse of 200-ps duration.

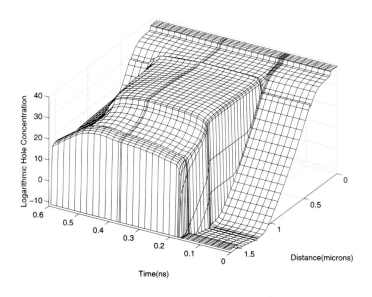

Figure 6.33: Variation of hole concentration in AlGaAs/GaAs p-i-n
for an optical pulse of 200-ps duration.

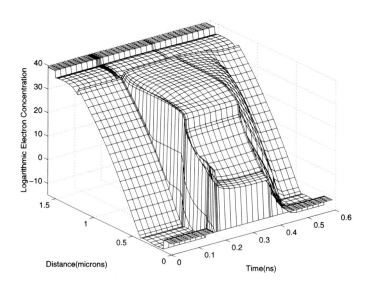

Figure 6.34: Variation of electron concentration in AlGaAs/GaAs p-i-n
for an optical pulse of 200-ps duration.

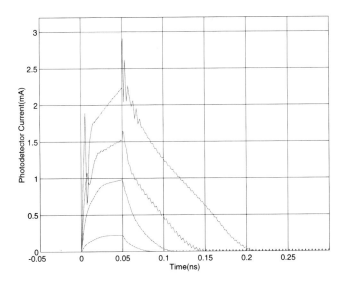

Figure 6.35: Photodiode current under increasing optical illumination.

example, the detailed device-level model of the photodiode is combined with the circuit-level lumped models for the rest of the photoreceiver circuit. Since the device-level simulation of the p-i-n photodiode is carried out using the circuit simulator, the link between the device- and circuit-level simulation results is automatically established. The time-independent carrier concentrations and the local potential yield the terminal current of the simulated photodiode. The transient current-voltage characteristics of the photodiode are combined with the external circuit, and the voltage boundary conditions are evaluated at every time point by solving the entire OEIC self-consistently. This self-consistent solution yields the true large-signal response of the device and the OEIC. Figure 6.37 shows the simulated photocurrent of the detector and the variation of the detector bias under a 50-ps optical pulse; approximately 0.5 mW is absorbed in the middle of the depletion region.

Figure 6.38 shows the transient variation of the base current and the collector voltage of the transistor of the photoreceiver circuit. The transient variation of the base current clearly shows a reversal in the bias current when the optical pulse arrives. The variation of the collector potential from saturation to nearly 2 V when the optical pulse strikes the detector demonstrates the large-signal capability of the model. Note that all the circuit variables are evaluated simultaneously with the local device variables; therefore, device performance can be tested under realistic bias conditions. Furthermore, this mixed-mode device/circuit simulation capability

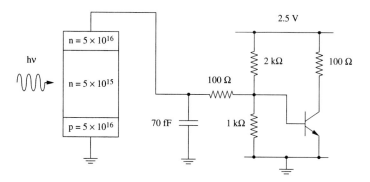

Figure 6.36: Photoreceiver circuit consisting of p-i-n detector and BJT amplifier.

enables the designer to investigate the effects of various subtle changes in device geometry and/or material parameters on circuit performance, a task which is difficult, if not impossible, to accomplish using only lumped circuit-level models.

6.6 Summary

In this chapter, a novel approach has been presented for incorporating the transient solution of the one-dimensional semiconductor equations within a general circuit simulation environment. This approach allows a simple representation of the localized carrier transport models through equivalent-circuit elements such as voltage-controlled current sources and capacitors.   The availability of local photogeneration capability at every grid point enables the simple simulation of various optoelectronic devices.

The model was utilized to calculate results supporting the theory of the RCE p-i-n and to compare it with its conventional counterpart. A threefold enhancement in the bandwidth-quantum efficiency product was predicted. Large-signal capability was demonstrated via the simulation of a heterojunction photodiode.

Since the device-level simulation is carried out using an equivalent-circuit representation, this approach lends itself directly to the mixed-mode simulation of devices and circuits with a single simulation tool. The mixed-mode simulation capability of the new approach was demonstrated for the time-domain analysis of a photoreceiver circuit.   It was shown that the transient characteristics of one-dimensional device structures, operating within a realistic circuit environment, can be simulated without requiring dedicated software tools. The device-level modeling approach introduced can be easily extended for mixed two-dimensional device and circuit simulation. The device simulation model presented in this chapter has been implemented and tested in SPICE 3F2 as well as on two SPICE-like general-purpose circuit simulation programs.   The implementation of the model requires no

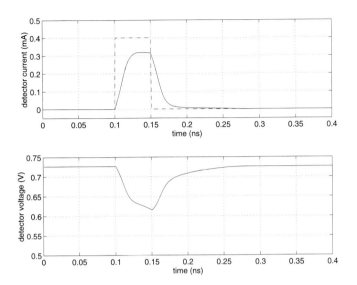

Figure 6.37: Transient response of OEIC photoreceiver using mixed-mode device/circuit simulation.
Top: Diode current (solid line) and photogeneration (dashed line, arbitrary units).
Bottom: Transient variation of the photodiode terminal potential.

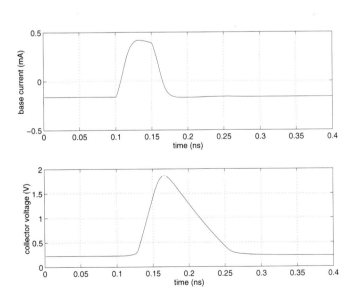

Figure 6.38: Transient response of OEIC photoreceiver using mixed-mode device/circuit simulation.
Top: Base current. Bottom: Collector potential.

modifications to the internal structure of the circuit simulator; thus, it can be implemented quickly and easily on a number of different computational platforms. The versatility of the equivalent-circuit representation of the device equations should enable the incorporation of additional differential equations describing other physical phenomena.

6.7 Appendix — Normalization Factors

Normalization of the semiconductor equations is performed largely to simplify numerical calculations [6.9]. The table below shows the normalization factors used for the simulations presented in this chapter.

Variable	Normalization Factor
Position x, h	$L_{Di} = \dfrac{\varepsilon_{s,ref}\varepsilon_o kT}{q^2 n_{i,ref}}$
Potential V, ψ_n, ψ_p	$V_{nor} = \dfrac{kT}{q}$
Carrier concentration n, p, n_i, N_A, N_D	$n_{i,ref}$
Dielectric constant ε	$\varepsilon_{s,ref}$
Diffusion coefficient D_o	$D_o = 1 \dfrac{\text{cm}^2}{\text{s}}$
Carrier mobility μ_n, μ_p	$\dfrac{D_o}{V_{nor}}$
Current density J_n, J_p	$\dfrac{qD_o n_{i,ref}}{L_{Di}}$
Generation, recombination G, R	$\dfrac{D_o n_{i,ref}}{L_{Di}^2}$
Carrier lifetime τ_n, τ_p	$\dfrac{L_{Di}^2}{D_o}$

In this table, $\varepsilon_{s,ref}$ and $n_{i,ref}$ are the relative permittivity and the intrinsic carrier concentration, respectively, of the reference semiconductor material. In the simulations presented in this chapter, the reference has been GaAs.

6.8 References

[6.1] E. M. Buturla, P. E. Cottrell, B. M. Grossman, and K. A. Salsburg, "Finite element analysis of semiconductor devices: the FIELDAY program," *IBM Journal of Research and Development*, vol. 25, pp. 218-231, 1981.

[6.2] P. C. H. Chan and C. T. Sah, "Exact equivalent circuit model for steady-state characterization of semiconductor devices with multiple energy-level recombination centers," *IEEE Transactions on Electron Devices*, vol. ED-26, pp. 924-936, 1979.

[6.3] R. G. Hunsperger, *Integrated Optics: Theory and Technology*. New York: Springer Verlag, 1991.

[6.4] K. Kani, "A survey of semiconductor device analysis in Japan," *Proceedings of the NASEC-ODE I Conference*, pp. 104-119, 1979.

[6.5] R. Kielkowski, *Inside SPICE*. New York: McGraw-Hill, Inc. 1994.

[6.6] K. Kishino, M. S. Ünlü, J. I. Chyi, J. Reed, L. Arsenault, and H. Morkoç, "Resonant cavity enhanced (RCE) photodetectors," *IEEE Journal of Quantum Electronics*, vol. 27, no. 8, pp. 2025-2034, 1991.

[6.7] Y. Leblebici, M. S. Ünlü, H. Morkoç, and S. M. Kang, "One-dimensional transient device simulation using a direct method circuit simulator," *Proceedings of the IEEE International Symposium on Circuits and Systems*, pp. 895-898, 1992.

[6.8] Y. Leblebici, M. S. Ünlü, S. M. Kang, and B. M. Onat, "Transient simulation of heterojunction photodiodes — Part I: Computational methods," *IEEE Journal of Lightwave Technology*, vol. 13, no. 3, pp. 396-405, 1995.

[6.9] M. S. Lundstrom, S. Datta, R. J. Schuelka, S. Bondyopadhyey, and P. Sorlie, "Physics and Modeling of Heterostructure Semiconductor Devices," Purdue University Technical Report, no. TR-EE 84-35, Purdue University, September 1984.

[6.10] O. Madelung and M. Schulz, *Numerical Data and Functional Relationships in Science and Technology, Volume 22: Semiconductors*. New York: Springer Verlag, 1990.

[6.11] J. C. Palais, *Fiber-Optic Communications*. Englewood Cliffs, NJ: Prentice Hall, 1992.

[6.12]   R. F. Pierret, *Advanced Semiconductor Fundamentals, Volume 6: Modular Series on Solid-State Devices*. Reading, MA: Addison-Wesley, 1987.

[6.13]   M. R. Pinto, C. S. Rafferty, and R. W. Dutton, "PISCES II – Poisson and continuity equation solver," Stanford Electronics Laboratory Technical Report, Stanford University, September 1984.

[6.14]   C. T. Sah, "New integral representations of circuit models and elements for the circuit technique for semiconductor analysis," *Solid-State Electronics*, vol. 30, pp. 1277-1281, 1987.

[6.15]   B. E. A. Saleh and M. C. Teich, *Fundamentals of Photonics*. New York: John Wiley & Sons, 1991.

[6.16]   S. Selberherr, W. Fichtner, and H. W. Poetzl, "MINIMOS — a two-dimensional MOS transistor analyzer," *IEEE Transactions on Electron Devices*, vol. ED-27, pp. 1540-1550, 1980.

[6.17]   S. Selberherr, *Analysis and Simulation of Semiconductor Devices*. New York: Springer Verlag, 1984.

[6.18]   M. S. Ünlü, K. Kishino, J. I. Chyi, J. Reed, S. Noor Mohammad, and H. Morkoç, "Resonant cavity enhanced AlGaAs/GaAs heterojunction phototransistors with an intermediate InGaAs region in the collector," *Applied Physics Letters*, vol. 57, no. 8, pp. 750-752, 1990.

[6.19]   M. S. Ünlü, K. Kishino, J. I. Chyi, L. Arsenault, J. Reed, and H. Morkoç, "Wavelength demultiplexing heterojunction phototransistors," *Electronics Letters*, vol. 26, no. 22, pp. 2857-2858, 1990.

[6.20]   M. S. Ünlü, B. Onat, and Y. Leblebici, "Large-signal transient simulation of optoelectronic devices," *Proceedings of the International Semiconductor Device Research Symposium*, Charlottesville, VA, 1993.

[6.21]   M. S. Ünlü, B. Onat, and Y. Leblebici, "Transient simulation of optoelectronic integrated circuits using 'SPICE'," *Proceedings of the OSA Integrated Photonics Research Symposium*, pp. 228-229, 1994.

[6.22]   M. S. Ünlü, B. Onat, and Y. Leblebici, "Transient simulation of heterojunction photodiodes — Part II: Analysis of resonant cavity enhanced photodetectors," *IEEE Journal of Lightwave Technology*, vol. 13, no. 3, pp. 406-415, 1995.

[6.23]   J. T. Verdeyen, *Laser Electronics*. Englewood Cliffs, NJ: Prentice Hall, 1989.

[6.24]   M. S. Ünlü, "Resonant cavity enhanced photodetectors and optoelectronic switches," Ph.D. dissertation, University of Illinois at Urbana-Champaign, 1992.

7

TABLE-BASED SIMULATION

*B*oth the necessity and the usefulness of optoelectronic circuit simulation have been demonstrated throughout the previous chapters. As illustrated in Chapters 3 and 5, however, device modeling for circuit simulation usually requires extensive knowledge of the device's physical behavior, both electrical and optical, under all regimes of possible operation (AC, DC, transient). While simulators such as iSMILE can ease the process of *implementing* an existing model, *development* of the model itself often requires significant amounts of time and research into the device physics. The problem is made even worse in the exploratory research and development environment, where a device is often successfully fabricated, but the device physics governing its operation are not yet well understood. Circuit design and integration in this situation cannot proceed until extensive research is conducted on the underlying physical mechanisms involved and a new circuit model is created. This process can often consume an unacceptable amount of time; in fact, in the development of new devices, unfavorable attributes are often discovered that prove the device to be useless before the new model can even be completed. Clearly, an alternate method for fast prototyping of new devices is necessary [7.1] – [7.7].

The table-based approach to circuit simulation relieves the designer of the task of creating a physics-based device model. Under normal operation, a circuit simulator presents a set of currents and voltages to a device model. The model then predicts the device's response to these stimuli by evaluating the physical equations contained within, then passes the results back to the circuit simulator. With the table-based method, devices are modeled not by physical equations, but rather by tables of data containing information about the device's behavior over various regions of operation. As before, circuit simulation is initiated by presenting a set of currents and voltages to the table-based device model. However, the table-based model does not use physical equations to predict the device's response; rather, it searches the relevant data tables for an entry that matches the input currents and voltages, then retrieves the device's response from the table. When the exact values of currents and voltages are not contained in the table, the table-based model uses various forms of interpolation to estimate what the device's response will be.

The biggest requirement for table-based simulation is the availability of large amounts of device data; indeed, the more data that are available, the less frequently interpolation will be necessary, and the more accurate the model will be. Clearly, the table-based approach is purely numerical; it is an empirical fit of mathematical equations to measured data. In general, this type of model is not as flexible as a physically based model. However, with sufficient data, table-based simulation can be extremely effective in the quick prototyping of new devices. This allows for the evaluation of the feasibility and performance of circuits and systems composed of such devices, without the time-consuming task of researching the physical mechanisms involved and creating a new model. If the new device is later determined to be a success, with the potential for a long future, *then* the necessary research can be conducted to develop a more accurate physical model.

One potential drawback of table-based simulation is that it is not possible to predict the behavior of the device with different physical attributes (length, width, etc.) [7.1]. That is, since the data are measured for *one particular* device, it is not possible to predict the effects of physically modifying (such as scaling) the device. It is only possible to predict the device's response to a given current or voltage input that falls within the range over which the measurements were taken. However, this problem can be circumvented by measuring many tables of data, each set corresponding to a physically different device. Appropriate numerical schemes can then be used to interpolate not only between currents and voltages, but also between different device attributes. For example, measurement of the entire set of current and voltage data for a transistor can be replicated for several different channel widths.

While the data are usually measured from actual, fabricated devices, they can also be obtained from the results of very detailed device simulations. Since device simulation involves the discretization and solution of the fundamental semiconductor equations, such results are often nearly as accurate as actual measurements. This approach is especially useful in the design of a new device when the required fabrication facilities are either not available or are behind schedule.

In this chapter, several examples of table-based modeling will be presented for devices of interest to optoelectronic integration, such as the laser diode, the HBT, and the HFET. Case studies using both actual measured data and simulation data as the table inputs are considered. As with several of the previous chapters, iSMILE will be used as an illustrative vehicle for demonstrating the principles of table-based simulations; however, it is to be emphasized that these principles are general, and can be applied to any suitable tool.

7.1 The Table-Based Approach

To perform table-based simulation over all of the ranges of interest (AC, DC, transient), two sets of data are necessary. The first set, which is referred to as the *DC table*, contains tabulated data measured at steady state [7.1]. The DC table can be used to predict the characteristics of resistive components in the device. The second set of data is referred to as the *dynamic table*; this table contains data necessary to predict the performance of storage elements, such as capacitors and inductors. The dynamic table can be prepared from either time-domain or frequency-domain measurements. For the former, the data are often measured from the step response of the device, while for the latter, the data are usually measured as S parameters over the appropriate region of operation. The DC table is created by making measurements of the current and voltage responses over the entire DC range of interest. The dynamic table is then constructed by measuring the dynamic response of the device, either in the time domain or the frequency domain, at each DC bias point.

In the measurement of data for table-based simulation, it is crucial to have sufficient samples to ensure interpolation accuracy. In addition, it is highly desirable to include more data points over regions of operation under which the device undergoes sudden or drastic changes. For example, the DC behavior of a diode could be sampled sparsely above and below threshold, but aggressively at or about the turn-on point. In the time domain, it would be prudent to sample the step response heavily right at the step transition and less intensely at points before and after the turn on.

Once the tables have been created, the next step is to construct a simple equivalent circuit to *qualitatively* replicate the behavior represented by the measured data. This differs from the physically based circuit modeling approaches of the previous chapters, as will be demonstrated throughout this chapter. While use of an equivalent circuit is certainly not the only method available, it is chosen in this section as the most general approach. That is, while other, purely numerical, methods do exist, the equivalent-circuit method tends to lend itself most easily to the implementation of new, arbitrary models.

For example, consider the arbitrary time-domain step response depicted in Figure 7.1. In this particular example, no assumption is made about the nature of this response; it could be electrical, optical, thermal, or mechanical. It could even represent the growth of the economy in response to a sudden infusion of dollars by the government. By examination, the response of this data *qualitatively* resembles that of a series RC circuit, as depicted in Figure 7.2. The values of the resistor R_0 and the capacitor C_0 must be chosen so that the voltage across the capacitor matches

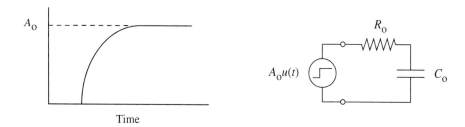

Figure 7.1: Arbitrary step response.   **Figure 7.2:** Qualitative equivalent circuit.

that of Figure 7.1. In this simple case, the elements are linear, and their values can be determined more or less by inspection; however, in general, the elements are nonlinear and more involved algorithms are employed to extract their characteristic functions from the data table. Extraction of these characteristic functions results in a numerical method for predicting the response of each of the circuit elements. The general procedure will be illustrated by example in the next section.

The next step in the process is the development of an algorithm to interpolate between data points when necessary. There are many methods of performing interpolation, such as curve fitting via linear [7.3], piecewise polynomial [7.6], or spline [7.5] algorithms, or even multidimensional techniques such as fitting to shape functions. Simple curve fitting techniques are often adequate for two-terminal devices, as will be illustrated in the next section. Multiterminal devices, however, often require more sophisticated approaches as detailed in [7.7]. Examples of interpolation using shape functions will be depicted in Sections 7.3 and 7.4.

The final task in the development of a table-based model is the implementation of an algorithm for differentiation. As detailed in previous chapters, not only must a device model return a set of currents and voltages to the circuit simulator, it must also return a set of derivatives. Again, many different methods exist for numerical differentiation, ranging from the very simple (such as the divided difference method [7.8]) to the complex (such as isoparametric shape functions [7.1], [7.7]). Examples will be presented of both in this chapter. Once these steps have been completed, the model can be verified and executed.

7.2 Laser Diode

In this section, the table-based laser diode model of Pang et al. [7.1] will be presented. Pang's method is based on the equivalent-circuit approach detailed in the previous section. In this case, rather than using measured data as the basis for the table construction, [7.1] uses the analytical circuit model presented in Chapter 3 as proof that the table-based approach is valid. However, examples using measured data and device simulation data will follow in subsequent sections.

7.2.1 Laser diode electrical properties

Since the laser diode is an optoelectronic device, it is necessary in circuit simulation to model both its optical and electrical properties. This was demonstrated in the laser diode models of Chapters 3 and 5 as well as the transmitter simulations of Chapter 4. To create the electrical portion of the table-based model, both the DC and dynamic tables are constructed from current-voltage data. As mentioned previously, for this example these data are taken from an analytical model in order to show the validity of the table-based approach. Figure 7.3 depicts the circuit setup and the resulting simulation output used to construct the DC table, while Figure 7.4 illustrates the corresponding scenario for creation of the dynamic table. In order to increase interpolation accuracy, data are sampled at a higher rate in the DC table at points greater than $V_{in} = 0.5$ V since there is little or no activity before that point. The range of inputs has been arbitrarily chosen to be ± 1.5 V, as this was the assumed region of operation. A subset of the transient curves necessary for construction of the dynamic table is depicted in Figure 7.4. It should be emphasized that a separate step response is necessary for each bias point in the DC table.

As described in the last section, the next step in the creation of a table-based laser diode model is the formulation of an equivalent circuit that *mimics* the I-V and transient operation represented by the DC and dynamic tables. In constructing such an equivalent circuit, two underlying assumptions are made [7.1]. The first is that capacitors and inductors appear as open circuits and short circuits, respectively, in

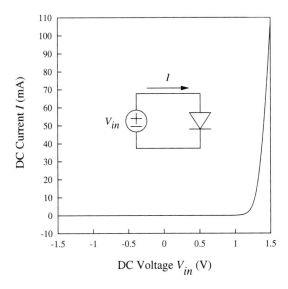

Figure 7.3: Circuit and data used to construct DC I-V table.

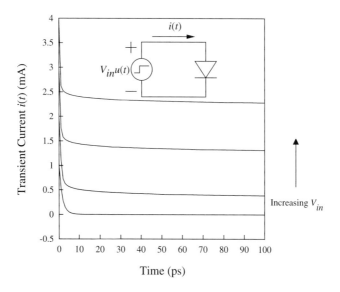

Figure 7.4: Circuit and data used to construct dynamic I-V table.

steady state. The second assumption is that the circuit is relaxed before the input is applied at $t = 0$, where t is time; that is, $i(t = 0^-) = 0 = v(t = 0^-)$.

The equivalent circuit used to model the DC and dynamic electrical responses is depicted in Figure 7.5. In this figure, the current through the nonlinear resistor R_s is used to model the information contained in the dynamic table. The operation of this circuit can be understood by examining both Figure 7.4 and Figure 7.5. As assumed, the node voltages are all zero before any input is applied. At $t = 0$, a step voltage is input and the capacitor C_p begins to charge, a process which continues until steady state, after which C_p behaves as an open circuit. In steady state, the only current flow is DC, as modeled by R_s and R_p. Thus, the laser's I-V characteristic in this state can be represented by the information contained in the DC table. This operation qualitatively matches the responses of Figures 7.3 and 7.4 quite well.

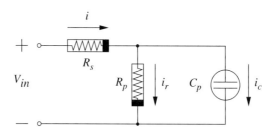

Figure 7.5: Equivalent circuit for laser diode electrical characteristics.

The next step is to determine the characteristic functions of each of the components in the equivalent circuit. For the nonlinear resistors, the characteristic function will consist of an I-V relationship, while for the nonlinear capacitor, a Q-V (charge-voltage) relationship must be determined. To construct the characteristic functions, it is useful to map the equivalent circuit of Figure 7.5 into two regimes. Figure 7.6a depicts the equivalent circuit at $t = 0^+$, that is, immediately after the voltage is applied to the input. At $t = 0^+$, there is not yet any voltage across the capacitor; it thus appears as a short circuit. Since the only component through which current flows in this situation is R_s, the characteristic function for R_s is constructed from the I-V relationship $i(0^+)$ versus V_{in}, which can be deduced from Figure 7.4.

To determine the I-V relationship for R_p, consider that in steady state, since capacitors behave as open circuits, Figure 7.5 can be represented by Figure 7.6b; the steady state current through the laser is labeled I_{ss}. If the voltages across R_s and R_p are denoted as V_s and V_p, respectively, then by Kirchhoff's law, $V_p = V_{in} - V_s$. The input voltage V_{in} is a known quantity, and the voltage V_s can be deduced from the characteristic I-V function of R_s determined previously. The steady-state current I_{ss} as a function of V_{in} (and hence V_s) is known both from the DC table and the dynamic table. Thus, the characteristic function for the resistor R_p is determined by the data I_{ss} versus V_p or, equivalently, I_{ss} versus $V_{in} - I_{ss}R_s$. It is clear that the I-V relations for both resistors can be extracted from the data in the DC and dynamic tables.

Finally, the characteristic function for the capacitor C_p must be extracted. As stated previously, this takes the form of a charge-voltage expression. Referring to Figure 7.5, the charge through the capacitor can be expressed as

$$Q = \int_0^\infty i_c(t)\, dt \tag{7.1}$$

where i_c is the capacitor current and is related to the other currents by $i_c = i - i_r$ (i is known from the dynamic table at any given time). The determination of i_r is clearly necessary to determine i_c. This can be accomplished through (7.2) – (7.4).

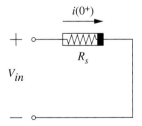

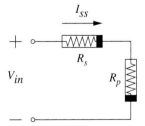

Figure 7.6a: Equivalent circuit at $t = 0^+$.

Figure 7.6b: Equivalent circuit in steady state.

$$V_s = iR_s \tag{7.2}$$

$$V_p = V_{in} - V_s \tag{7.3}$$

$$i_r = \frac{V_p}{R_p} \tag{7.4}$$

In (7.2), V_s is known for a given i because the I-V characteristic of R_s is known. In (7.4), i_r is known because the I-V characteristic of R_p is known. Thus, (7.2) – (7.4) are summarized as

$$i_r(t) = \frac{V_{in} - i(t)R_s}{R_p} \tag{7.5}$$

Note that this analysis began by extracting i from the dynamic table at a given t. This process (7.2) – (7.5) must be repeated for all times represented in the dynamic table in order to establish the behavior of $i_c(t)$. Once $i_c(t)$ is known for all time t, it can be numerically integrated as required by (7.1). Since each set of data in the dynamic table is measured at a given input voltage, this procedure must be performed for each DC bias point.

Curve fitting of tabular data for numerical integration can be performed in many different ways. In the approach of Pang's model, the data are first fitted to a cubic-spline function [7.9]. For a better fit using the cubic spline algorithm, the first and second derivatives are calculated at the end points in [7.1] through the divided difference method [7.8] shown in (7.6) and (7.7).

$$\frac{dx}{dt} \approx \frac{x_{n+1} - x_n}{t_{n+1} - t_n} \tag{7.6}$$

$$\frac{d^2x}{dt^2} \approx 2 \cdot \frac{\dfrac{x_{n+1} - x_n}{t_{n+1} - t_n} - \dfrac{x_{n+2} - x_{n+1}}{t_{n+2} - t_{n+1}}}{t_{n+2} - t_n} \tag{7.7}$$

Once the tabular data have been curve fitted, evaluation and then integration of the resulting spline function can take place (7.1).

In the preceding paragraphs, the characteristic functions of each of the equivalent-circuit elements were numerically constructed from data. The next issue to be addressed is the actual operation of the model during simulation. As with an analytical model, when called by the simulation program, the table-based model must return the values of currents or voltages, as well as the necessary derivatives. It is crucial, then, to be able to extract this information from the characteristic functions. Since the inputs from the simulator to the table-based model will be continuous, it is necessary to develop a strategy for interpolating between data points stored in the DC and dynamic tables. As mentioned previously, various interpolation methods exist. For the two-terminal laser diode model, Pang uses piecewise linear interpolation, which is shown in [7.1] to be more robust in this particular situation than the cubic spline algorithm. In this particular scenario, if the voltage V_{in} is applied across the two terminals of the laser, then the table model must return the resulting current I. Since $V_{in} = V_s + V_p$, the current can be determined through piecewise linear interpolation of the two resistor (R_s and R_p) I-V characteristics.

In addition, given the input V_{in}, the appropriate derivatives must also be calculated. For differentiation, Pang uses the cubic spline again, with endpoint derivatives calculated through (7.6) and (7.7). The first quantity to be calculated is the derivative of the current with respect to the voltage (i.e., the conductance g):

$$g = \left. \frac{dI}{dV} \right|_{V_{in}} \tag{7.8}$$

In addition, the charge derivative (i.e., the capacitance) must also be extracted. This is accomplished through interpolation of the capacitor's Q-V characteristic:

$$Q = \int_0^\infty i_c(t)\, dt \rightarrow C = \left. \frac{dQ}{dV} \right|_{V_{in}} \tag{7.9}$$

A comparison of the table-based model with data obtained from circuit simulation of the analytical model of Gao et al. presented previously is depicted in Figures 7.7 and 7.8. Figure 7.9 depicts a simulation of a laser transmitter using the analytical HEMT model of Cioffi et al. and the table-based laser model of Pang et al. As seen in all of these examples, the table-based model matches the analytical model quite well. Considering the complexity of the analytical model, the advantages of the table-based approach are clear.

7.2.2 Laser diode optical properties

Table-based modeling of the laser's output power characteristic can be accomplished by applying the same procedure used for generation of the I-V characteristics. The data used to construct the DC and dynamic tables are depicted in Figures 7.10 and 7.11, respectively. As before, the first step is to qualitatively map the behavior of the laser output into simple circuit elements. As in the example of Section 7.1, the behavior of the dynamic table (Figure 7.11) closely resembles that of a charging capacitive circuit; that is, it starts at a low value and then rises until it reaches a steady-state value. In this particular case, the steady-state capacitive

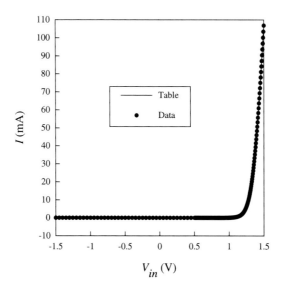

Figure 7.7: DC simulation of laser diode electrical (I-V) characteristics.

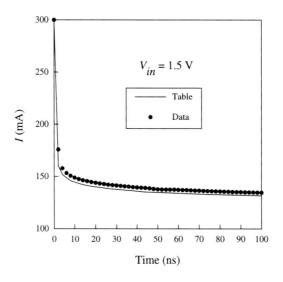

Figure 7.8: Transient simulation of laser diode electrical (I-V) characteristics.

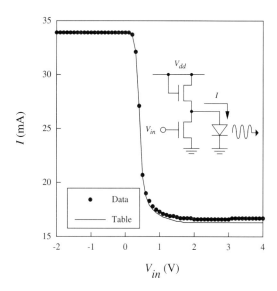

Figure 7.9: Laser transmitter electrical simulation.

voltage must be mapped into the steady-state optical power, denoted P_{ss}. This behavior can be represented by the middle portion of the equivalent circuit of Figure 7.12. (The left and right sides of the circuit will be explained shortly.) Based on this circuit, the voltage charging the capacitor must have a value equal to P_{ss}, a quantity which is extracted from the V_{in}-P_{ss} DC table.

In the construction of the equivalent circuit, it would seem natural to model the $(V_{in}$-$P_{ss})$ DC table by a nonlinear resistor as in the modeling of the laser's electrical characteristics. However, as Pang explains in [7.1], this is not possible. The reason is that to use this resistive representation, P_{ss} would have to be a current, which it cannot be since it was already defined as the capacitor charging *voltage*. The remedy to this problem is to define P_{ss} as a current and to use a CCVS (current-controlled voltage source) with a voltage equal to the current P_{ss} multiplied by a constant of 1 V/A. This explains the dependent source in the middle of Figure 7.12. With P_{ss} defined as a current, the resistor R_{in} can be used to model the V_{in}-P_{ss} relation contained in the DC table. Since power is, in reality, optical emission and is *not* an electrical current (it is only being *modeled* as a current), it is important to ensure that the resistor modeling the V_{in}-P_{ss} relationship not be allowed to connect electrically to external parts of the circuit. Thus, the VCVS (voltage-controlled voltage source) on the left side of the equivalent circuit serves to isolate the P_{ss} current that models the power. As can be deduced from Figure 7.12, the multiplier on the VCVS is unity. For this same reason, the capacitive voltage V_c (that models the output power through the dynamic table) is isolated through another unity VCVS on the right side

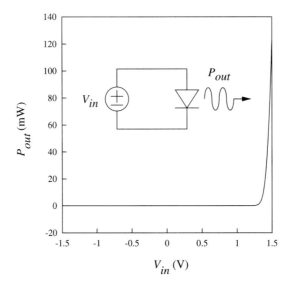

Figure 7.10: Circuit and data used to construct DC output power table.

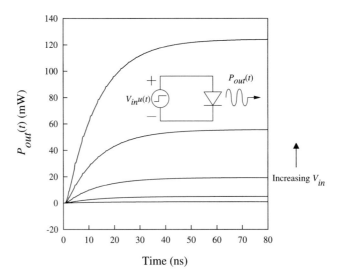

Figure 7.11: Circuit and data used to construct dynamic output power table.

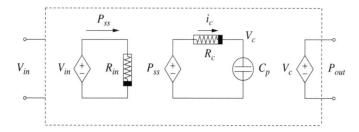

Figure 7.12: Equivalent circuit for modeling laser output power.

of the circuit. Thus, the input to the circuit is the voltage V_{in}, and the output is the optical power P_{out}.

Finally, characteristic functions must, as in the electrical case, be assigned to each of the equivalent circuit elements in Figure 7.12. As stated, the characteristic function of R_{in} is the V_{in}-P_{ss} DC table. The situation for R_c and C_p, however, is not as straightforward. At $t = 0^+$, the capacitor appears as a short circuit; thus, the only element connected to the CCVS is the resistor R_c. Unfortunately, there is no information contained in either the DC or the dynamic table about the current through R_c at $t = 0^+$. In steady state, the current through R_c is zero; again, no information can be extracted. Thus, it would seem impossible to extract the characteristic function for R_c. In [7.1], Pang takes an iterative approach. Under this technique, the nonlinear resistor R_c is linearized and all other characteristic functions are determined based on this linear assumption:

$$P_{ss} = f(i_c) + V_c \rightarrow P_{ss} = i_c R_c + V_c \qquad (7.10)$$

Once this process is complete, comparison with the data is performed to adjust the value of R_c. This, in turn, requires recalculation of the other characteristic functions, a process that continues until the value of R_c agrees with both the DC and the dynamic tables [7.1].

The dynamic table contains information about $P_{out}(t)$ and V_{in} versus time. Since this behavior is modeled by the capacitor, the characteristic Q-V function of C_p is again determined by

$$Q = \int_0^\infty i_c\, dt = \frac{1}{R_c} \int_0^\infty [P_{ss}(t) - V_c(t)]\, dt \qquad (7.11)$$

where

$$P_{ss}(t) = i_c(t) R_c + P_{out}(t) \qquad (7.12)$$

Once the characteristic functions have been defined, interpolation and differentiation occur in a manner identical to that used in the electrical case. That is, piecewise linear interpolation and cubic spline differentiation are used to determine values not directly contained in the DC and dynamic tables. Comparisons of the power characteristic of the table-based model with the analytical model are shown in Figures 7.13 and 7.14, and an optical simulation of Figure 7.9 is depicted in Figure 7.15. As before, the matches are quite good.

7.2.3 Table-based model using device simulation

As shown in the previous sections, simulation output can often be used in place of measured data in the construction of a table-based laser diode model. This is particularly helpful when measured laboratory data are not available. Sections 7.2.1 and 7.2.2 showed how well the table-based model fits the analytical laser diode model; however, as far as the user is concerned, the data from the analytical model could just as well have been measured data. This is especially true for device simulation, which involves the very detailed solution of the equations of fundamental semiconductor and quantum-mechanical physics. In this section, an example of using device simulation output, in place of measured data, to construct a table-based laser model will be illustrated.

In order to avoid replication of the simulation results presented previously, the example presented in this section will focus on modeling the relaxation oscillation

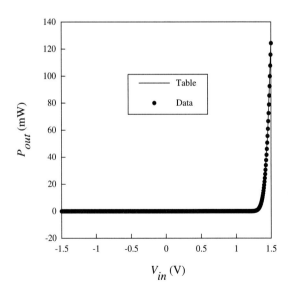

Figure 7.13: DC simulation of laser diode optical characteristics.

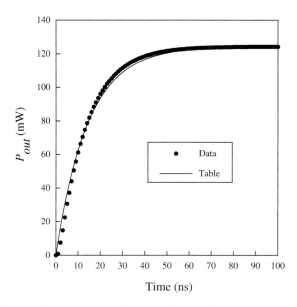

Figure 7.14: Transient simulation of laser diode optical characteristics.

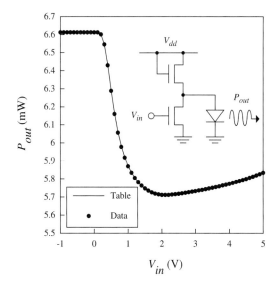

Figure 7.15: Laser transmitter optical simulation.

portion of the laser's transient response. The tool to be used in this section is MINILASE [7.10], a two-dimensional quantum-well laser device simulator.

As a review, the relaxation oscillation phenomenon occurs at the beginning of the laser transient when current is suddenly injected. A good illustration of this is Figure 3.7. When this happens, the electron concentration rises abruptly. Normally, the laser handles this excess electron population through spontaneous emission or nonradiative recombination; that is, the excess electrons are depleted through various physical mechanisms. However, if the laser is turned on very quickly, the electron population builds up and reaches the threshold density more quickly than the equilibration process can occur. When recombination finally begins to take place, since the electron population is so large, a huge number of photons is emitted through stimulated emission, which quickly drives the electron population below the threshold density. The resulting dip in the number of electrons causes the system to try to equilibrate through a rapid buildup of electrons, which then results in a large number of stimulated photon emissions. This process repeats itself, over and over, the effect becoming less profound at each iteration. Eventually, the system equilibrates to some steady state as seen, again, in Figure 3.7. Figure 7.16 depicts a MINILASE device simulation of a laser diode turn-on transient. As explained, the current is initially high, the power is initially low, and both undergo damped oscillations. As expected, the electrical and optical powers are "90° out of phase" [7.1] indicating an exchange of energy between the electrical and optical powers.

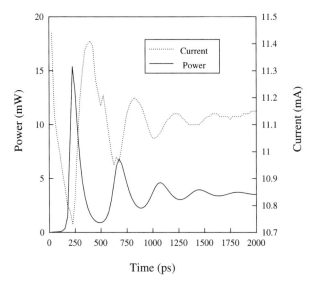

Figure 7.16: MINILASE simulation of laser transient.

   The table-based approach can be used to model the decaying oscillatory behavior of the turn-on transient by mapping it into an RLC circuit (Figure 7.17). In general, an LC circuit will exhibit oscillatory behavior as energy is dynamically exchanged between the inductor and the capacitor. Addition of a resistor to the circuit will dampen the oscillations by dissipating energy. Since the instantaneous current flow across an inductor is zero, the laser's injected current cannot change instantaneously; this models the behavior of the pump source attempting to maintain a steady flow of charge [7.1]. Stimulated emission is modeled in Pang's approach by the capacitor, which uses up charge from the pump source. Since the storage elements are relaxed at $t = 0$, (i.e., the inductor acts as an open circuit and the capacitor as a short circuit), the series resistor R_s is necessary to ensure a finite amount of current flow into the laser. The resistor R_p is used to model the steady-state value of the current since, in steady state, the capacitor is open circuited and the inductor is short circuited.

   By analyzing Figure 7.17, the current resulting from a step input V_{in} can be expressed as (7.13), where the time derivative of V_{in} is zero for $t \geq 0^+$.

$$\frac{d^2 i}{dt^2} + \left(\frac{L + R_s R_p C}{R_s L C}\right) \cdot \frac{di}{dt} + \left(\frac{R_s + R_p}{R_s L C}\right) \cdot i = \frac{V_{in}}{R_s L C} \tag{7.13}$$

This can be recast into the form of

$$\frac{d^2 i}{dt^2} + 2\alpha \frac{di}{dt} + \omega_0^2 i = \kappa \tag{7.14}$$

where α and ω_0 are referred to, in standard circuit theory, as the exponential damping coefficient and the undamped resonant frequency, respectively [7.11]. The homogeneous solution of (7.14) is

$$i(t) = K \exp(st) \tag{7.15}$$

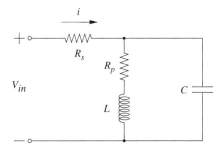

Figure 7.17: RLC circuit to model current relaxation oscillations.

where

$$s = -\alpha \pm \sqrt{\alpha^2 - \omega_0^2} \tag{7.16}$$

and K is an arbitrary constant. For the damped oscillatory response exhibited by an RLC circuit, the quantity $\alpha^2 - \omega_0^2$ is negative, making s a complex number. If the damping frequency β is defined as $\beta^2 = \omega_0^2 - \alpha^2$, then (7.15) can be rewritten as

$$i(t) = \exp(-\alpha t) [A_1 \cos(\beta t) + A_2 \sin(\beta t)] \tag{7.17}$$

The values of the components in the equivalent circuit of Figure 7.17 must, then, be chosen so that the damped oscillatory response of Figure 7.16 matches (7.17).

The data used to construct one entry of the dynamic table, corresponding to $V_{in} = 1.7u(t)$, are graphed in Figure 7.18. (As a review, the dynamic table consists of the transient response to *all* step inputs V_{in}). From the equivalent circuit of Figure 7.17, when the input is applied at $t = 0^+$, the capacitor is short circuited; the inductor L and the resistor R_p are, thus, removed from consideration. Therefore, the series resistor R_s can be expressed as

$$R_s = \frac{V_{in}}{i(t=0^+)} \tag{7.18}$$

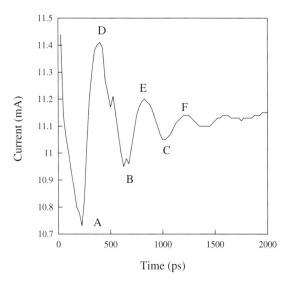

Figure 7.18: MINILASE transient current response.

In steady state, the capacitor is open circuited and the inductor is short circuited; thus, the equivalent circuit of Figure 7.17 is reduced to the series combination of R_s, R_p, and V_{in}. Since R_s and V_{in} are known, R_p can be expressed as

$$R_p = \frac{V_{in}}{I_{ss}} - R_s \tag{7.19}$$

where the steady-state current I_{ss} for a given V_{in} is extracted either from the DC table or from the steady-state portion of the dynamic table.

The damping coefficient α can be determined from Figure 7.18 by estimating the envelope of the relaxation oscillations as a decaying exponential with an offset:

$$i_{envelope}(t) = I_{ss} \pm I_o \exp(\gamma t) \tag{7.20}$$

In Pang's approach, I_o and γ are estimated by simple approximation from the first two peaks and the first two valleys of Figure 7.18 (points D & E and A & B, respectively). This results in two sets of I_o and γ; the damping coefficient α (7.17) is taken to be the average of the two values of γ. The damping frequency β can be estimated in a similar manner.; the amount of time between peaks and valleys is sampled (i.e., the amount of time between points A & B, B & C, D & E and E & F in Figure 7.18). The average of these times is taken to be the damping period τ, from which β is determined by

$$\beta = \frac{2\pi}{\tau} \tag{7.21}$$

Finally, ω_o^2 is calculated through $\omega_o^2 = \beta^2 + \alpha^2$.

With the determination of α and ω_o^2, the quantities R_s, R_p, L, and C can be determined from (7.13) and (7.14):

$$2\alpha = -\frac{L + R_s R_p C}{R_s L C} \tag{7.22}$$

$$\omega_o^2 = \frac{R_s + R_p}{R_s L C} \tag{7.23}$$

Solution of these relationships results in $R_s = 144.80\ \Omega$, $R_p = 8.07\ \Omega$, $C = 5.17$ pF, and $L = 0.85$ nH. A comparison of a table-based simulation with the original MINILASE data is shown in Figure 7.19. While the match obtained in this particular case is not perfect, this example is an excellent illustration of the use of device simulation output in place of actual, measured data. The deficiency in this approach is most likely due to the idealized linear assumptions made in Pang's technique [7.1].

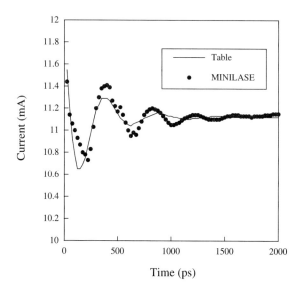

Figure 7.19: Comparison of table-based model with device simulation output.

A similar approach is taken in [7.1] for modeling the relaxation oscillations in the laser's optical output. The damped oscillatory behavior can, again, be modeled by a series RLC circuit, as shown in Figure 7.20. The inductor prevents instantaneous current flow into the laser diode and the capacitor prevents instantaneous voltage build-up. If the capacitor voltage is used to mimic optical power, then the build-up of charge on the capacitor can be used to represent the build-up and generation of photons during stimulated emission. The resistor models the effect of absorption in the laser cavity, which eventually brings the system to a steady-state equilibrium. The equivalent-circuit parameters R, L, and C are determined in a manner similar to that used to determine the circuit parameters for the current model. The capacitor voltage, which is used to model optical power, can be expressed as (7.24).

$$\frac{d^2V_c}{dt^2} + \frac{R}{L} \cdot \frac{dV_c}{dt} + \frac{V_c}{LC} = \frac{P_{ss}}{LC} \tag{7.24}$$

Following the method used to model the laser current, (7.24) is rewritten as

$$\frac{d^2V_c}{dt^2} + 2\alpha\frac{dV_c}{dt} + \omega_o^2 V_c = \kappa \tag{7.25}$$

where β is again defined as $\beta^2 = \omega_o^2 - \alpha^2$. The values of L and C are calculated from (7.26) and (7.27).

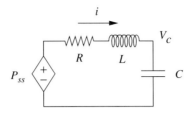

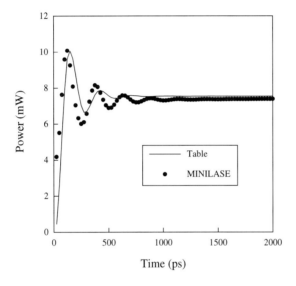

Figure 7.20: Equivalent circuit for modeling power oscillations (top) and comparison of table-based model with device simulation output (bottom).

$$2\alpha = \frac{R}{C} \tag{7.26}$$

$$\omega_o^2 = \frac{1}{LC} = \beta^2 + \alpha^2 \tag{7.27}$$

The extracted circuit parameters for an input voltage of 1.8 V are $R = 6.19$ mΩ, $L = 365.90$ fH, and $C = 4.32$ nF. A comparison is illustrated in Figure 7.20. Again, the match is not perfect, but this exercise remains a good example of the concept of combining device simulation with table-based circuit simulation.

7.3 Heterojunction Bipolar Transistor (HBT)

Several popular models exist for the circuit simulation of bipolar junction transistors (BJTs), such as the Ebers-Moll [7.12] and Gummel-Poon models [7.13]. However, as was the case with the laser diode, no such *standard* models exist for the heterojunction bipolar transistor (HBT). The lack of a suitable circuit model for the HBT is a considerable deficiency considering the popularity of HBT use in OEIC receiver and transmitter design. Acceptance of a standard HBT model has been hindered by the considerable complexity of the physics involved; this problem has been exacerbated by constantly evolving device structures and processing techniques. As mentioned throughout the chapter, this is exactly the type of situation in which table-based modeling is useful. While a universal HBT structure has not yet been adopted, there still exists a need for simulating HBTs in optoelectronic integrated circuits. As the HBT technology matures, a more accurate analytical circuit model will undoubtedly be developed.

In this section, Pang's table-based HBT model will be demonstrated [7.1]. Because the HBT is a multiterminal device, more involved numerical interpolation techniques are required than those used in the modeling of the (two-terminal) laser diode. The approach taken in this model is the use of multidimensional linear isoparametric shape functions for both numerical interpolation and differentiation. This algorithm will be elaborated upon shortly.

As with the table-based laser diode model, two tables are required for simulation: a DC table and a dynamic table. The DC table is generated by sweeping the voltages across the base-emitter (V_{be}) and collector-emitter (V_{ce}) junctions, then measuring and tabulating the resulting base (I_b) and collector (I_c) currents, as shown in Figure 7.21a. The dynamic table is generated from small-signal S parameters measured at each of the DC biases, V_{be} and V_{ce}. The resulting S parameters, their corresponding frequencies, and the DC bias points constitute the dynamic table.

The first step in the development of the table-based laser diode model was the determination of an equivalent circuit that mimicked the behavior of the measured data. For the DC HBT model, Pang uses the equivalent circuit of Figure 7.21b. The characteristic functions of R_{be} and R_{ce} are readily determined directly from the DC table. For the dynamic model, Pang uses the popular *hybrid-π* model (Figure 7.22). The feature which differentiates this approach from the standard hybrid-π model is the fitting of the HBT to the model and the use of table-based numerical techniques for interpolation and differentiation. In Figure 7.22, the contact resistors r_b, r_c, and r_e are assumed to be linear, while the resistors r_{be} and r_{ce} are nonlinear. The base-emitter (C_{be}) and base-collector (C_{bc}) capacitances are also taken to be nonlinear. Once the decision is made to use the hybrid-π model, the next task is to determine the values of the circuit elements that best fit the device behavior exhibited by the measured S parameters. In this particular example, a commercial AC parameter extraction tool [7.14] is used. To expedite the extraction process, it is helpful to have some idea of the ranges of the elements before beginning. Since r_b and r_c are

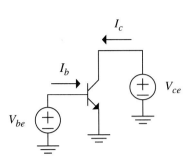

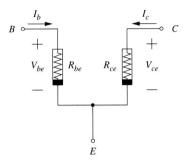

Figure 7.21a: Setup for DC measurements.

Figure 7.21b: DC equivalent circuit for table model.

assumed to be linear, their values can be estimated from the DC table as the inverse slopes of the base current and collector current I-V characteristics, respectively, above threshold. A priori knowledge of as many element values as possible significantly increases the extraction accuracy of the other components. The DC values of r_b and r_c are, thus, used as initial guesses for the AC parameter extraction. In the extraction process, not only is r_e determined, but r_b and r_c are each slightly optimized (the inverse slope calculation was only an estimate). Once the values of the three linear resistors have been determined, the voltages v_π and v_o can be calculated through Kirchhoff's laws:

$$i_e = i_b + i_c \tag{7.28}$$

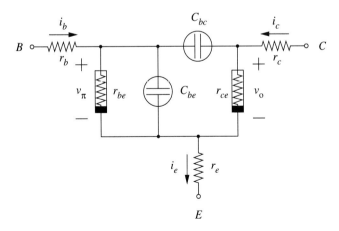

Figure 7.22: Hybrid π small-signal bipolar model.

$$v_\pi = v_{be} - i_b r_b - i_e r_e \tag{7.29}$$

$$v_o = v_{ce} - i_c r_c - i_e r_e \tag{7.30}$$

From these equations, the characteristic functions for r_{be} and r_{ce} are, thus, i_b versus (v_π, v_o) and i_c versus (v_π, v_o), respectively.

The same AC parameter extraction tool is used in [7.1] to determine the capacitor values from the S parameter data at each DC bias point. Once the capacitance-voltage (C-V) relationship has been determined, it can be integrated to obtain the Q-V characteristic of each capacitor.

In the laser diode model, the I-V characteristics were only single-variable functions. In the HBT model, however, the I-V characteristics of r_{be} and r_{ce} are multivariable functions. Thus, simple piecewise linear and cubic spline interpolation and differentiation algorithms cannot be applied. For these operations, Pang uses the multidimensional linear isoparametric shape functions. These functions not only have superior convergence characteristics but also reduce the Newton-Raphson iteration time [7.1]. In [7.7], it is shown that use of isoparametric shape functions in MOSFET modeling can reduce the overall simulation time by 10%. In the same reference, successful interpolation and differentiation were demonstrated for as few as 30 points in the data tables. For these reasons, Pang uses 2-D isoparametric shape functions to model the behavior of the three-terminal HBT.

To provide some background, the term *shape functions* refers to a group of functions used to interpolate a given function in space through linear combinations. In mathematical terms, a set of shape functions forms a *basis* of the space. (A detailed discussion about *bases* can be found in references such as [7.15].) If tuning parameters are included to permit interpolation over a wider space, the shape functions are termed *parametric*. The term *isoparametric* refers to the situation in which both the function and the coordinate system can be mapped using the same set of shape functions [7.16] – [7.20]. This definition will become clear as the discussion progresses.

As mentioned previously, the problem of interpolation becomes considerably more complex when a function of more than one variable is involved. The two-dimensional interpolation problem can be expressed in Figure 7.23. In this figure, f is a function of two variables $f(x,y)$, and the value of f at four different (x,y) coordinates is known, as would be the case in a table of measured data. The difficulty is in trying to predict the value of the function for an (x,y) that falls between the four points. In the case of the HBT model, x would represent the voltage v_π, and y the voltage v_o.

The technique begins by assuming that the interpolation can be expressed in a form such as

$$f(x, y) = a + bx + cy + dxy \tag{7.31}$$

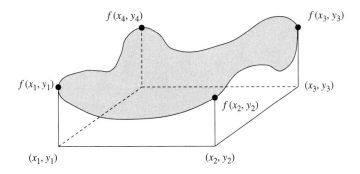

Figure 7.23: Function f of two variables x and y.

where a-d are constants. Since four points of the function are known, the constants a-d can be evaluated according to

$$
\begin{aligned}
f(x_1, y_1) &\equiv f_1 = a + bx_1 + cy_1 + dx_1y_1 \\
f(x_2, y_2) &\equiv f_2 = a + bx_2 + cy_2 + dx_2y_2 \\
f(x_3, y_3) &\equiv f_3 = a + bx_3 + cy_3 + dx_3y_3 \\
f(x_4, y_4) &\equiv f_4 = a + bx_4 + cy_4 + dx_4y_4
\end{aligned}
\tag{7.32}
$$

It can be shown that once a-d are determined, (7.31) can be expressed as

$$
f(x, y) = \sum_{i=1}^{4} f_i S_i
\tag{7.33}
$$

where the functions f_i are as defined in (7.32) and the S_i terms are called *shape functions* and are expressed as [7.17]

$$
S_1 = \frac{(x_2 - x)(y_3 - y)}{(x_2 - x_1)(y_3 - y_1)}
\tag{7.34}
$$

$$
S_2 = \frac{(x_1 - x)(y_3 - y)}{(x_1 - x_2)(y_3 - y_2)}
\tag{7.35}
$$

$$
S_3 = \frac{(x_1 - x)(y_1 - y)}{(x_1 - x_3)(y_1 - y_3)}
\tag{7.36}
$$

$$S_4 = \frac{(x - x_2)\,(y_1 - y)}{(x_4 - x_2)\,(y_1 - y_4)} \tag{7.37}$$

In addition, the x and y coordinates can also be expressed in terms of the shape functions as shown in (7.38). This was the situation alluded to previously in which both the function and the coordinate system can be mapped using the same set of shape functions.

$$x = \sum_{i=1}^{4} x_i S_i \qquad\qquad y = \sum_{i=1}^{4} y_i S_i \tag{7.38}$$

Consider the situation in which the x and y coordinates are uniformly distributed on a two-dimensional grid. The next step is to express the system in terms of *local coordinates* (Figure 7.24). In this case, the local coordinates are ξ and η and the mapping between (x,y) and (ξ,η) is deduced from Figure 7.24 to be

$$(x, y) \rightarrow (\xi, \eta) = \begin{cases} (x_1, y_1) \rightarrow (-p, -q) \\ (x_2, y_2) \rightarrow (p, -q) \\ (x_3, y_3) \rightarrow (p, q) \\ (x_4, y_4) \rightarrow (-p, q) \end{cases} \tag{7.39}$$

With these definitions, the shape functions can be rewritten as [7.17]

$$S_1(\xi, \eta) = \frac{(p - \xi)\,(q - \eta)}{4pq} \tag{7.40}$$

$$S_2(\xi, \eta) = \frac{(p + \xi)\,(q - \eta)}{4pq} \tag{7.41}$$

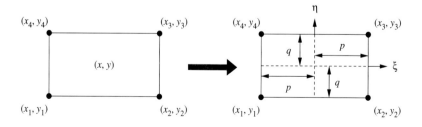

Figure 7.24: Transformation of coordinates.

$$S_3(\xi, \eta) = \frac{(p+\xi)(q+\eta)}{4pq} \tag{7.42}$$

$$S_4(\xi, \eta) = \frac{(p-\xi)(q+\eta)}{4pq} \tag{7.43}$$

Finally, the shape functions are normalized with the nondimensional relations

$$r = \frac{\xi}{p} \qquad\qquad s = \frac{\eta}{q} \tag{7.44}$$

The nondimensional shape functions are thus

$$
\begin{aligned}
S_1 &= 0.25\,(1-r)\,(1-s) \\
S_2 &= 0.25\,(1+r)\,(1-s) \\
S_3 &= 0.25\,(1+r)\,(1+s) \\
S_4 &= 0.25\,(1-r)\,(1+s)
\end{aligned}
\tag{7.45}
$$

The quantities r and s are termed the local coordinates; by their normalization, their ranges are restricted to $-1 \le r \le 1$, $-1 \le s \le 1$. The 2-D isoparametric shape functions are graphed in Figure 7.25. Modeling through use of isoparametric shape functions requires, then, that a function in an arbitrary two-dimensional coordinate space (such as (x,y)) be mapped into the (r,s) coordinate space. In this particular case, a function $f(x,y)$ must be mapped into $g(r,s)$.

While the mapping in this example was made for a rectangular shape, in [7.17] it is shown that this same set of isoparametric shape functions (7.45) can be used to map an arbitrary quadrilateral into the (r,s) domain (Figure 7.26). It is natural to inquire about the advantages of this technique, as compared to other numerical interpolation methods. As discussed previously in this section, interpolation via isoparametric shape functions improves convergence time and often requires fewer data points when compared to methods such as spline and polynomial interpolation. Furthermore, the shape functions are of a form which provides continuity at the boundaries between quadrilateral regions as well as monotonicity [7.7]. The shape functions described thus far have represented four-node elements. To describe more complex shapes and functions of greater dimensionality, higher-order isoparametric shape functions can be used; an example of the use of a twentieth order isoparametric interpolation will be demonstrated in the modeling of the HFET in Section 7.4.

Finally, the shape function derivatives must be computed; recall that in circuit simulation, not only the currents and voltages, but also their derivatives, must be returned at the end of each simulation iteration. The derivatives are calculated in (7.46), where the partial derivatives of S_i are given by (7.47) and (7.48) [7.1].

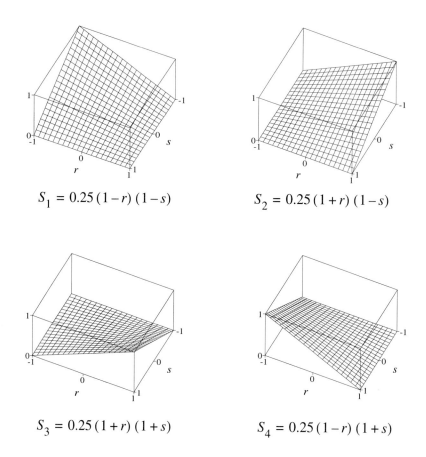

$$S_1 = 0.25\,(1-r)\,(1-s)$$ $$S_2 = 0.25\,(1+r)\,(1-s)$$

$$S_3 = 0.25\,(1+r)\,(1+s)$$ $$S_4 = 0.25\,(1-r)\,(1+s)$$

Figure 7.25: 2-D isoparametric shape functions.

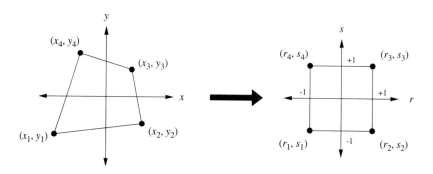

Figure 7.26: Isoparametric mapping of quadrilateral.

$$\frac{\partial f(x,y)}{\partial x} = \sum_{i=1}^{4} \frac{\partial S_i}{\partial x} \cdot f_i \qquad\qquad \frac{\partial f(x,y)}{\partial y} = \sum_{i=1}^{4} \frac{\partial S_i}{\partial y} \cdot f_i \qquad (7.46)$$

$$\frac{\partial S_i}{\partial x} = \frac{\partial S_i}{\partial r} \cdot \frac{\partial r}{\partial x} + \frac{\partial S_i}{\partial s} \cdot \frac{\partial s}{\partial x} \qquad (7.47)$$

$$\frac{\partial S_i}{\partial y} = \frac{\partial S_i}{\partial r} \cdot \frac{\partial r}{\partial y} + \frac{\partial S_i}{\partial s} \cdot \frac{\partial s}{\partial y} \qquad (7.48)$$

Incidentally, the partial derivatives of r and s with respect to x and y in these equations are the same ones that would be used in the Jacobian matrix for transformation between coordinate systems [7.17], [7.20].

$$\bar{J} = \begin{bmatrix} \dfrac{\partial x}{\partial r} & \dfrac{\partial y}{\partial r} \\[2ex] \dfrac{\partial x}{\partial s} & \dfrac{\partial y}{\partial s} \end{bmatrix} \qquad (7.49)$$

In Pang's approach, the first step is the construction of the data tables; as alluded to previously, these consist of v_π and v_0 as the (x,y) coordinates and i_c and i_b as the functions for the dynamic table, and V_{be} and V_{ce} as the (x,y) coordinates and I_b and I_c as the functions for the DC table. The next step in the method of [7.1] is to determine an appropriate set of (r,s)-domain coordinates for each set of (x,y). These are denoted in [7.1] as $(v_{\pi,i}, v_{0,i})$ and $(V_{be,i}, V_{ce,i})$ for the dynamic and DC cases, respectively. The coordinate system mapping is accomplished through isoparametric shape functions, as expressed in (7.38). Once the local coordinates (r,s) are determined from (x,y), the values of the currents are interpolated from the shape functions (7.33) and their derivatives are determined from (7.46). A DC simulation of a TTL inverter is presented in Figure 7.27, while a transient simulation is presented in Figure 7.28.

7.4 Heterojunction Field-Effect Transistor (HFET)

As the last example of this chapter, the table-based HFET model of Cho et al. [7.21], [7.22] will be presented. This model is essentially an extension of Pang's HBT model in that it, too, uses isoparametric shape functions for interpolation and differentiation. However, while the HBT model was two-dimensional (V_{be} and V_{ce}), Cho's HFET model is three-dimensional. The three dimensions arise from the

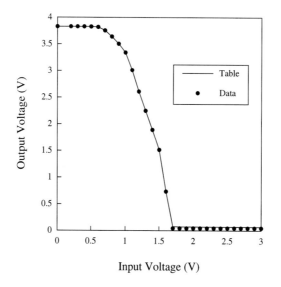

Figure 7.27: DC simulation of TTL inverter.

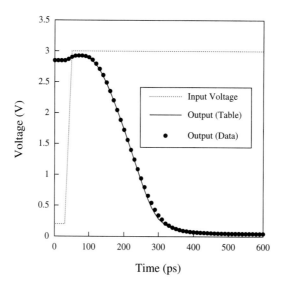

Figure 7.28: Transient simulation of TTL inverter.

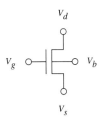

Figure 7.29: Node voltages of the HFET.

gate-source, drain-source, and bulk-source voltages (V_{gs}, V_{ds}, and V_{bs}, respectively), as depicted in Figure 7.29.

It was possible to solve the two-dimensional problem with a set of four isoparametric shape functions via the mapping of the original surface onto a square (Figure 7.26). However, the three-dimensional case requires mapping onto not a square, but a cube; this requires 8 nodes. In Cho's HFET model, additional accuracy is achieved through the addition of an extra node at the midpoint of each cube edge, increasing the number of nodes from eight to twenty. The 20-node element allows the original boundary to be curved, which increases interpolation accuracy [7.19]. In addition, the use of extra nodes necessitates a polynomial fit, rather than linear one, which also increases numerical accuracy. The mapping of a twenty-node element from global (x,y,z) coordinates to local (r,s,t) coordinates is depicted in Figure 7.30. Following the example in the previous section, the shape functions necessary to interpolate the twenty-node element are given by (7.50) [7.21], [7.22], where as before, the original function $f(x,y,z)$ is expressed as (7.51). In (7.50) and (7.51), the points (x_i,y_i,z_i) and (r_i,s_i,t_i) are taken to be the node points in the global and local

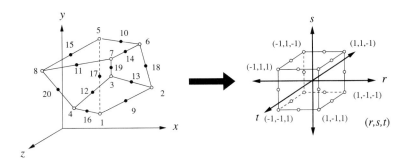

Figure 7.30: Isoparametric mapping of the 20-node element.

coordinate systems, respectively. Also, as before, since the shape functions are isoparametric, the global and local coordinates are related by (7.52).

$$S_i = 0.125\,(1 + rr_i)\,(1 + ss_i)\,(1 + tt_i) \quad \text{for } i = 1 \text{ to } 8$$

$$S_i = 0.25\,(1 - r^2)\,(1 + ss_i)\,(1 + tt_i) \quad \text{for } i = 9 \text{ to } 12$$

$$S_i = 0.25\,(1 + rr_i)\,(1 - s^2)\,(1 + tt_i) \quad \text{for } i = 13 \text{ to } 16 \tag{7.50}$$

$$S_i = 0.25\,(1 + rr_i)\,(1 + ss_i)\,(1 - t^2) \quad \text{for } i = 16 \text{ to } 20$$

$$f(x, y, z) = \sum_{i=1}^{20} f_i S_i \qquad f_i = f(x_i, y_i, z_i) \tag{7.51}$$

$$x = \sum_{i=1}^{20} x_i S_i \qquad y = \sum_{i=1}^{20} y_i S_i \qquad z = \sum_{i=1}^{20} z_i S_i \tag{7.52}$$

The partial derivatives can, again, be expressed as

$$\frac{\partial f}{\partial x} = \sum_{i=1}^{20} \frac{\partial S_i}{\partial x} \cdot f_i \qquad \frac{\partial f}{\partial y} = \sum_{i=1}^{20} \frac{\partial S_i}{\partial y} \cdot f_i \qquad \frac{\partial f}{\partial z} = \sum_{i=1}^{20} \frac{\partial S_i}{\partial z} \cdot f_i \tag{7.53}$$

where [7.21], [7.22]

$$\frac{\partial S_i}{\partial x} = \frac{\partial S_i}{\partial r} \cdot \frac{\partial r}{\partial x} + \frac{\partial S_i}{\partial s} \cdot \frac{\partial s}{\partial x} + \frac{\partial S_i}{\partial t} \cdot \frac{\partial t}{\partial x}$$

$$\frac{\partial S_i}{\partial y} = \frac{\partial S_i}{\partial r} \cdot \frac{\partial r}{\partial y} + \frac{\partial S_i}{\partial s} \cdot \frac{\partial s}{\partial y} + \frac{\partial S_i}{\partial t} \cdot \frac{\partial t}{\partial y} \tag{7.54}$$

$$\frac{\partial S_i}{\partial z} = \frac{\partial S_i}{\partial r} \cdot \frac{\partial r}{\partial z} + \frac{\partial S_i}{\partial s} \cdot \frac{\partial s}{\partial z} + \frac{\partial S_i}{\partial t} \cdot \frac{\partial t}{\partial z}$$

For accurate simulation of AC characteristics, accurate information about the terminal capacitances is required. Since capacitance is related to charge by (7.55) (where j represents a FET terminal), [7.21] bases its AC model on two tables of data: one for the gate charge Q_g and one for the bulk charge Q_b. Accordingly, interpolation between data table points is achieved through (7.56).

$$C_j = \frac{\partial Q_j}{\partial V_j} \qquad (7.55)$$

$$Q_g = \sum_{i=1}^{20} S_i(r, s, t) \cdot Q_{g,i} \qquad Q_b = \sum_{i=1}^{20} S_i(r, s, t) \cdot Q_{b,i} \qquad (7.56)$$

The corresponding capacitances are computed by [7.21]

$$C_{gj} = \sum_{i=1}^{20} \frac{\partial S_i(r, s, t)}{\partial V_{gj}} \cdot Q_{g,i} \qquad j = d, s, b$$

$$= \frac{\partial Q_g}{\partial V_{gj}} = \frac{\partial Q_g}{\partial r} \cdot \frac{\partial r}{\partial V_{gj}} + \frac{\partial Q_g}{\partial s} \cdot \frac{\partial s}{\partial V_{gj}} + \frac{\partial Q_g}{\partial t} \cdot \frac{\partial t}{\partial V_{gj}} \qquad (7.57)$$

$$C_{bj} = \sum_{i=1}^{20} \frac{\partial S_i(r, s, t)}{\partial V_{bj}} \cdot Q_{b,i} \qquad j = g, d, s$$

$$= \frac{\partial Q_b}{\partial V_{bj}} = \frac{\partial Q_b}{\partial r} \cdot \frac{\partial r}{\partial V_{bj}} + \frac{\partial Q_b}{\partial s} \cdot \frac{\partial s}{\partial V_{bj}} + \frac{\partial Q_b}{\partial t} \cdot \frac{\partial t}{\partial V_{bj}} \qquad (7.58)$$

In these equations, g, d, s, and b are the FET gate, drain, source, and bulk nodes, respectively, and $Q_{g,i}$ and $Q_{b,i}$ are the tabulated values of the gate and bulk charges, respectively. In [7.21], Cho calculates Q_d and Q_s through the charge-partitioning method:

$$Q_c = Q_d + Q_s = -(Q_g + Q_b) \qquad (7.59)$$

$$Q_d = X_{QC} Q_c \qquad (7.60)$$

$$Q_s = (1 - X_{QC}) Q_c \qquad (7.61)$$

Here Q_c refers to the channel charge and X_{QC} is the coefficient of channel charge share, which is used in the SPICE Level 2 Ward & Dutton charge model to subdivide the channel charge between the source and the drain [7.23].

In contrast to the charge model, for table-based DC analysis, Cho uses only an eight-node isoparametric interpolation because of the lower degree of complexity involved [7.7], [7.21], [7.22]. The isoparametric shape functions required for the eight-node interpolation are given by (7.62).

$$S_1 = 0.125 \, (1-r) \, (1-s) \, (1-t)$$

$$S_2 = 0.125 \, (1+r) \, (1-s) \, (1-t)$$

$$S_3 = 0.125 \, (1-r) \, (1+s) \, (1-t)$$

$$S_4 = 0.125 \, (1+r) \, (1+s) \, (1-t)$$

$$S_5 = 0.125 \, (1-r) \, (1-s) \, (1+t)$$ (7.62)

$$S_6 = 0.125 \, (1+r) \, (1-s) \, (1+t)$$

$$S_7 = 0.125 \, (1-r) \, (1+s) \, (1+t)$$

$$S_8 = 0.125 \, (1+r) \, (1+s) \, (1+t)$$

Graphically, the eight-node mapping is identical to the twenty-node case in Figure 7.30, without the mid-edge nodes.

For the DC table, values of current as a function of node voltages $I_{ds}(V_{ds}, V_{gs}, V_{bs})$ are compiled. As with the previous cases, the current can be interpolated by [7.7]

$$I_{ds}(V_{ds}, V_{gs}, V_{bs}) = \sum_{i=1}^{8} S_i(r, s, t) \cdot I_{ds,i} \qquad (7.63)$$

where $I_{ds,i}$ is the value of the current in the table. The local and global coordinates are related by

$$V_{gs} = \sum_{i=1}^{8} S_i \cdot V_{gs,i} \qquad V_{ds} = \sum_{i=1}^{8} S_i \cdot V_{ds,i} \qquad V_{bs} = \sum_{i=1}^{8} S_i \cdot V_{bs,i} \qquad (7.64)$$

where voltage values subscripted with an i correspond to table entries. The derivatives, i.e., the conductances, are similarly expressed in (7.65) – (7.67).

$$g_{ds} = \frac{\partial I_{ds}}{\partial V_{ds}} = \frac{\partial I_{ds}}{\partial r} \cdot \frac{\partial r}{\partial V_{ds}} + \frac{\partial I_{ds}}{\partial s} \cdot \frac{\partial s}{\partial V_{ds}} + \frac{\partial I_{ds}}{\partial t} \cdot \frac{\partial t}{\partial V_{ds}}$$

$$= \sum_{i=1}^{8} \frac{\partial S_i(r, s, t)}{\partial V_{ds}} \cdot I_{ds,i} \qquad (7.65)$$

$$g_m = \frac{\partial I_{ds}}{\partial V_{gs}} = \frac{\partial I_{ds}}{\partial r} \cdot \frac{\partial r}{\partial V_{gs}} + \frac{\partial I_{ds}}{\partial s} \cdot \frac{\partial s}{\partial V_{gs}} + \frac{\partial I_{ds}}{\partial t} \cdot \frac{\partial t}{\partial V_{gs}}$$

$$= \sum_{i=1}^{8} \frac{\partial S_i(r,s,t)}{\partial V_{gs}} \cdot I_{ds,i}$$

(7.66)

$$g_{bs} = \frac{\partial I_{ds}}{\partial V_{bs}} = \frac{\partial I_{ds}}{\partial r} \cdot \frac{\partial r}{\partial V_{bs}} + \frac{\partial I_{ds}}{\partial s} \cdot \frac{\partial s}{\partial V_{bs}} + \frac{\partial I_{ds}}{\partial t} \cdot \frac{\partial t}{\partial V_{bs}}$$

$$= \sum_{i=1}^{8} \frac{\partial S_i(r,s,t)}{\partial V_{bs}} \cdot I_{ds,i}$$

(7.67)

Thus, the eight-node mapping accomplishes both current interpolation and differentiation [7.7]. Application of the table-based model to an AlSbAs/InAs HFET [7.21] is depicted in Figure 7.31. The DC simulation of this HFET is significant because it accurately represents data for a device for which no analytical model exists.

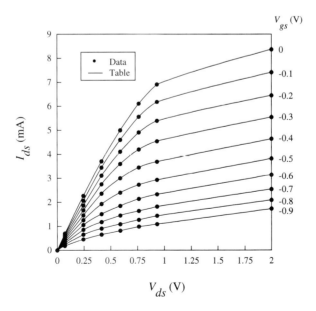

Figure 7.31: DC simulation of AlSbAs/InAs HFET.

7.5 Summary

In this chapter, several different approaches were demonstrated for the table-based simulation of optoelectronic devices, based on the work of Pang et al. and Cho et al. Numerical methods, such as table-based simulation, are advantageous because of the considerable time and effort involved in the development of a full analytical model. In many cases, modeling capability is needed sooner than an analytical model can be developed; an example of this is the need for an equivalent-circuit model to allow inclusion of a new device in an experimental integrated circuit. Often, for one reason or another, the device proves itself unworthy of further pursuit. When analytical models are the only option, the device then becomes obsolete before the model can even be finished. Thus, rapid prototyping is a key motivation for table-based simulation.

Table-based modeling is also useful when the physics of a device are not yet fully understood. A device can often be fabricated in the lab simply by scaling dimensions or varying epitaxial growth layers. While the fabrication *process* can be well understood, it can lead to devices whose behaviors are not, such as ultra-small feature size transistors that invoke quantum-size effects, or novel laser structures.

This chapter demonstrated the use of table-based models for the rapid prototyping of optoelectronic devices, such as the laser diode, and corresponding electronic devices used in optoelectronic systems, such as the HFET and HBT. The methods used varied from simple linear interpolation to four-, eight-, and twenty-node isoparametric shape function interpolation and differentiation. The choice of the interpolation method was shown to be dependent on the complexity of the device behavior. A noteworthy procedure demonstrated in this chapter was the use of device simulation output as a substitute for measured data. This technique allows for optimization of the circuit-level model without dependence on the schedule or resources of an associated fabrication facility.

7.6 References

[7.1]   J. C. Pang, "Fast prototyping of device models using table-lookup methods in computer simulation," M. S. thesis, University of Illinois at Urbana-Champaign, 1990.

[7.2]   T. Shima, H. Yamada, and R. Dang, "Table look-up MOSFET modeling system using a 2-D device simulator and monotonic piecewise cubic interpolation," *IEEE Transactions on Computer-Aided Design of Integrated Circuits and Systems*, vol. CAD-2, no. 2, pp. 121-126, 1983.

[7.3]   K. Sakui, T. Shima, Y. Hayashi, F. Horiguchi, and M. Ogura, "A simplified accurate three-dimensional table lookup MOSFET model for VLSI circuit simulation," *Proceedings of the IEEE Custom Integrated Circuit Conference*, pp. 347-351, 1985.

[7.4]   P. E. Allen and K. S. Yoon, "A table look-up model for analog applications," *Proceedings of the IEEE International Conference on Computer-Aided Design*, pp. 124-127, 1988.

[7.5]   J. Baby, J. Vlach, and K. Singhal, "Tableau formation for optimized polynomial spline MOSFET model approximations," *Proceedings of the IEEE International Symposium on Circuits and Systems*, pp. 1127-1138, 1986.

[7.6]   Y. H. Jun and S. B. Park, "Piecewise polynomial models for MOSFET DC characteristics with continuous first order derivative," *Proceedings of the IEEE International Symposium on Circuits and Systems*, pp. 2589-2592, 1988.

[7.7]   D. H. Cho, T. H. Kim, and J. T. Kong, "A table-lookup model using a 3-D isoparametric shape function with improved convergence," *IEEE International Conference on Computer-Aided Design*, pp. 244-247, 1989.

[7.8]   M. J. Maron, *Numerical Analysis: A Practical Approach*. New York: Macmillan Publishing Company, 1987.

[7.9]   W. H. Press, S. A. Teukolsky, W. T. Vetterling, and B. P. Flannery, *Numerical Recipes in C*. New York: Cambridge University Press, 1992.

[7.10] G. H. Song, K. Hess, T. Kerkhoven, and U. Ravaioli, "Two-dimensional simulation of quantum well lasers," *European Transactions on Telecommunications and Related Technologies*, vol. I, no. 4, pp. 375-381, 1990.

[7.11] J. D. Irwin, *Basic Engineering Circuit Analysis*. New York: Macmillan Publishing Company, 1987.

[7.12] J. J. Ebers and J. L. Moll, "Large-signal behavior of junction transistors," *Proceedings of the IRE*, vol. 42, no. 12, pp. 1761-1772, 1954.

[7.13] H. K Gummel and H. C. Poon, "An integral charge control model of bipolar transistors," *Bell System Technical Journal*, vol. 49, no. 5, pp. 827-852, 1970.

[7.14] J. Guild, *Touchstone & Libra: Reference Manual*. Westlake Village, CA: EEsof, Inc., 1990.

[7.15] S. J. Leon, *Linear Algebra with Applications*. New York: MacMillan Publishing Company, 1986.

[7.16] H. C. Martin and G. F. Carey, *Introduction to Finite Element Analysis: Theory and Application*. New York: McGraw-Hill Book Company, 1973.

[7.17] W. B. Bickford, *A First Course in the Finite Element Method*. Boston, MA: Richard D. Irwin, Inc., 1990.

[7.18] B. Irons and S. Ahmad, *Techniques of Finite Elements*. New York: John Wiley & Sons, 1980.

[7.19] K.-J. Bathe, *Finite Element Procedures in Engineering Analysis*. Englewood Cliffs, NJ: Prentice Hall, Inc., 1982.

[7.20] W. Kaplan, *Advanced Calculus*. Reading, MA: Addison-Wesley Publishing Company, 1984.

[7.21] D. H. Cho and S. M. Kang, "An accurate charge-oriented table look-up model for circuit simulation," Internal memorandum, University of Illinois at Urbana-Champaign, 1994.

[7.22] D. H. Cho and S. M. Kang, "An accurate AC characteristic table look-up model for VLSI analog circuit simulation applications," *IEEE International Conference on Computer-Aided Design*, pp. 1531-1534, 1993.

[7.23] G. Massobrio and P. Antognetti, *Semiconductor Device Modeling with SPICE*. New York: McGraw-Hill, Inc., 1993.

8

SYSTEM AND MIXED-LEVEL SIMULATION

The previous chapters dealt primarily with the computer-aided design of optoelectronic devices, integrated circuits, and subsystems. It is also desirable to analyze all of these from the *systems* perspective. When constructing systems, such as fiber-optic links or optical interconnects, it is crucial to understand not only the behavior of the individual devices themselves, but also the impact that they have on the system as a whole. In the design of a fiber-optic system, for example, it is desirable to minimize the number of repeaters or optical amplifiers required and to optimize their spacing, based on the attributes of the specific devices involved. The power budget, both electrical and optical, is another important system characteristic that is determined by the performance of the individual devices. Other examples are the system bit-error rate (BER), signal integrity, eye diagram, and overall bit rate.

Optoelectronic system design is often done from the "bottom up"; that is, design of the system frequently begins by designing and optimizing individual circuits and devices. The system is then built around that subset of devices that not only has sufficient performance but also proves to be reliable, economical, and manufacturable. Once the devices are obtained, the system is often assembled by trial and error or breadboarding; that is, although simulation may have been used to design the individual components, it is often not used in the systems integration.

When system design is performed in the "bottom-up" manner, the system designer often has no choice but to accept whatever devices are available, then "piece together" the system. By making system-level simulation of an optoelectronic link the *first* step in system design, the system designer can determine, beforehand, what types of devices and device performance will be necessary to construct the desired system. This "top-down" approach allows the system designer to drive the device technology, rather than be a passive customer. Since system simulation attempts to model the overall performance of the optoelectronic link, it is even possible to modify specific system-level attributes of individual devices to determine what effect those modifications will have on the system behavior. In this manner, "top-down" design can be used not only for designing entirely new systems, but also for improving existing designs by pinpointing those device attributes that contribute most critically to system performance.

System simulation involves a higher level of abstraction in the description of the optoelectronic link than those approaches discussed in previous chapters. As is the case when determining whether to perform device or circuit simulation, the choice between system and circuit simulation involves a series of trade-offs between simulation time, accuracy, flexibility, and usefulness. However, as seen previously, such trade-offs need not always be absolute. The mixed-mode approach depicted in Chapter 6 illustrated a situation in which those portions of the circuit that required a higher degree of accuracy were simulated at the device level, while those portions for which accurate compact models existed were simulated at the circuit level. The simulations were performed dynamically, yielding both time-efficient and accurate results.

In this chapter, the illinois FibeR-optic and Optoelectronic Systems Toolkit (iFROST), developed by Whitlock et al. [8.1] – [8.3], [8.69], [8.70] will be discussed as an approach to the system-level simulation of optoelectronic links. An overview of this tool will be presented and examples will be given illustrating its use in the computation of system bandwidth, maximum bit rate, rise time, bit-error rate, and eye diagram based on the performance characteristics of the individual components comprising the link. In this chapter, a mixed-level simulation environment will also be discussed as an approach to dealing with particularly troublesome portions of the system which require a higher degree of accuracy than that afforded by system-level simulation. This environment was created by combining the system-level tool iFROST with the circuit simulator iSMILE, which was discussed in Chapters 3 and 4. When system components require a high level of detail, iFROST can invoke, or call, iSMILE to simulate those individual parts at the circuit level. Because the circuit and system simulations are performed sequentially, rather than simultaneously, the environment is referred to as "mixed-level," rather than "mixed-mode."

In the "bottom-up" approach to system design, components are often obtained commercially, or "off-the-shelf," the actual design of the components being beyond the control of the system designer. In this scenario, it is desirable to use data sheets provided by the device manufacturer as inputs to a system simulation. In the case where such data are not available, the mixed-level iFROST-iSMILE simulation method can be used to generate such system-level information from a circuit-level simulation. This information can then be used as the input to the system simulation. However, even when device data sheets *are* available, they are often incomplete, frequently listing data for only a single bias point, temperature, or other set of conditions. The use of circuit simulation is ideal for generating a more comprehensive set of system inputs than that which is available from the provider of the device. In "top-down" system design, the circuit simulation portion of this mixed-level environment can be used to specify the particulars of the device (geometry, material, etc.) which must be improved or met in order to satisfy system requirements.

8.1 The iFROST System Simulator

While VHDL has had phenomenal success in the description of digital electronic systems, tools appropriate for the description of optoelectronic systems are rare. iFROST is a software tool, written in the ANSI C programming language, that has been designed to perform system-level simulation of optoelectronic links based on the system-level attributes of the individual devices comprising the link. The simulation modes supported include the high-level analytic mode, which computes a link's bandwidth and maximum bit rate, and the waveform simulation mode, which simulates the system's waveforms to determine the bit-error rate and to generate eye diagrams. To clarify what is meant by "system-level attributes," consider the circuit-level models for the laser diode presented previously. The input parameters to those models included geometrical information such as cavity width, length, and thickness, as well as physical information such as recombination and coupling coefficients, lifetimes, and confinement factors. While this information is quite useful in the design of a laser diode, it is of little value in the design of an optical link. Of greater importance are *system-level attributes*, such as the laser's threshold current, spectral width, differential quantum efficiency, and bandwidth. System simulators, such as iFROST, require parameters such as these, rather than the physical, geometrical, and material parameters that are required by device and circuit simulators.

iFROST is a data-driven simulator which traces signals from the input node(s) all the way through the system to the output node(s). The high level at which the simulation is performed allows for lower CPU times and better convergence than lower-level tools, such as circuit or device simulators. The modular nature of the program allows for the easy inclusion of new system-level models for components that are either under development or are yet to be developed. For example, while iFROST does not currently support erbium- or praseodymium-doped fibers, the addition of such a model is straightforward. It is also possible to add simulation capability for wavelength-division multiplexed (WDM) systems; in this fashion, iFROST can support both time- and wavelength-domain models.

8.1.1 High-level analytic optoelectronic models

In order to perform optoelectronic system simulation, it is necessary to have suitable models. In this version of iFROST [8.1] – [8.3], a library has been incorporated for the high-level analytic mode that includes models for optical sources, optical fibers, and optical receivers. The models are based on Gaussian approximations; that is, they assume that the input signals can be modeled as Gaussian in shape and that each of the components can be modeled as linear time-invariant (LTI) Gaussian systems.

As Brown showed in [8.4], when a system with a Gaussian impulse response $h(t)$ with RMS width σ

$$h(t) = \frac{1}{\sigma\sqrt{2\pi}}\exp\left(\frac{-t^2}{2\sigma^2}\right) \tag{8.1}$$

is excited by a step input function, the resulting output pulse has a 10% to 90% rise time given by

$$t_r = \sigma[\text{erf}^{-1}(0.9) - \text{erf}^{-1}(0.1)] \tag{8.2}$$

where the error function is defined as

$$\text{erf}(x) = \frac{2}{\sqrt{\pi}}\int_0^x \exp(-t^2)\,dt \tag{8.3}$$

By taking the Fourier transform of the impulse response, the -3 dB and -6 dB bandwidths are easily determined to be

$$BW_{-3\text{ dB}} = \frac{\sqrt{0.3\ln 10}}{2\pi\sigma} \tag{8.4}$$

$$BW_{-6\text{ dB}} = \frac{\sqrt{0.6\ln 10}}{2\pi\sigma} \tag{8.5}$$

Finally, it is well known that when several such Gaussian components are cascaded, the overall system RMS width σ_{sys}, rise time t_{sys}, and bandwidth BW_{sys} can be expressed in terms of the attributes of the individual components through

$$\sigma_{sys}^2 = \sigma_1^2 + \sigma_2^2 + \sigma_3^2 + \dots \tag{8.6}$$

$$t_{sys}^2 = t_1^2 + t_2^2 + t_3^2 + \dots \tag{8.7}$$

$$BW_{sys}^{-2} = BW_1^{-2} + BW_2^{-2} + BW_3^{-2} + \dots \tag{8.8}$$

Those elements in the digital optical link which are not Gaussian take advantage of the central limit theorem, which dictates that for sufficiently large numbers of events, the sum of a large number of random variables converges to a Gaussian distribution [8.5]. Because the component impulse responses can be treated as probability density functions, simple relationships for the overall system rise time and bandwidth, similar to (8.6) – (8.8), can be derived even for non-Gaussian systems, based on the attributes of the individual components [8.4]. The high-level analytic iFROST models and analyses used by Whitlock et al. and presented in this section are heavily based on these principles.

LED

Light emitting diodes have been used as sources in optical links, primarily when low cost is a concern and when speed is not. While LEDs can be fabricated very economically, their long rise time limits the system bandwidth and maximum bit rate, and their broad spectral width increases dispersion in fibers. The impulse response of the LED takes an exponential form, and is defined in [8.4] as

$$h_s(t) = \frac{1}{\tau_s} \exp\left(-\frac{t}{\tau_s}\right) \tag{8.9}$$

where $t \geq 0$ and τ_s is the time constant of the exponential which is also equal to the RMS width of the impulse response. Under an electrical step input to the LED, the 10 to 90 percent rise time of the optical output can be computed through convolution as [8.4]

$$t_s = \tau_s \ln 9 \tag{8.10}$$

These relations can be used to derive the -3 dB and -6 dB bandwidths by Fourier transforming (8.9):

$$BW_{s,\,-3\ \mathrm{dB}} = \frac{0.3489}{t_s} \tag{8.11}$$

$$BW_{s,\,-6\ \mathrm{dB}} = \frac{0.6037}{t_s} \tag{8.12}$$

The RMS spectral width of the LED is arrived at from the FWHM spectral width according to (8.13) [8.1].

$$\sigma_{\lambda} \approx \frac{\Delta\lambda_{FWHM}}{1.177} \qquad (8.13)$$

Thus, the main parameters of interest are the rise time of the LED, its bandwidths, and its RMS spectral width. Since the impulse response was not Gaussian, but exponential, further system analyses will take advantage of the Central limit theorem.

Laser diode

Both single-mode and multimode lasers, over a wide range of wavelengths, are used in optical links. It is well known that long-wavelength lasers are of particular value in fiber-optic links due to the low loss and low dispersion minima in silica optical fibers at 1.55 μm and 1.31 μm, respectively [8.6]. In fact, the presence of these minima actually *drove* laser research at these wavelengths, a good example of system considerations influencing device development. Short-wavelength lasers (usually 670 nm ~ 850 nm) can be used in fiber-optic links, as well, but because their center wavelength falls well short of the minimum-dispersion wavelength in optical fiber, such systems usually suffer a severe dispersion penalty. However, these devices usually have a significant cost advantage over long-wavelength lasers because they are able to utilize the lower cost GaAs, rather than the InP, materials system. Since the amount of dispersion in fiber-optic links increases with distance, short-wavelength laser diodes are usually used in short-haul links, while long-wavelength lasers are used in long-distance links. Single-mode lasers obviously offer significantly lower dispersion than multimode lasers, but usually at a higher cost; system simulation can be used to determine the various trade-offs involved in selecting among several different laser diodes.

The laser diode has much better performance than an LED; it has a faster rise time, a narrower spectral width, and a narrower emission profile (which facilitates coupling of the output power to an optical fiber). When the laser diode is biased above threshold and then digitally modulated, bit rates in the several tens of Gb/s can be achieved.

It is widely recognized, however, that because of the tremendous amount of device physics involved, the laser diode is the most difficult element to model in an optical link. The rise time and spectral/RMS width of the laser output are not as easily derived as the LED. Since iFROST is a *system* simulator, however, it is not overly concerned with *determining* these system-level parameters (τ_s and σ_{λ}); rather, the iFROST laser diode model expects to *receive* these parameters as inputs to the simulation. In Section 8.2, the mixed-level simulation environment alluded to previously will be presented. This environment creates a link with the iSMILE laser diode model presented in Chapter 3 and provides an interface that transforms detailed circuit simulation outputs into parameters that iFROST expects to receive as inputs. Alternatively, information from manufacturer data sheets can be used in place of circuit-simulation results, though such information is often inadequate for full system simulation. Other parameters of interest in the system simulation of laser diodes are the center wavelength of the laser, the maximum amount of frequency and wavelength chirp, the RMS relative intensity noise (RIN), the threshold current, and the slope efficiency. As with the LED, iFROST's high-level analytic model assumes that the laser output is a Gaussian pulse.

Optical fiber

When used to transmit data digitally, optical fibers are limited by three fundamental dispersion mechanisms: first-order chromatic dispersion, second-order chromatic dispersion, and modal dispersion. Chromatic dispersion is basically the result of the fact that different wavelengths of light travel at different group velocities through the fiber [8.6]. Modal dispersion, on the other hand, is caused by the fact that different modes travel at different group velocities. As detailed by Whitlock in [8.1] and Brown in [8.4], first-order chromatic dispersion dominates second-order chromatic dispersion for wavelengths far away from the zero-dispersion wavelength, while second-order chromatic dispersion dominates for wavelengths near the zero-dispersion wavelength. Similar to the previous elaborations, the Gaussian impulse response for first-order chromatic dispersion can be expressed as

$$h_{cd1}(t) = \frac{1}{\sigma_\lambda DL \sqrt{2\pi}} \exp\left[\frac{-(t-t_o)^2}{2(\sigma_\lambda DL)^2}\right] \qquad (8.14)$$

From (8.14), the RMS pulse width of the output signal is derived to be [8.4]

$$\sigma_{cd1}^2 = (LD\sigma_\lambda)^2 \qquad (8.15)$$

where L is the length of the fiber in km, D is the chromatic dispersion in units of ns of dispersion (pulse spreading) per nm of source spectral width per km of optical fiber (ns/nm·km), and σ_λ is the RMS spectral width of the source in nm. Using [8.4], again, as a reference, the -3 dB and -6 dB bandwidths that correspond to first-order chromatic dispersion are determined by the Fourier transform of (8.14):

$$BW_{cd1, -3\ dB} \approx \frac{0.1323}{\sigma_{cd1}} \qquad (8.16)$$

$$BW_{cd1, -6\ dB} \approx \frac{0.1871}{\sigma_{cd1}} \qquad (8.17)$$

Similarly, the impulse response, the RMS pulse width, and the -3 dB and -6 dB bandwidths for second-order chromatic dispersion are expressed in (8.18) – (8.21) [8.4].

$$h_{cd2}(t) = \frac{1}{\sqrt{\pi \left(S + \frac{2D}{\lambda_c} \right) L \sigma_\lambda^2 t}} \exp \left[\frac{-t}{\left(S + \frac{2D}{\lambda_c} \right) L \sigma_\lambda^2} \right] \tag{8.18}$$

$$\sigma_{cd2} = \frac{\left(S + \frac{2D}{\lambda_c} \right) L \sigma_\lambda^2}{\sqrt{2}} \tag{8.19}$$

$$BW_{cd2,\,-3\text{ dB}} \approx \frac{0.2748}{\sigma_{cd2}} \tag{8.20}$$

$$BW_{cd2,\,-6\text{ dB}} \approx \frac{0.6133}{\sigma_{cd2}} \tag{8.21}$$

In these equations, S is the slope of the dispersion in ns/km·nm^2 at the source centroid wavelength and λ_c is the source centroid wavelength in nm. Since the source spectrum can sometimes be asymmetric, Whitlock defines the centroid wavelength λ_c as the wavelength at peak optical power, rather than the wavelength at the center of the source spectrum.

There are several methods among which iFROST chooses when computing D and S at λ_c, depending on the type of fiber, the type of optical source, and the available component parameters. If multimode fiber is being used and the zero-dispersion wavelength λ_o is known, the chromatic dispersion and the dispersion slope at the source centroid wavelength λ_c are determined by the Sellmier approximations [8.1], [8.4]

$$D = \frac{S_o}{4} \lambda_c \left(1 - \frac{\lambda_o^4}{\lambda_c^4} \right) \tag{8.22}$$

$$S = \frac{S_o}{4} \left(1 + 3\frac{\lambda_o^4}{\lambda_c^4} \right) \tag{8.23}$$

$$S_o = 0.1064 \times 10^{-3} \ \frac{\text{ns}}{\text{nm}^2 \cdot \text{km}} \tag{8.24}$$

If λ_o is not known, then D and S are calculated according to [8.7], [8.8] as given by (8.25) and (8.26)

$$D = 10^{-3} \left[\left(0.0266 - \frac{10^{-3}}{4\pi^2 n_c ca^2} \right) \lambda_c - \frac{6.985 \times 10^{10}}{\lambda_c^3} \right] \tag{8.25}$$

$$S = 10^{-3} \left(0.0266 - \frac{10^{-3}}{4\pi^2 n_c ca^2} + \frac{20.95 \times 10^{10}}{\lambda_c^4} \right) \tag{8.26}$$

where n_c is the index of refraction of the fiber cladding, c is the speed of light in vacuum (km/ns), and a is the radius of the fiber core in μm. For a single-mode fiber for which λ_o is known, an effective value of the fiber core can be used in (8.25) and (8.26) [8.1], [8.8]

$$a = \sqrt{\frac{1}{1.8\pi^2 (0.0266 - 6.985 \times 10^{10} \lambda_o^{-4})}} \tag{8.27}$$

If the fiber is dispersion shifted, and λ_c is in the vicinity of 1550 nm, then D and S can be estimated by [8.8]

$$D = 10^{-3} [2 + 0.08 (\lambda_c - 1550)] \tag{8.28}$$

$$S = 0.08 \times 10^{-3} \frac{\text{ns}}{\text{nm}^2 \cdot \text{km}} \tag{8.29}$$

As with chromatic dispersion, modal dispersion in multimode fibers can also be modeled by an RMS impulse response width σ_m. Following the previous elaboration, the -3 dB and -6 dB bandwidths resulting from modal dispersion are approximated by [8.1], [8.4]

$$BW_{m, -3 \text{ dB}} \approx \frac{0.2524}{\sigma_m} \tag{8.30}$$

$$BW_{m, -6 \text{ dB}} \approx \frac{0.1478}{\sigma_m} \tag{8.31}$$

Optical receiver

The rise time and bandwidth of a specific photoreceiver, consisting of a photodetector and an electrical amplifier, can be calculated without much effort. This was demonstrated for the MSM-HEMT photoreceivers illustrated in Chapter 4.

If the (electrical) bandwidth of such a photoreceiver is denoted by BW_e, and if the receiver is assumed, for simplicity, to have a raised cosine frequency dependence, the impulse response can be approximated as [8.4]

$$h_r(t) \approx \frac{\sin(5.5 \pi t BW_e)}{2\pi t [1 - (5.5 t BW_e)^2]} \tag{8.32}$$

The 10% – 90% rise time can be calculated from the assumed BW_e as [8.1], [8.4]

$$t_r \approx \frac{0.3529}{BW_e} \tag{8.33}$$

Alternatively, the rise time can be determined from the circuit elements that comprise the photoreceiver, or it can be obtained from a data sheet. For completeness, [8.1] and [8.4] calculate the -3 dB and -6 dB bandwidths of the receiver as

$$BW_{r,-3 \text{ dB}} = BW_e \tag{8.34}$$

$$BW_{r,-6 \text{ dB}} = \frac{11}{8} BW_e \tag{8.35}$$

8.1.2 High-level system analysis

With the high-level models described in the previous sections, it is possible to determine the overall system bandwidth, rise time, and maximum bit rate in terms of the attributes of the individual components. By utilizing (8.7), iFROST determines the overall rise time of a fiber-optic link consisting of an optical source, an optical fiber, and a photoreceiver as [8.1], [8.4]

$$t_{sys} \approx \sqrt{(1.17 t_s)^2 + (2.56 \sigma_{cd1})^2 + (1.81 \sigma_{cd2})^2 + (2.56 \sigma_m)^2 + (0.94 t_r)^2} \tag{8.36}$$

where the parameters assume their previously defined meanings.

Similarly, in [8.1] and [8.4], Whitlock and Brown model the overall system -3 dB and -6 dB bandwidths as (8.37) and (8.38).

$$BW_{sys,-3 \text{ dB}} \approx \frac{1}{\sqrt{\left(\dfrac{t_s}{0.291}\right)^2 + \left(\dfrac{\sigma_{cd1}}{0.132}\right)^2 + \left(\dfrac{\sigma_{cd2}}{0.187}\right)^2 + \left(\dfrac{\sigma_m}{0.1323}\right)^2 + \left(\dfrac{t_r}{0.361}\right)^2}} \tag{8.37}$$

$$BW_{sys, -6\, dB} \approx \frac{1}{\sqrt{\left(\frac{t_s}{0.411}\right)^2 + \left(\frac{\sigma_{cd1}}{0.187}\right)^2 + \left(\frac{\sigma_{cd2}}{0.265}\right)^2 + \left(\frac{\sigma_m}{0.187}\right)^2 + \left(\frac{t_r}{0.511}\right)^2}} \qquad (8.38)$$

The system rise time and bandwidth can be used to compute the maximum bit rate under different modulation schemes [8.6]. For return-to-zero modulation (RZ), the bit rate can be expressed as

$$\text{Bit Rate} = \frac{0.35}{t_{sys}} \qquad (8.39)$$

For nonreturn-to-zero (NRZ) modulation, the bit rate can be expressed as

$$\text{Bit Rate} = \frac{0.70}{t_{sys}} \qquad (8.40)$$

The overall system dispersion can also be determined based on individual component attributes. The dispersion penalty P_d when the optical source is a multimode laser is given by [8.8] – [8.13]

$$P_d = -5\left(\frac{x+2}{x+1}\right)\log\left\{1 - \frac{k^2 Q^2}{2}\left(\frac{\pi m_d L}{T}\right)^4\left[1 + 10.5\left(\frac{\sigma_\lambda}{\lambda_c - \lambda_o}\right)^2 + 3\left(\frac{\sigma_\lambda}{\lambda_c - \lambda_o}\right)^4\right]\right\} \qquad (8.41)$$

In (8.41), x is the excess noise factor if the photodetector is an APD ($x = \infty$ for a PIN detector), Q is related to the BER (for BER=10^{-9}, $Q=6$), $1/T$ is the bit rate in Gb/s, m_d is the chromatic dispersion of the fiber in ns/nm/km, σ_λ is the RMS laser linewidth, L is the length of the optical fiber link in km, λ_c is the laser centroid wavelength in nm, and λ_o is the zero-dispersion wavelength of the fiber in nm. The parameter k is the mode partition coefficient of the laser; mode partition noise in lasers occurs when the total energy in a multimode laser cavity fluctuates between the various longitudinal modes. The parameter k is a measure of the cross-correlation between the modes [8.6]. The chromatic dispersion of the fiber is given by Whitlock in [8.1] as (8.42), in which the first term accounts for dispersion at a given (centroid) wavelength and the second term accounts for dispersion due to the wavelength spread σ_λ.

$$m_d = \sqrt{D^2 + \left(\frac{S\sigma_\lambda}{2\sqrt{2}}\right)^2} \qquad (8.42)$$

The zero-dispersion wavelength λ_o is either known or approximated by (8.43).

$$\lambda_o \approx \sqrt[4]{\frac{6.985 \times 10^{10}}{0.0266 - \frac{10^{-3}}{4\pi^2 n_c ca^2}}} \tag{8.43}$$

In iFROST, Whitlock also accounts for optical chirp, which becomes significant in single-mode lasers. The dispersion penalty due to laser chirp is given by [8.8], [8.14] – [8.19]

$$P_d = -10\left(\frac{x+2}{x+1}\right)\log(1-\Omega) \tag{8.44}$$

Here, Ω is the eye diagram aperture factor which is calculated as

$$\Omega = \left(5.16 t_c m_d L \frac{\delta\lambda_c}{T^2}\right) \cdot \left[1 + \frac{2}{3T}(m_d L \delta\lambda_c - t_c)\right] \tag{8.45}$$

if the laser chirp time t_c is less than $0.5T$ and

$$\Omega = 1 - \frac{1}{\xi}\left[2\exp\left(\frac{-6}{\xi^2}\right) - 1\right] \tag{8.46}$$

if t_c is greater than $0.5\ T$. In these equations, ξ is defined in (8.47) and (8.48).

$$\xi^2 = 1 + 12\left(\frac{t_c}{\gamma^2}\right)X + 8\left(\frac{t_c}{\gamma^3}\right)(X - t_c)XT \tag{8.47}$$

$$X = m_d L \frac{\delta\lambda_c}{T^2} \tag{8.48}$$

Here, $\delta\lambda_c$ is the wavelength chirp in nm and γ is the duty ratio of the optical pulse.

8.1.3 High-level simulation examples

To demonstrate the capabilities of iFROST, simulations of the optical interconnect link iPOINT testbed (illinois Pulsar-based Optical INTerconnect), developed at the University of Illinois at Urbana-Champaign, will be illustrated in this section. The iPOINT network is a fiber-optic communication link that utilizes the $n \times n$ nonblocking Pulsar switch [8.20] to accomplish ATM (Asynchronous Transfer Mode) networking between Unix workstations [8.21]. As shown in Figure 8.1, iPOINT uses high-speed fiber-optic channels to provide bidirectional data transmission between the workstations and the electronic ATM switch. Because of the short-distance nature of the optical links in the iPOINT testbed, inexpensive multimode fiber is used to achieve the target bit rate of 1.25 Gb/s.

To demonstrate the use of iFROST in a top-down design, in [8.1], Whitlock analyzes the requirements for the laser diode in the iPOINT testbed. By using iFROST, the effect of the laser parameters on the overall system rise time and maximum attainable bit rate can be determined. Furthermore, it is shown in [8.1] that iFROST can also be used to determine an acceptable range of laser parameters for a given set of system specifications.

Figure 8.2 depicts a typical link in the iPOINT testbed. The photoreceiver has a bandwidth of 2 GHz. As previously mentioned, the optical fiber is multimode with a 50-μm core diameter and a zero-dispersion wavelength of 1300 nm. The laser diode is also multimode, with a center (centroid) wavelength of 850 nm. In the following simulations, the linewidth (FWHM) and rise time of the laser are varied to evaluate their effects on the overall system rise time, bandwidth, and maximum bit rate.

Figure 8.3 shows that with a rise time of 0.3 ns, as long as the laser linewidth is less than 5.0 nm, the system can meet the 1.25 Gb/s NRZ specified bit rate. Figure 8.4 shows that with a linewidth of 1.0 nm, as long as the rise time is kept below 0.4 ns, the speed specification can be met. Figure 8.5 is a three-dimensional combination of Figures 8.3 and 8.4, simultaneously showing the effects of linewidth and rise time on the bit rate. These figures also show that for laser rise times above ≈ 0.2 ns, the overall system bit rate is dominated by the rise time of the laser; thus, for optimum system performance, the laser rise time should be as fast as possible.

8.2 Circuit-System Mixed-Level Simulation

It was previously mentioned that for system-level simulation, physical, material, and device parameters are of little use. Fortunately, input parameters more suitable for system simulation can often be obtained from manufacturer data sheets or directly from the research laboratories where the devices were fabricated. Unfortunately, the information available is often incomplete; in fact, the parameters of interest usually have complex bias-, modulation-, frequency-, or temperature-dependent natures which are difficult to capture completely in a few data sheet entries. In situations such as these, circuit simulation can be used to generate very detailed information such as individual currents, voltages, powers, and the like. However, circuit simulation outputs are too detailed to be of direct use in system simulation. These outputs must first be manipulated and converted into a form suitable for use as system simulation inputs. In this section, a mixed-level circuit-system interface for laser diode simulation will be presented. This approach uses the iSMILE laser diode model of Chapter 3 to produce detailed circuit-level outputs, interprets these values, and converts them into a more usable form, then passes them on to iFROST as system-level input parameters.

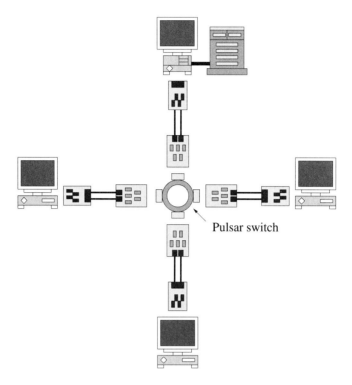

Figure 8.1: The iPOINT testbed.

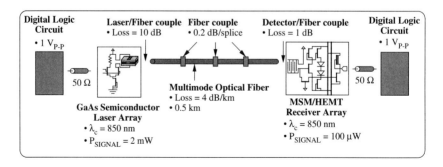

Figure 8.2: Typical link in iPOINT testbed.

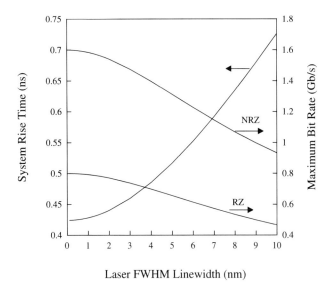

Figure 8.3: Overall system performance as a function of laser linewidth.
Laser rise time = 0.3 ns.

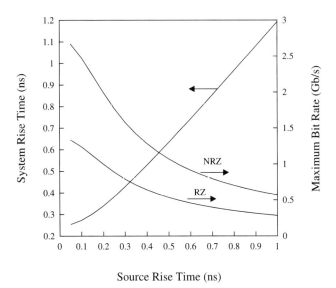

Figure 8.4: Overall system performance as a function of laser rise time.
Laser linewidth = 1.0 nm.

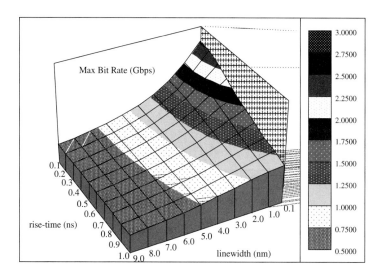

Figure 8.5: System bit rate as a function of both laser linewidth and rise time.

Nearly all of the parameters required to use the rate equation-based laser diode model are much too detailed for system simulation. For example, it is of no interest to iFROST how long or wide the laser is or how thick its quantum well is. Material parameters, such as recombination rates and lifetimes, are similarly of little use in system simulation. Typical inputs required for a system simulation are quantities such as the threshold current, differential quantum efficiency, rise time, center wavelength and spectral (line) width. Other parameters of interest are chirp, relative intensity noise (RIN), and the frequency and decay rate of relaxation oscillations. A major problem in optoelectronic design, then, is the determination of such system-level quantities for a given laser.

In this section, the iSMILE-iFROST vertically linked simulation environment will be described [8.2], [8.3]. The entire simulation is run with iFROST as the main user interface. The detailed rate-equation input parameters are fed to iFROST as inputs (Figure 8.6). These inputs are then passed from iFROST to iSMILE, upon which detailed circuit simulation of the laser diode occurs. Upon completion, iFROST takes the individual current and voltage outputs from the iSMILE simulation, converts them into the system-level quantities mentioned previously, then uses them as the laser diode inputs for its link simulation. Whereas quantities in Section 8.1, such as the laser linewidth and chirp, were either determined by iFROST through simplified calculations or read in from a data sheet, with this approach these quantities are determined by accurate circuit simulation. In addition, the bias dependence of these quantities can be accounted for with the iFROST-iSMILE mixed-level simulation environment, as will be shown in subsequent sections. Data

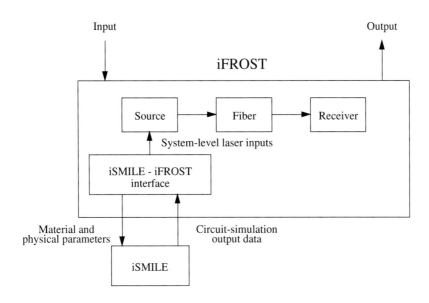

Figure 8.6: iSMILE-iFROST linkage.

sheets rarely contain such information. Since iFROST is written in C, this iSMILE-iFROST interface was also written in C so that it could be integrated seamlessly into iFROST. The interface was constructed to extract difficult-to-compute information from iSMILE simulations and then to use well-established theory to calculate relevant figures of merit from these data. It should be mentioned, in passing, that should adequate information be available (either from the manufacturer or the fabrication facility), this interface can always be bypassed; that is, the user has the choice of either providing the input parameters himself or running circuit simulation to determine the system-level inputs.

Since iFROST is essentially the simulation controller, the input to iFROST for laser diode simulations would be the parameters required by the iSMILE rate equation model. This input takes the form of a text file containing two columns: the first column is a list of the parameter names and the second column is a list of the corresponding parameter values (see Table 8.1). As seen in Figure 8.6, iFROST uses this information to run iSMILE. The iSMILE-iFROST interface actually constructs an iSMILE input file from the two-column text input file mentioned earlier, then invokes iSMILE via a C system call. The interface then interprets the iSMILE output, extracts and calculates the required quantities, and passes this information to iFROST. With knowledge of various detailed laser attributes from circuit simulation, for a given level of injected current or output power, pertinent system-level parameters can easily be determined. Once these parameters have been calculated, they are then passed on to iFROST as "system-level" inputs. At this stage, iFROST

Model Parameter	Symbol	Value
Length	L	500 µm
Width	W	4 µm
Quantum-Well Thickness	L_z	70 Å
Number of Quantum Wells	N	1
Facet Reflectivity	R	0.32
Photon Energy	E_{photon}	1.556 eV
Temperature	T	300 K
Optical Confinement Factor	Γ	0.3
Dielectric Constant	$\bar{n}$	3.55
Radiative Recombination Coefficient	B	1.6×10^{-11} cm^3/s
Spontaneous Emission Coupling Coefficient	β	5×10^{-4}
Intrinsic Cavity Loss	α_i	3 cm^{-1}
Saturation Photon Density	S_{sat}	5×10^{23} m^{-3}
Series Resistance	R_b	1 Ω
Diode Ideality Factor	nn	2.11
Nonradiative Recombination Current	I_o	0.84 µA
Depletion Capacitance	C_o	3.17 pF

Table 8.1 (Table 3.2): Input parameters for laser diode circuit simulation.

system simulation proceeds in a manner similar to that depicted in the previous section. The following sections will illustrate the calculation of system-level laser parameters using this environment.

8.2.1 Threshold current

One of the most useful system-level laser attributes is the threshold current, a figure of merit that is always included on data sheets of commercial laser diodes. However, when using custom-made devices, the threshold current is difficult to determine when attempting to do system design *before* the laser diode is even fabricated. Circuit simulation of the laser has been demonstrated previously; however, the standard outputs of the iSMILE laser model are current, voltage, and optical power. To determine the threshold current within a circuit-simulation framework, the onset of lasing is defined as the point at which the laser gain equals the total loss in the laser cavity. As a review, the gain was computed quantum mechanically in the laser model as

$$g(\omega) = \frac{q^2 m_r^*}{\varepsilon_0 m_0^2 c \bar{n} \hbar E L_z} |M|^2 [f_c(\omega) - f_v(\omega)] \tag{8.49}$$

while the loss was given by

$$\alpha = \Gamma \alpha_{int} - \frac{1}{L} \ln(R) \tag{8.50}$$

The gain in (8.49) is a function of the quasi-Fermi levels which are, in turn, functions of the injected current. Since the gain is computed within the model at every iteration, the threshold current can be determined by performing a DC L-I simulation in which the injected current is swept. The gain and the loss are monitored at each bias point; the unique injected current for which the gain equals the loss is, by definition, the threshold current.

The simulation flow occurs as follows. The data in Table 3.2 (Table 8.1) are input to iFROST. These parameters, which are the inputs for the rate-equation model, are passed by the simulation interface depicted in Figure 8.6 from iFROST to iSMILE along with a set of instructions for iSMILE to return the net laser gain (net = gain - loss) and the injected current. Upon completion of the DC circuit simulation, iSMILE delivers the requested information back to the simulation interface, upon which the interface determines the threshold current. For the laser of Table 8.1, the interface determined a threshold current of approximately 10 mA. This visually matches the turn-on point of the simulated L-I curve of Figure 3.6, which is repeated for convenience as Figure 8.7.

Since the quantum-mechanical gain, the electron density, and their dependence on the injected current are difficult to evaluate, circuit simulation is an excellent way to determine the threshold current. In fact, it was mentioned in Chapter 3 that the computation of the quantum-mechanical gain was one of the biggest contributions of the iSMILE laser diode model. While analytical, closed-form, expressions of the threshold current do exist, they are usually functions of either the gain g, the electron density n, or both; thus, while such seemingly simple expressions do exist, their *evaluation* under a given set of conditions is often quite difficult. With the mixed-level simulation approach, the user simply begins with a set of input parameters for the rate-equation model and ends up with the threshold current.

8.2.2 Quantum efficiency

Another figure of merit of use in system simulation is the laser diode efficiency. Fortunately, this can also be simply determined from circuit simulation. The

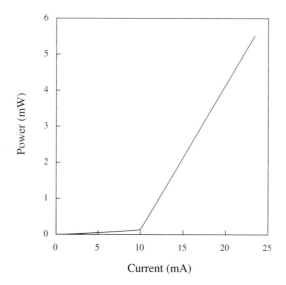

Figure 8.7: Simulated DC L-I characteristic.

differential quantum efficiency is defined to be the slope of the L-I curve (or more rigorously, the derivative of the output light with respect to the input current); thus, the rate-equation input parameters are again input to iFROST which, via the simulation interface, passes them on to iSMILE. The interface instructs iSMILE to perform a DC simulation, this time requesting that only the input current and output power for values of current greater than the threshold current be returned. The interface then computes the slope between each successive set of two (L-I) points and averages the slope over the total number of data point pairs. For the laser of Table 8.1, the iSMILE-iFROST interface calculated a slope efficiency of 0.40 mW/mA, which again matches the simulation output depicted in Figure 8.7.

8.2.3 Optical wavelength and frequency

The photon energy of the laser diode is an input parameter to the iSMILE laser model and is determined analytically by

$$E_{ph} = \frac{\hbar^2 \pi^2}{2L_z^2} \left(\frac{1}{m_c^*} + \frac{1}{m_v^*} \right) + E_g \qquad (8.51)$$

However, of more use in system simulation are the laser's center/centroid wavelength and frequency. The wavelength and frequency are computed by the iFROST-iSMILE interface through the well-known relations contained in (8.52).

$$\lambda = \frac{hc}{E_{ph}} \qquad\qquad f = \frac{c}{\lambda} \qquad\qquad (8.52)$$

In this case, the simulation interface does not require an actual iSMILE circuit simulation since the wavelength and frequency can be computed directly from the input parameters. Using a bulk bandgap of $E_g = 1.424$ eV and an input photon energy of 1.56 eV, the simulation interface computed a wavelength of 0.80 µm and an optical frequency of 106 THz for the laser of Table 8.1.

8.2.4 Noise and laser noise

The evaluation of noise is an important issue not only in assessing the quality of a laser diode, but also in determining the overall performance of *systems* using laser diodes. In this, and subsequent, sections, the simulation of various classes of laser noise and deviations from ideal laser performance will be investigated from a systems perspective. It was stated in Chapter 3 that the standard rate equations are inadequate for modeling phenomena that occur on atomic time scales. A rigorous treatment of laser noise requires the quantization of both the electromagnetic field and the atoms, an extremely involved process. A list of classical works on this subject is included in [8.22] – [8.31]. Fortunately, however, an alternate method that is a cross between the purely quantum approach and the standard rate equation representation exists. Before embarking on this topic, however, it is instructive to review a few basic noise concepts.

Noise is random in nature and can only be described in terms of stochastic variables; thus, it is impossible to predict the *instantaneous* value of noise. This is a considerable drawback of the rate equation model, which is purely deterministic. Many laser noise mechanisms can be described by Gaussian distributions; this considerably simplifies the analysis since a Gaussian random variable can be characterized by its mean and its variance. A noisy signal can be represented most simply as the sum of a noiseless deterministic signal and a random noisy component, as depicted in Figure 8.8. From this figure, it should be noted that in traditional (Gaussian) noise analysis, the mean of the noise itself is usually considered to be zero; that is, noise is treated as a set of random fluctuations, centered about zero. On the other hand, a deterministic (noiseless) signal is exactly known at all times; therefore, its variance (i.e., its deviation from its mean) is zero. If a composite signal is defined as a noiseless signal plus noise, the mean of the composite signal will be the original, noiseless, signal itself (since the mean of the noise is zero), while the variance of the composite signal will be equal to the variance of the noise (since the variance of the original signal is zero).

To illustrate how the mean and variance of a Gaussian random process can be used to model noise, consider Figure 8.9 in which a Gaussian distribution is rotated and superimposed on a composite signal similar to that of Figure 8.8. The dotted

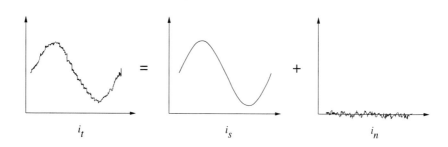

Figure 8.8: Composite signal consisting of the sum of a noiseless, deterministic signal and random (zero-mean) noise.

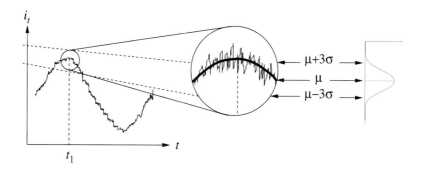

Figure 8.9: Description of noise through the variance of the Gaussian random variable that characterizes it.

lines in the expanded view indicate the extent of the noise and show how the noise can be represented by the variance of a Gaussian distribution. (In this figure a value of 3σ is illustrated because over 99% of a Gaussian distribution is captured within $\pm 3\sigma$ of its mean). Thus, if the variance of a Gaussian noise mechanism σ^2 can be determined, the "amount" of noise can be *statistically* quantified. Therefore, a major issue in the Gaussian-based simulation of laser noise is the determination of the variance σ^2 for various laser noise mechanisms.

If the composite signal is denoted as i, its noise can be expressed in the time domain in terms of its autocorrelation [8.32]:

$$R_i(\tau) = \langle i(t)\, i(t+\tau) \rangle \tag{8.53}$$

It is often convenient to represent noise in the frequency domain, as well. The frequency-domain representation of noise is called the spectral density, $S_i(f)$. According to the Wiener-Khintchine theorem [8.33], [8.34], the autocorrelation and the spectral density form a Fourier transform pair [8.34]:

$$S_i(f) = 2\int_{-\infty}^{\infty} R_i(\tau)e^{-j2\pi f\tau}d\tau \tag{8.54}$$

$$R_i(\tau) = \frac{1}{2}\int_{-\infty}^{\infty} S_i(f)e^{j2\pi f\tau}df \tag{8.55}$$

Since the variance of the signal itself is zero, the mean-square noise is defined as

$$\langle i^2(t)\rangle = R_i(0) = \int_0^{\infty} S_i(f)\,df \tag{8.56}$$

Noise in semiconductor laser diodes is caused primarily by intrinsic quantum fluctuations in the laser cavity. The biggest contribution to laser noise is due to the addition of incoherent power to the coherent lasing mode through spontaneous emission. Since the electromagnetic field which produces this power is not coherent with the lasing mode (i.e., different phase, frequency, and direction), random fluctuations in both amplitude and phase occur, resulting in nonideal laser performance [8.35]. Random amplitude fluctuations manifest themselves most commonly in intensity noise ($I \propto |E_0|^2$) in a manner similar to that depicted in Figure 8.8. Relative intensity noise (RIN), a common figure of merit, is defined to be the ratio of the intensity noise to the output power. Random phase fluctuations (or *phase noise*) manifest themselves in many different ways, the most common being linewidth broadening. An ideal laser produces output power at a single frequency; however, phase noise causes lasing to occur over a *range* of frequencies, spread about the center/centroid frequency. Linewidth broadening has significant implications on communication systems as it determines the maximum speed at which data can be transmitted. Phase noise can also result in *chirp* which is a shift (rather than a spreading/broadening) of the center wavelength. To further complicate matters, the amplitude noise and the phase noise are not independent but are correlated. Shot noise in the injected carriers also affects laser performance, as fluctuations in the carrier density will cause fluctuations in the gain, which will in turn cause fluctuations in the amplitude and the phase. When lasing does not occur solely in a single (longitudinal) mode, *mode partition noise* (MPN) can occur. In multimode lasers, the intensities at each frequency vary randomly, as energy is interchanged between the modes. MPN has been recognized as a serious problem when lasers are used in communication systems since it can result in modal noise and dispersion in optical fibers, limiting transmission rates [8.36] – [8.40]. As with the work presented thus far, however, the simulation of laser noise in this chapter will focus mainly on single-mode lasers. Detailed descriptions of these noise mechanisms will be presented in later sections. In the next section, a mathematical treatment of laser noise will be presented based on the hybrid combination of the rate equations and the strictly quantum approach.

8.2.5 Mathematical formulation of laser noise

As previously stated, the standard rate equations, being derived from semiclassical considerations, are inadequate for modeling laser noise; instead, a fully quantum-mechanical representation must be used. Such a derivation, however, is not trivial. In fact, the theory is so complex that it often obscures the desired results. Thus, in this section "noise-driven" rate equations are presented, based on the full quantum-mechanical approach in conjunction with the standard rate equations.

In [8.22] – [8.31], [8.41], and [8.42], laser noise is derived using the standard Langevin approach detailed in [8.33] and [8.43]. These results can be used to develop a set of single-mode, "noise-driven" rate equations [8.42] that are based on the standard rate-equation model presented in Chapter 3. As seen in (8.57) – (8.59), the noise-driven rate equations consist of the original three rate equations with the addition of one extra term. The added terms F_n, F_s, and F_ϕ are all stochastic and are referred to as the Langevin noise sources.

$$\frac{dn}{dt} = \frac{J}{qNL_z} - Bn^2 - \Gamma gc'S + F_n(t) \tag{8.57}$$

$$\frac{dS}{dt} = \beta Bn^2 + \Gamma gc'S - \frac{S}{\tau_{ph}} + F_S(t) \tag{8.58}$$

$$\frac{d\phi}{dt} = \frac{\alpha}{2}\left(\Gamma gc' - \frac{1}{\tau_{ph}}\right) + F_\phi(t) \tag{8.59}$$

In the Langevin approach, noise is expressed by first defining a set of physical differential equations describing the noiseless system under consideration. For lasers, the differential equations are the original rate equations. Next, the noise is modeled through a "random source function" containing all of the noise properties of the system; in the noise-driven rate equations (8.57) – (8.59), $F_n(t)$, $F_s(t)$, and $F_\phi(t)$ are the Langevin noise forces for the electron density, the photon density, and the phase, respectively. Since the Langevin sources are stochastic processes, their incorporation causes the rate equations to become nondeterministic. Following the derivations of [8.22] – [8.31], [8.41], and [8.42], the moments of the Langevin forces can be expressed as

$$\langle F_n(t) F_n(t+\tau) \rangle = (2\beta B n_o^2 S_o V^2 + 2 B n_o^2 V)\, \delta(\tau) \tag{8.60}$$

$$\langle F_S(t) F_S(t+\tau) \rangle = (2\beta B n_o^2 S_o V^2)\, \delta(\tau) \tag{8.61}$$

$$\langle F_\phi(t) F_\phi(t+\tau) \rangle = \left(\frac{\beta B n_o^2}{2 S_o} \right) \delta(\tau) \tag{8.62}$$

$$\langle F_n(t) F_S(t+\tau) \rangle = (-2\beta B n_o^2 S_o V^2)\, \delta(\tau) \tag{8.63}$$

$$\langle F_n(t) F_\phi(t+\tau) \rangle = 0 \tag{8.64}$$

$$\langle F_S(t) F_\phi(t+\tau) \rangle = 0 \tag{8.65}$$

In the Langevin moments, n_o and S_o represent the steady-state, average values of the electron and photon densities, respectively, while $\delta(\tau)$ is the delta function and V is the cavity volume. The Langevin force for the electron density has its origins in carrier shot noise, while the forces for the photon density and the phase result from spontaneous emission [8.42]. The electron and photon density Langevin forces are not independent; their relationship is expressed through their cross-correlation in (8.63). However, as seen in (8.64) and (8.65), the Langevin force for the phase is independent of both the electron and photon densities. The Langevin forces were derived in [8.22] – [8.31], [8.41], and [8.42] on the Markovian assumption that the processes of interest take place on time scales much greater than those required for atomic processes. The noise-driven rate equation formulation is quite convenient since it encompasses both semiclassical and quantum effects. It will serve as the basis for the laser noise simulations presented in this chapter.

The Langevin noise sources were presented in the previous section without derivation; however, it is obviously desirable to have *some* physical understanding of their origins. The quantum-mechanical description of a laser, as well as the quantum theory behind laser noise, is extremely complicated, at most points being more mathematical than physical. Many researchers have recognized this problem and have presented alternate derivations of laser noise which arrive at the same final result [8.44] – [8.46]. The most intuitive, however, was presented by Henry in [8.47] – [8.50], in which spontaneous emission is depicted in the form of a phasor diagram. Since the focus of these sections is on the simulation of laser noise, rather than its derivation, the reader is referred to these references for a better understanding of the laser noise mechanisms themselves.

In [8.42], the noise-driven rate equations are linearized, then solved analytically in the small-signal regime. The first step in Agrawal and Dutta's approach in [8.42] is to split the various quantities into two components

$$n = n_0 + \delta n \qquad S = S_0 + \delta S \qquad \phi = \phi_0 + \delta\phi \qquad (8.66)$$

where n_0, S_0, and ϕ_0 are the steady-state values of the electron and photon densities and the phase, respectively, and δn, δS, and $\delta\phi$ are the small-signal fluctuations. The noise-driven rate equations are then found to be

$$\frac{d(\delta n)}{dt} = -\Gamma_n \delta n - \Gamma c' \left(g + \frac{\partial g}{\partial S} S_0 \right) \delta S + \frac{d(\delta J)}{qNL_z} + F_n(t) \qquad (8.67)$$

$$\frac{d(\delta S)}{dt} = -\Gamma_S \delta S + \left(\Gamma c' S_0 \frac{\partial g}{\partial n} + 2\beta B n_0 \right) \delta n + F_S(t) \qquad (8.68)$$

$$\frac{d(\delta\phi)}{dt} = \left(\frac{\alpha}{2} \right) \frac{\partial g}{\partial n} \delta n + F_\phi(t) \qquad (8.69)$$

where Γ_n and Γ_S are defined as

$$\Gamma_n \equiv 2Bn_0 + \Gamma c' \frac{\partial g}{\partial n} S_0 \qquad (8.70)$$

$$\Gamma_S \equiv \frac{\beta B n_0^2}{S_0} - \frac{\partial g}{\partial S} S_0 \qquad (8.71)$$

and are interpreted in [8.42] as the small-signal decay rates of the electron and photon densities, respectively. Notice that in (8.67) – (8.69), the Langevin noise sources are left in their original form since they are inherently "small-signal" fluctuations. The gain derivatives are evaluated at the steady-state electron and photon densities n_0 and S_0. In [8.42], it is assumed that the injected current J is constant (i.e, $\delta J = 0$). This has important implications as it limits the analysis to unmodulated lasers; that is, it allows prediction of only the *steady-state* noise characteristics of laser diodes. Modeling laser noise under modulation presents additional complexity, as shown in [8.51]. In order to demonstrate the simulation of laser noise and its place within a mixed-level simulation framework, the simpler task of modeling steady-state laser noise will be demonstrated in this chapter. The techniques illustrated here can be expanded to encompass the noise behavior of modulated laser diodes.

The steady-state densities, n_0 and S_0, are determined by setting the time derivatives equal to zero in the rate equations and solving for n and S respectively. Since n_0 and S_0 can be determined so simply, it is natural to inquire why simulation is necessary at all. While many relevant quantities can be determined analytically, their *computation* is not simple. For example, the quantum-mechanical gain equation (8.49) is a function of the Fermi-level difference, a quantity that is not trivial to evaluate. While the derivatives of the gain with respect to the electron and photon densities appear symbolically (analytically) in the rate equations, their computation is also quite difficult. Determination of these quantities is further complicated when a drive circuit is attached to the laser; when designing transmitters, it is necessary to determine the change in the electron and photon densities, gain, and the gain derivatives as a function of inputs to the drive circuit. Matters only become worse when the characteristics of arrays of lasers or arrays of transmitters must be analyzed. In a manner consistent with the mixed-level simulation approach presented in previous sections, the simulation of laser noise will use iSMILE and Gao's laser diode model to compute difficult-to-evaluate quantities such as the steady-state electron and photon densities, the gain, and the derivatives of the gain with respect to the electron and photon densities. It will then use analytical techniques to determine other quantities of interest. The simulation interface will, again, control and coordinate the activity.

As mentioned previously, the analysis of noise quantities is often simpler in the frequency domain. By defining the Fourier transform as

$$\tilde{f}(\omega) = \int_{-\infty}^{\infty} f(t) e^{j\omega t} dt \tag{8.72}$$

and making use of the fact that time derivatives map into $j\omega$ in the frequency domain, following the approach of Agrawal and Dutta in [8.42], the rate equations are transformed into the frequency domain according to (8.73) – (8.75).

$$j\omega \delta \tilde{n} = -\Gamma_n \delta \tilde{n} - \Gamma c' \left(g + \frac{\partial g}{\partial S} S_o \right) \delta \tilde{S} + \tilde{F}_n \tag{8.73}$$

$$j\omega \delta \tilde{S} = \Gamma_S \delta \tilde{S} + \left(\Gamma c' S_o \frac{\partial g}{\partial n} + 2\beta B n_o \right) \delta \tilde{n} + \tilde{F}_S \tag{8.74}$$

$$j\omega \delta \tilde{\phi} = \left(\frac{\alpha}{2} \right) \frac{\partial g}{\partial n} \delta \tilde{n} + \tilde{F}_\phi \tag{8.75}$$

In the frequency domain, the rate equations form a set of linear algebraic equations which can be solved by conventional methods. The Fourier transforms of the moments of the Langevin noise sources take advantage of the fact that delta functions in the time domain map into constants in the frequency domain, and, using the methods detailed in [8.42], are given by (8.76) – (8.79).

$$\langle \tilde{F}_n(\omega)\, \tilde{F}_n(\omega) \rangle = 2\beta Bn_o^2 S_o V^2 + 2Bn_o^2 V \tag{8.76}$$

$$\langle \tilde{F}_S(\omega)\, \tilde{F}_S(\omega) \rangle = 2\beta Bn_o^2 S_o V^2 \tag{8.77}$$

$$\langle \tilde{F}_\phi(\omega)\, \tilde{F}_\phi(\omega) \rangle = \frac{\beta Bn_o^2}{2S_o} \tag{8.78}$$

$$\langle \tilde{F}_n(\omega)\, \tilde{F}_S(\omega) \rangle = -2\beta Bn_o^2 S_o V^2 \tag{8.79}$$

Algebraically solving these equations results in closed-form expressions for the electron density, the photon density, and the phase (8.80) – (8.82).

$$\delta \tilde{n}(\omega) = \frac{(\Gamma_S + j\omega)\, \tilde{F}_n - \Gamma c'\left(g + \frac{\partial g}{\partial S} S_o\right)\tilde{F}_S}{(\Omega_R + \omega - j\Gamma_R)\,(\Omega_R - \omega + j\Gamma_R)} \tag{8.80}$$

$$\delta \tilde{S}(\omega) = \frac{(\Gamma_n + j\omega)\, \tilde{F}_S + \left(\Gamma c' S_o \frac{\partial g}{\partial n} + 2\beta Bn_o\right)\tilde{F}_n}{(\Omega_R + \omega - j\Gamma_R)\,(\Omega_R - \omega + j\Gamma_R)} \tag{8.81}$$

$$\delta \tilde{\phi}(\omega) = \frac{1}{j\omega}\left\{\left(\frac{\alpha}{2}\right)\frac{\partial g}{\partial n}\left[\frac{(\Gamma_S + j\omega)\, \tilde{F}_n - \Gamma c'\left(g + \frac{\partial g}{\partial S} S_o\right)\tilde{F}_S}{(\Omega_R + \omega - j\Gamma_R)\,(\Omega_R - \omega + j\Gamma_R)}\right] + \tilde{F}_\phi\right\} \tag{8.82}$$

The quantities Γ_R and Ω_R are the decay rate and frequency of the laser relaxation oscillations. The simulation of these phenomena will be presented in the next section; however, their values are presented here as

$$\Gamma_R = \frac{(\Gamma_n + \Gamma_S)}{2} \tag{8.83}$$

$$\Omega_R = \sqrt{\Gamma c'\left(g + \frac{\partial g}{\partial S} S_o\right)\left(\Gamma c' \frac{\partial g}{\partial n} S_o + 2\beta Bn_o\right) - \left(\frac{\Gamma_n + \Gamma_S}{2}\right)^2} \tag{8.84}$$

where Γ_n and Γ_S are the electron and photon density decay rates as given in (8.70) and (8.71).

Equations (8.80) – (8.82) could, in principle, be inverse Fourier transformed back into the time domain in order to determine the transient characteristics of the electron density, photon density, and phase. Alternatively, the noise-driven rate equations could have been solved directly in the time domain using numerical or

Monte Carlo techniques, a process further complicated by the stochastic nature of the noise sources. In any event, it should be noted that simulation of the transient behavior *is* possible. For the analyses of laser noise presented in the mixed-level framework of this chapter, $\delta n(t)$ and $\delta S(t)$, *themselves*, are not of interest, as will be detailed shortly. The time-domain properties of the phase will, however, be exploited in later sections. Thus, for the analyses to be considered in this chapter, it is simpler to leave these quantities in the frequency domain.

An important point that has been continuously stressed throughout this chapter is that while closed-form analytical expressions do exist (8.80) – (8.82), their evaluation is difficult since quantities such as the quantum-mechanical gain, the gain partial derivatives, and the electron and photon densities must be computed numerically.

8.2.6 Relaxation oscillations

As is well known from circuit and system theory, the "natural" response of a system can be deduced by first removing all sources, storing some initial energy in the system, then determining how the system equilibrates when the stored energy is released [8.52]. The classic example is a parallel RLC circuit with its current source removed. Energy is initially stored in the inductor and the capacitor, in the form of initial currents and voltages, then released; the differential equation describing this system is called the characteristic equation. The characteristic equation is then solved for the currents and voltages resulting in either an underdamped, overdamped, or critically damped response, depending on whether the roots of the characteristic equation are real and distinct, complex, or real and equal, respectively.

In [8.42], Agrawal and Dutta apply this same principle to the small-signal rate equations. After all sources, including the Langevin noise sources, are removed and initial conditions δn_0 and δS_0 are established, the rate equations constitute a simple eigenvalue equation:

$$\begin{bmatrix} \dfrac{d(\delta n)}{dt} \\[2ex] \dfrac{d(\delta S)}{dt} \end{bmatrix} = \begin{bmatrix} -\Gamma_n & -\Gamma c'\left(g + \dfrac{\partial g}{\partial S}S_0\right) \\[2ex] \Gamma c'S_0\dfrac{\partial g}{\partial n} + 2\beta Bn_0 & -\Gamma_S \end{bmatrix}\begin{bmatrix} \delta n \\[2ex] \delta S \end{bmatrix} \tag{8.85}$$

The solution to this system is known to be

$$\delta n(t) = \delta n_0 e^{-\lambda t} \qquad \delta S(t) = \delta S_0 e^{-\lambda t} \tag{8.86}$$

The eigenvalue λ is given by

$$\lambda = \Gamma_R \pm j\Omega_R \tag{8.87}$$

where the quantities Γ_R and Ω_R are as given in (8.83) and (8.84).   Since the eigenvalues are complex, the system (laser) is overdamped. Substitution of λ into the solutions for the small-signal electron and photon densities (8.86) will result in an exponentially decaying sinusoidal characteristic. Thus, Γ_R can be interpreted as the decay rate of the sinusoidal oscillations and Ω_R as the frequency. Examination of the turn-on characteristic of the laser diode (which is repeated here as Figure 8.10 for convenience) shows that the laser does in fact exhibit exponentially decaying oscillations. This phenomenon is called relaxation oscillation and has its origins in the dynamic exchange of energy between the electron and photon densities in the laser cavity during laser turn on. This is analogous to the exchange of energy between the inductor and capacitor in the RLC circuit. As detailed in previous chapters, turning on the laser suddenly creates an abrupt rise in the electron concentration. The electron population usually equilibrates, or relaxes, through recombination, producing a photon. However, if the current pulse is sharp enough, the electron density can rise so quickly that it exceeds the threshold density before such equilibration can take place. When recombination finally begins to take place, since the electron population is so large, a huge number of photons are emitted (i.e., a huge number of recombination events takes place). This drives the electron population below threshold. The resulting dip in the electron population causes a rapid buildup of electrons which then results in a large number of photon emissions. This process repeats itself, over and over, the effect becoming less profound at each

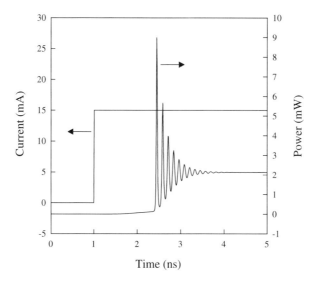

Figure 8.10: Laser turn-on transient exhibiting relaxation oscillations.

iteration. Eventually, the system equilibrates to some steady state as seen in Figure 8.10.

In a manner consistent with previous simulations using the mixed-level circuit-system interface, circuit simulation is used to numerically determine the electron and photon densities and gain related quantities in (8.83) and (8.84). The simulation interface then uses (8.83) and (8.84) to calculate the frequency and decay rates of the relaxation oscillations. Since all of the quantities extracted from circuit simulation depend on the input current (n_0, S_0, g, $\partial g/\partial n$, $\partial g/\partial S$), the dependence of Γ_R and Ω_R on the input current (or equivalently, on the bias point) can be determined, as in Figure 8.11 ($f_R = \Omega_R/2\pi$). Under this interpretation, the independent variable is injected current; however, since output power is a unique function of injected current (Figure 8.7), power is used as the independent variable in Figure 8.11. This representation is used mainly because it appears to be the most prevalent approach in the literature. While the frequency and decay rate of the oscillations could be determined by analyzing the circuit simulation output (Figure 8.10) directly and extracting Γ_R and Ω_R, the mixed-level simulation approach is much more efficient. Extraction of Γ_R and Ω_R directly from circuit simulation would require numerical optimization in a manner similar to that taken by Pang in his table-based laser model of Chapter 7.

The analysis of relaxation oscillations is important because, as seen in Figure 8.10, this phenomenon places an upper limit on the modulation and turn-on responses. The frequency of the oscillations is usually quite high (2 – 13 GHz in

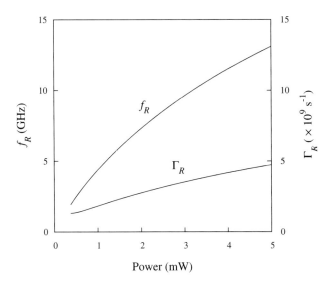

Figure 8.11: Frequency and decay rate of relaxation oscillations.

Figure 8.11); thus, the oscillations often do not appear during regular measurement [8.53]. High-speed oscilloscopes are usually necessary to observe this phenomenon. It is of interest to note that the curves in Figure 8.11 do not extend all the way down to 0 mW. The reason for this is that the laser of Table 8.1 produces only spontaneous emission until about 0.3 mW; that is, below 0.3 mW, lasing does not occur. Since the laser is not yet on, there can be no relaxation oscillations.

8.2.7 Relative intensity noise (RIN)

The most easily understood type of laser noise is relative intensity noise, or RIN. Intensity noise, as the name implies, is the random fluctuation of intensity in the laser output power (Figure 8.12). *Relative* intensity noise is defined to be the ratio of the mean-square optical noise to the square of the average, steady-state power.

$$\text{RIN}(t) = \frac{\langle \delta P^2(t) \rangle}{\langle P \rangle^2} \tag{8.88}$$

In (8.88), power is denoted by P and power fluctuations by δP. While the time-domain representation of Figure 8.12 is represented by (8.88), as mentioned in previous sections, it is much easier to determine noise, such as RIN, in the frequency domain [8.34]:

$$\text{RIN}(\omega) = \frac{S_P(\omega)}{\langle P \rangle^2} = \frac{\langle |\delta \tilde{P}(\omega)|^2 \rangle}{\langle P \rangle^2} = \frac{\langle |\delta \tilde{S}(\omega)|^2 \rangle}{\langle S \rangle^2} \tag{8.89}$$

In (8.89), S_P is the power spectral density (frequency domain ω), S denotes the photon density, and the third equality is made by canceling the scaling factor between photon density and power in both the numerator and the denominator.

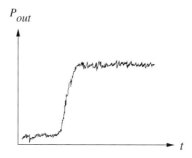

Figure 8.12: Illustration of intensity noise.

Using the three relations for the Langevin noise sources in the frequency domain (8.76), (8.77), (8.79), as well as the expression for the small-signal frequency-domain photon density (8.81), (8.89) can be used to determine RIN in the frequency-domain through (8.90) [8.42]. As with the modeling and simulation of relaxation oscillations, Gao's iSMILE laser diode model is used to compute the gain, the gain partial derivatives, and the steady-state electron and photon densities, then the mixed-level simulation interface extracts these values from the circuit simulation output and computes the RIN (8.90) as a function of frequency. The results of these simulations are shown in Figure 8.13.

$$\text{RIN}(\omega) = \frac{2\beta B n_o^2}{S_o} \cdot \frac{1}{[\Gamma_R^2 + (\Omega_R + \omega)^2][\Gamma_R^2 + (\Omega_R - \omega)^2]} \cdot \qquad (8.90)$$

$$\left\{ (\Gamma_n^2 + \omega^2) + \left(\frac{\partial g}{\partial n} S_o + 2\beta B n_o \right) \left[\left(\frac{\partial g}{\partial n} S_o + 2\beta B n_o \right) \left(1 + \frac{1}{\beta S_o V} \right) - 2\Gamma_n \right] \right\}$$

Typically, researchers measure the RIN as a function of frequency for several different levels of output power. In Figure 8.13, this approach is extended to simulate the RIN as a continuous function of both frequency and power. This figure shows that the RIN peaks at about the relaxation oscillation frequency Ω_R; this is also seen by examining the denominator of (8.90). From this figure it can also be

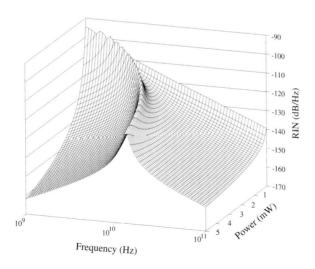

Figure 8.13: Modeling of RIN in the frequency domain.

seen that the peak of the RIN not only decreases but also shifts to greater frequencies for higher powers. The decrease in the RIN peak is caused by the S_0 term in the denominator of (8.90) and can be qualitatively understood by recalling that RIN is defined to be the intensity noise divided by the square of the output power. The frequency shift of the RIN peak is caused by the S_0-dependence of Ω_R (i.e., through negative values of the partial derivative of the gain with respect to S_0). Finally, it should be noted that the RIN in Figure 8.13 is not plotted all the way down to 0 mW — this would not make sense in view of the facts that, physically, RIN cannot be defined before the onset of lasing and, mathematically, the power appears in the denominator of the RIN expression. (Before lasing, only spontaneous emission emerges from the laser. Since spontaneously emitted photons are incoherent, all of the output power is noise).

While the frequency-domain solution of the RIN is quite useful, it cannot be used directly to represent the transient intensity noise depicted in Figure 8.12. The transient intensity noise is of much greater interest (Figure 8.9) since most fiber-optic systems employ digital pulses rather than small-signal sinusoidal modulation. Using the Wiener-Khintchine Theorem, the spectral density in the frequency-domain definition of the RIN can be mapped into the mean-square intensity noise, according to (8.56):

$$\mathrm{RIN}(t) = \int_0^\infty \mathrm{RIN}(f)\,df = \frac{1}{2\pi}\int_0^\infty \mathrm{RIN}(\omega)\,d\omega \tag{8.91}$$

Unfortunately, the integration of $\mathrm{RIN}(\omega)$ (8.91) is not trivial; however, by using complex contour integration, an analytical expression *can* be derived in the form of (8.92).

$$\int_0^\infty \frac{ax^2+b}{\left[c^2+(x+d)^2\right]\left[c^2+(x-d)^2\right]}\,dx = \frac{\pi}{4c}\left[a+\frac{b}{(c^2+d^2)}\right] \tag{8.92}$$

Using (8.92), the time-domain (mean-square) RIN can be expressed as (8.93).

$$\mathrm{RIN}(t) = \frac{\pi}{4\Gamma_R}\cdot\frac{2\beta B n_0^2}{S_0}\cdot$$

$$\left\{1+\frac{\Gamma_n^2+\left(\frac{\partial g}{\partial n}S_0+2\beta B n_0\right)\left[\left(\frac{\partial g}{\partial n}S_0+2\beta B n_0\right)\left(1+\frac{1}{\beta S_0 V}\right)-2\Gamma_n\right]}{(\Gamma_R^2+\Omega_R^2)}\right\} \tag{8.93}$$

Since output power is a unique function of the steady-state injected current, as with the relaxation oscillations, it is common to plot the time-domain RIN as a function of output power (Figure 8.14). In order to produce the data of Figure 8.14, as with the frequency-domain RIN, iSMILE was used to compute the relevant quantities and the mixed-level interface was used to numerically calculate the mean-square RIN in the time domain. This value was then fed to iFROST as an input parameter. Figure 8.14 illustrates the ratio of the Gaussian variance (mean-square value) of the intensity noise to the square of the output power. As expected, this figure shows an inverse dependence on output power, approaching infinity for zero power and zero (or negatively infinite dB) for infinite power.

8.2.8 Linewidth broadening

Although single (longitudinal) mode lasing has been assumed thus far, in reality, lasing does not occur at a single wavelength (frequency). While the laser ideally exhibits a delta function-like spectrum, noise mechanisms cause a spreading, or broadening, of the laser line, as in Figure 8.15. In this figure, the center frequency is denoted by f_o and is determined in a semiconductor laser by the separation of energy levels. The linewidth Δf is sometimes referred to as the full width at half-maximum (FWHM) since it is defined to be the amount of linewidth broadening that occurs when the laser field is at one-half its peak value.

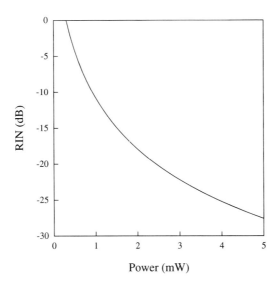

Figure 8.14: Modeling of mean-square RIN in the time domain.

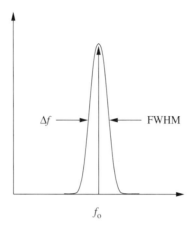

Figure 8.15: Laser linewidth broadening.

Since an optical signal experiences spreading, or dispersion, as it propagates through an optical fiber or waveguide, it is important to minimize the laser linewidth. In fact, the maximum bit rate achievable for an optical link is limited by the linewidth of the source, as evidenced by the iFROST simulations of previous sections. For a general overview of linewidth broadening, the reader is referred to [8.54] and [8.55].

The fundamental source of linewidth broadening is phase noise. While this is to be contrasted with intensity noise, which was discussed in the previous section, a coupling between intensity noise and phase noise does exist. Henry's description of laser noise [8.47] – [8.50] best explains the origins of phase noise. Essentially, phase noise originates from spontaneously emitted photons. Each of these photons has a random phase which causes an instantaneous change in the phase of the lasing field. The phase noise as expressed by the Langevin noise source in the third (phase) rate equation is the summation of these random phase additions to the lasing field, over all spontaneous emission events.

In addition, inspection of the expression for the phase, as calculated from the noise-driven rate equations, shows that the small-signal phase depends directly on the small-signal electron density; that is, (8.82) can be cast into the form of

$$\delta\tilde{\phi}(\omega) \ = \ \frac{1}{j\omega}\left[\left(\frac{\alpha}{2}\right)\frac{\partial g}{\partial n}\delta\tilde{n}(\omega) + \tilde{F}_{\phi}\right] \qquad (8.94)$$

This implies that changes in the electron density cause changes in the phase. Since changes in the number of electrons change the photon density, gain, refractive index and, most importantly, the optical intensity, this results in a coupling between the intensity and phase noises. This fact is encompassed in the linewidth enhancement factor α, previously introduced in the rate equation for the phase. This electron fluctuation-induced phase noise constitutes a second mechanism to the total phase noise, and is seen as the first component of (8.94). The physics behind α will be discussed shortly.

Since a change in phase will directly induce a change in frequency,

$$\delta f = \frac{1}{2\pi} \frac{d(\delta\phi)}{dt} \tag{8.95}$$

the linewidth broadening, or "frequency noise," can be determined from the spectral density of the time derivative of the phase in a manner similar to that used to compute the intensity noise from the spectral density of the photon density in the previous section [8.42]:

$$S_{\overset{\bullet}{\phi}}(\omega) = \langle |j\omega\delta\tilde{\phi}(\omega)|^2 \rangle = \langle |\omega\delta\tilde{\phi}(\omega)|^2 \rangle \tag{8.96}$$

In (8.96) use is made of that fact that time derivatives map into a multiplication by $j\omega$ in the frequency domain. Omitting the lengthy derivation found in [8.42] and [8.57], the linewidth Δf is defined to be

$$\Delta f = \frac{S_{\overset{\bullet}{\phi}}(0)}{2\pi} = \frac{1}{4\pi} \cdot \frac{\beta B n_o^2}{S_o}(1 + \alpha^2) \tag{8.97}$$

where α is called the linewidth enhancement factor.

Since the quantum-mechanical gain equation is a function of the Fermi-level difference, it depends directly on the injected carrier concentration. As a result, changes in the electron density will cause changes in the gain. Since the gain and refractive index are linked through the Kramers-Kronig relations, changes in the electron density will also cause changes in the refractive index. According to Henry in [8.47], there is an instantaneous change in field intensity brought about by each spontaneous emission event. In response to this perturbation, the laser tries to restore equilibrium through a change in the electron density. This, in turn, forces a change in the gain and, correspondingly, a change in the refractive index. Since a change in the refractive index will shift the longitudinal-mode frequency [8.42], the phase will change as well. A change in intensity will therefore result in a change in the phase. Thus, there are two sources of phase noise; the noise caused by the coupling between intensity noise and phase, and the noise caused by the addition of random phase to the lasing field from spontaneous emission.

Henry's linewidth enhancement factor α describes the proportionality between the gain and the index changes, as shown in (8.98). In this equation, $\bar{n}$ is the refractive index and g is the gain. In fact, because α essentially describes the coupling between the intensity noise and the phase noise, it can be calculated by measuring the correlation between the AM and FM noises [8.57]. While α can range from about 4 ~ 8 in a bulk semiconductor laser, it is significantly lower (in the 1 ~ 2 range) in quantum-well lasers [8.58] – [8.60]. It has even been shown that α can be further reduced through use of strain in quantum-well lasers [8.61].

$$\alpha = -2\left(\frac{2\pi}{\lambda_o}\right) \cdot \frac{\left(\frac{d\bar{n}}{dn}\right)}{\left(\frac{dg}{dn}\right)} \qquad (8.98)$$

The linewidth enhancement factor is commonly assumed to be constant. However, it has been shown [8.62] that not only does α change with both the carrier density and the intensity $|E|^2$, it is also time dependent whenever the carrier density and the intensity are. Despite this variability, it is shown in [8.62] that α can be treated as a constant not only when the laser is in CW operation, but also when the laser is directly modulated, as long as the carrier densities do not change appreciably.

Finally, the linewidth enhancement factor has been shown to have a dependence on gain nonlinearities, mainly those due to gain saturation [8.63], [8.64]. As the intracavity intensity approaches the saturation intensity, the linewidth enhancement factor begins to increase from its value when gain nonlinearities are not considered. That is, while the linewidth initially decreases with increasing output power, for high levels of output power, α actually begins to increase.

As with the intensity noise, the linewidth broadening can be simulated by first using circuit simulation and the laser diode model to calculate relevant quantities, then using the mixed-level simulation interface to compute (8.97). So far, only the *frequency* spread in the linewidth has been considered; it is often of interest to obtain the spread of the *wavelength*. While the frequency and the wavelength are related simply by

$$f\lambda = c \qquad (8.99)$$

in order to determine the relationship between the incremental wavelength and frequency, it is necessary to differentiate (8.99):

$$f\Delta\lambda + \lambda\Delta f = 0 \qquad (8.100)$$

The wavelength spread is then calculated to be

$$\Delta\lambda = -\frac{\lambda}{f}\Delta f \qquad (8.101)$$

where the negative sign is usually dropped to consider absolute changes. Thus, with knowledge of Δf, (8.97) can be used to calculate $\Delta\lambda$, as well. In Figures 8.16 and 8.17 the linewidth for the laser of Table 8.1 in terms of both wavelength and frequency is depicted for $\alpha = 2$. As seen in Figure 8.16, the linewidth decreases with increasing power. In Figure 8.17, the linewidth is depicted versus the inverse power.

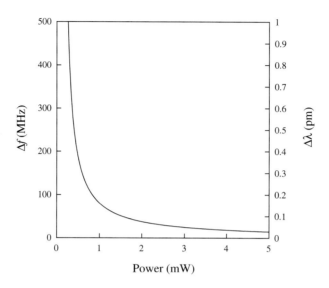

Figure 8.16: Mixed-level simulation of linewidth versus power.

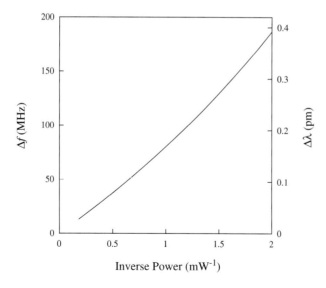

Figure 8.17: Mixed-level simulation of linewidth versus inverse power.

This representation is quite popular in the literature and stems from the fact that in a standard laser, with no gain saturation effects, the linewidth should be linear with respect to inverse power. This is clear from (8.97) which shows that since the output power is directly proportional to the photon density, the linewidth versus power relationship should be hyperbolic and the linewidth versus inverse power relationship should be linear. The slight nonlinearity in Figure 8.17 is attributed to gain saturation effects that were implemented in the laser diode model. Since the linewidth simulations of Figures 8.16 and 8.17 were performed on single-mode lasers, the frequency and wavelength spread were on the order of MHz and pm $(10^{-12}$ m), respectively. Multimode laser linewidth simulation should result in quantities on the order of GHz and nm.

8.2.9 Chirp

While the previous section showed how the linewidth enhancement factor can lead to a broadening of the frequency, α can also cause a transient *shift* in the center frequency. This phenomenon, which can occur during laser modulation, is known as chirp and is often the limiting mechanism in the speed of optical links. This is mainly due to the fact that shifts in the center wavelength will result in chromatic dispersion, a problem which is especially apparent when the original center wavelength is positioned so that it coincides with the fiber's zero-dispersion wavelength.

Physically, chirp is brought about by a shift in the phase. In fact, the transient chirp is defined to be the instantaneous frequency shift of the lasing mode from its steady-state value [8.65]:

$$\delta f(t) = \frac{1}{2\pi} \frac{d(\delta\phi)}{dt} \tag{8.102}$$

From examination of the rate equation for the phase (8.59), (8.69), as well as its solution (8.82), it is evident that the linewidth enhancement factor plays a significant role in determining the chirp. Actually, from (8.59) it is clear that if α were zero, the only contribution to the phase variation would be the Langevin phase noise.

While most of the analysis, so far, has been in the small-signal frequency domain, (8.102) shows one clear example where transient analysis is required. In addition, while previous analyses were concerned only with the steady-state noise characteristics of lasers, chirp, by definition, occurs only under modulation. Thus, it will not be possible to follow the same straightforward approach taken in previous sections. For simplicity, in this section noise due to the Langevin phase source will be neglected in order to concentrate on the intrinsic phase-shifting mechanisms present in laser diodes; however, determination of the phase derivative with the Langevin source *can* be accomplished, as shown in [8.51]. In most cases, however, the chirp that results from α effects will be much greater than the chirp caused by the

Langevin phase term. In order to determine chirp, the modulation current (δJ or δI, as the case may be) can no longer be assumed to be zero; the general approach taken in this section will closely parallel that of [8.42] and [8.65].

Since the modulation response of the laser must now be considered, δI must now be allowed to be nonzero. If the frequency-domain representation of δI is defined by its Fourier transform

$$\tilde{\delta I}(\omega) = \int_{-\infty}^{\infty} \delta I(t)\, e^{-j\omega t} dt \tag{8.103}$$

then the small-signal, frequency-domain rate equations can be used to solve for the small-signal phase variation [8.65]:

$$\tilde{\delta\phi}(\omega) = -\frac{\alpha \frac{\partial g}{\partial n} \tilde{\delta I}(\omega)(\omega - j\Gamma_S)}{2q\omega\,[\omega - (\Omega_R + j\Gamma_R)]\,[\omega - (-\Omega_R + j\Gamma_R)]} \tag{8.104}$$

However, this is not enough, as it is necessary to determine the time-domain phase derivative. Using the fact that differentiation in the time domain can be mapped into multiplication by $j\omega$ in the frequency domain, as well as the definition of the (inverse) Fourier transform, (8.102) can be used to express the chirp analytically:

$$\delta f(t) = \int_{-\infty}^{\infty} \left\{ \frac{\frac{1}{(2\pi)^2} \cdot \alpha \frac{\partial g}{\partial n} \tilde{\delta I}(\omega)(\omega - j\Gamma_S)}{2jq\,[\omega - (\Omega_R + j\Gamma_R)]\,[\omega - (-\Omega_R + j\Gamma_R)]} \right\} e^{j\omega t} d\omega \tag{8.105}$$

This expression is, however, a complex integral and can only be evaluated analytically through contour integration. Furthermore, in order to analytically integrate (8.105), the modulation current must be known analytically, as well.

With the expression for the chirp as a function of the modulation current (δI), the electron and photon densities, the gain, and the gain derivatives (via Ω_R, Γ_R and Γ_S), it is necessary to determine how to numerically evaluate (8.105) for an arbitrary modulation current. The approach is to, again, use iSMILE and the laser diode model to compute relevant quantities and to use the mixed-level interface to compute the rest. In order to compute the integral of (8.105) it should first be shown that (8.105) is a valid inverse Fourier transform. This should be apparent since it was arrived at through Fourier transform and linear frequency-domain techniques (8.103), (8.104). This problem can also be viewed as a convolution integral if the frequency-domain transfer function $H(\omega)$ is defined by (8.106).

$$H(\omega) = \left(\frac{\alpha \frac{\partial g}{\partial n}}{4\pi q}\right) \cdot \frac{\omega - j\Gamma_S}{j\left[\omega - (\Omega_R + j\Gamma_R)\right]\left[\omega - (-\Omega_R + j\Gamma_R)\right]} \qquad (8.106)$$

By using standard Fourier transform tables, the impulse response can be determined as

$$h(t) = \left(\frac{\alpha \frac{\partial g}{\partial n}}{4\pi q}\right) \cdot \frac{e^{-\Gamma_R t}}{\Omega_R} \left[\Omega_R \cos(\Omega_R t) + (\Gamma_S - \Gamma_R)\sin(\Omega_R t)\right] u(t) \qquad (8.107)$$

where $u(t)$ is the Heaviside step function. Since Ω_R, Γ_R, and Γ_S are all constant with respect to ω, it is clear that the impulse response $h(t)$ is the sum of two exponentially decaying sinusoids. This result is important because it shows that the impulse response is bounded. With this result, the chirp can be represented in the time domain as a convolution:

$$\delta f(t) = \delta I(t) * h(t) \qquad (8.108)$$

Now that the integral of (8.105) has been established as a valid inverse Fourier transform, it can be solved numerically. In order to make this mixed-level interface as general as possible, it will be assumed that the input modulation current is specified as a one-dimensional array of current versus time. In order to determine $I(\omega)$ the Brenner Fast Fourier Transform (FFT) algorithm is employed [8.66], an algorithm which is based on the Danielson-Lanczos lemma [8.66]. Two typical examples will be considered, closely following the analysis of Shen and Agrawal in [8.65]: square-wave and sinusoidal modulation.

Square-wave modulation

The ideal digital input pulse to an optical system is a square wave. Following the analysis of [8.65], this can be modeled mathematically by

$$I(t) = I_0\left\{[1 - e^{-t/\tau}]u(t) - [1 - e^{-(t-T)/\tau}]u(t-T)\right\} \qquad (8.109)$$

where I_0 is the peak value of the current, T is the pulse duration, τ is related to the rise and fall times, and $u(t)$ is the step function. In [8.65], the Fourier transform of $I(t)$ is analytically derived, substituted into (8.105), and then analytically integrated, resulting in an analytical expression for the time-domain chirp. By following the approach described in the previous section, the chirp of the laser of Table 8.1 that results from a square-wave input was simulated (Figures 8.18 and 8.19). In these simulations, it was assumed that the linewidth enhancement factor was two. To

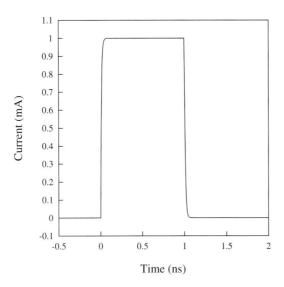

Figure 8.18: Square-wave pulse input.

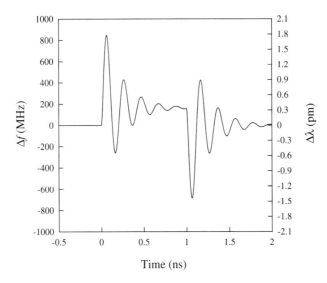

Figure 8.19: Chirp resulting from pulse input.

model the input pulse, (8.109) was discretized into a one-dimensional array of 2048 elements since the Brenner FFT requires that the number of elements in the input array be an integer power of two. A peak current of 1 mA was chosen, as well as a pulse width of 1 ns and a time constant of 0.1 ps. Once the input pulse was discretized, the Brenner FFT was applied to obtain the frequency-domain equivalent of $I(t)$, which was then multiplied term by term with the transfer function $H(\omega)$ of (8.106). This resulted in a numerical representation of the argument of the inverse Fourier transform in (8.109). Finally, the Brenner FFT was applied to this numerical result to transform it back into the time domain. These results matched the analytical solution of [8.65] very well (Figure 8.19).

From Figure 8.19, it is apparent that both the turn-on and turn-off transients cause chirping. For both square-wave modulation, as well as the sinusoidal modulation that will be considered in the next section, the bias conditions were such that the output power was approximately 1 mW. This is an important fact since the chirp depends directly on the bias conditions, as will be shown in a later section.

Sinusoidal modulation

A similar type of analysis can be performed using sinusoidal current modulation at the input. In order to make a comparison with the rectangular pulse, in [8.65] a raised cosine pulse is chosen according to

$$I(t) = \frac{I_o}{2}\left[1 - \cos\left(\frac{\pi t}{T}\right)\right] \tag{8.110}$$

In [8.65], the Fourier transform of (8.110) is analytically derived, upon which (8.105) is integrated analytically. By following the same steps as for the square-wave input, the simulated transient chirp for sinusoidal modulation, using the iSMILE-iFROST interface is depicted in Figures 8.20 and 8.21. Comparison with the analytical expressions for the transient chirp in [8.65] shows an excellent match. The scale of Figure 8.21 shows that the amount of chirp that results from a sinusoidal input pulse can be much smaller than that which occurs under square-wave modulation. However, it has been shown that the chirp that results from sinusoidal modulation is a strong function of the bit rate/period T, exhibiting a sharp peak in the vicinity of the frequency of relaxation oscillations.

While chirp simulation for both square-wave and sinusoidal input currents was demonstrated, since the approach was to FFT the input current, multiply it term by term with the transfer function, and then inverse FFT it back to the time domain, the approach is general and can be applied to any input waveform. The only condition is that the waveform be discretized into an array with a number of elements equal to an integral power of two.

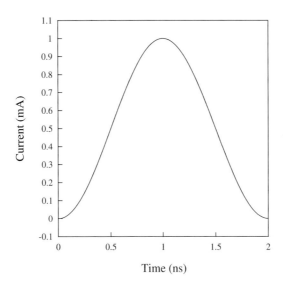

Figure 8.20: Raised cosine input pulse.

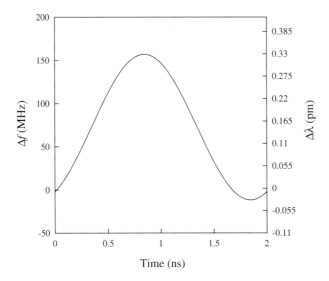

Figure 8.21: Chirp resulting from sinusoidal input.

Maximum chirp

Finally, while the transient chirp is a useful quantity for device analysis, of more importance for system simulation is the *maximum* amount of transient chirp that will occur for a given laser, under a given input. For example, the exact waveform of Figure 8.19 is of little interest; of greater interest is the fact that for a bias that results in 1-mW output, the *peak* chirp is about 850 MHz. By performing the chirp simulations discussed in previous sections for many different levels of current and power, the *maximum* chirp for both the square and sinusoidal input waveforms can be simulated, as seen in Figures 8.22 and 8.23. With knowledge of the maximum chirp for a given bias level, it is possible to do a systems analysis based on the worst-case chirp conditions. As with the evaluation of the transient chirp, the only restriction on the shape of the input waveform is that it is discretizable into 2^n elements, where n is an integer. It is important to bear in mind that the analysis of [8.65] considered only relatively small fluctuations of the current and power about some steady-state bias condition. This is more or less valid considering the magnitude of practical signals in optical communications. Large-signal analysis of the chirp is considerably more involved and is left as an issue for future investigation [8.67], [8.68].

8.3 Mixed-Level Waveform Simulation

Using the approach of the previous section, iFROST can be further combined with circuit-level models to perform data-driven waveform simulation using its waveform simulation mode [8.69], [8.70]. Typically, a digital optical link begins with the generation of a binary waveform which is usually the output of a digital circuit. This waveform is then used as the input to a laser driver which, in turn, controls the laser diode. The resultant optical output signal travels through a fiber, after which it is detected, amplified, and converted to a usable output voltage. The iFROST analyses of Section 8.1 enabled determination of overall system characteristics (rise time, bandwidth/bit rate) as functions of attributes of the individual components comprising the link. In this section, the extension of these capabilities to include data-driven simulation of the actual waveforms, using the quasianalytical technique for the treatment of noise, is depicted. The simulation begins with a simple binary waveform, and traces the changes in this waveform as it traverses the optical link. By including noise and the capability for generating pseudorandom bit stream inputs, the waveform eye pattern at the output of the system can be simulated. The BER can then be computed through the use of the quasianalytical technique. The eye diagram, which is essentially a superposition of many signal waveforms at the output of a system, is an important measure of signal integrity.   By using a pseudorandom bit stream as the input, all possible combinations of signals can be simulated. If logic 1's and logic 0's can still be distinguished at the system output, the eye is said to be "open." If not, the eye is

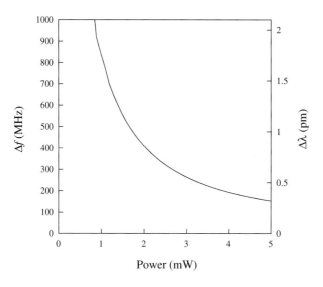

Figure 8.22: Maximum chirp — square-wave input.

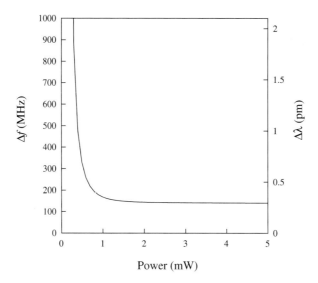

Figure 8.23: Maximum chirp — sinusoidal input.

"closed." Strict telecommunications standards exist to determine exactly "how open" an eye must be in order to meet specifications. System-level requirements such as these place restrictions on the noise, dispersion, and speed of the individual components that make up the link. The use of a mixed-level simulation environment which includes waveform simulation capability can help to facilitate the evaluation of the various trade-offs between system and device performance. The term "data driven" is used in this context to emphasize the sequential nature of the waveform simulation, as each individual component is simulated when the input signal data for that component is ready.

8.3.1 Component models

As shown in previous sections, in systems analysis, lower levels of simulation can be used to model particularly complex devices or subsystems. While the waveform simulation presented here uses the typical rate-equation model for the laser diode, several other models are required, as well.

Binary sequence generator

First, iFROST uses the deBruijn binary sequence [8.69] – [8.71] to generate a pseudorandom bit stream as the system input. While most pseudorandom algorithms produce sequences of length $2^n - 1$, the deBruijn sequence used by Whitlock creates a bit sequence of length 2^n. This is accomplished by adding the one n-bit pattern that does not occur naturally, namely a string of all zeros [8.71]. By definition, all possible n bit combinations in the sequence appear once, and only once; thus, by using this as the input, all possible memory and intersymbol interference effects can be observed. A further advantage of the deBruijn sequence is that it is equally balanced; that is, equal numbers of ones and zeros occur. While an extensive discussion of the generation of random sequences is beyond the scope of this book, an excellent review of this topic can be found in [8.71].

Manchester encoder

As detailed in [8.72], while straight binary signaling (such as NRZ) can be used in communication systems, such methods can often lead to long strings of 1's or 0's. In a serial digital link (i.e., one in which the clock is embedded with the data), long periods without data transitions make clock recovery very difficult. In parallel systems, a lack of transitions makes it difficult for the receiver to determine the beginning and end of each bit. To address these problems, data encoding is often used to translate or map bit streams into predetermined patterns that have a guaranteed number of transitions embedded in them. For example, the 8B/10B protocol maps an 8-bit byte into a 10-bit sequence that contains a predetermined number of transitions; as evidenced by the use of two extra bits, such encoding schemes obviously constitute some degree of overhead. In iFROST, a model that carries out the popular Manchester encoding scheme is included after the deBruijn binary generator. The Manchester encoding scheme embeds a transition after every bit (Figure 8.24); a logic one is signaled by a binary 1-0 sequence, while a logic zero is denoted by a binary 0-1 sequence. While the presence of a transition at the middle of each bit makes it easier for synchronization, the overhead paid is a halving of the available bandwidth since every bit is mapped into two bits. An advantage to using the Manchester encoding scheme is that it guarantees that the data are evenly balanced between 1's and 0's. This allows the use of a simpler AC-coupled receiver, rather than a more complicated DC-coupled receiver. The iFROST model accepts a binary signal as the input and outputs the corresponding Manchester-encoded binary signal.

Binary to voltage converter

In order to interface with models that use electrical signals, a method must be available to convert the simple digital (binary) data waveforms used thus far into usable electrical signals. iFROST also includes a model that accepts a binary waveform as the input and produces a valid electrical signal as the output. In addition to the binary signal, the model requires a set of characteristics describing the shape of the desired electrical waveform. Among these characteristics are the

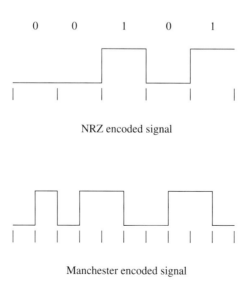

Figure 8.24: Manchester and straight binary (NRZ) data encoding schemes.

rise and fall times of the desired signal, the voltage used to denote a logic high, and the voltage used to denote a logic low. This model is of general use when connecting logic-level components with electrical components.

Laser driver

To model the laser driver, iFROST uses a behavioral description which represents the laser driver as a lowpass Butterworth filter. The voltage waveform that results from the signal converter of the previous paragraph is taken as the input to the laser driver model, which converts it into a current signal. Other inputs to the model include the gain and current offset of the output signal as well as various filter parameters. In general, iFROST has the capability to behaviorally model various types of digital IIR filters including Butterworth, Chebyshev, and elliptical filters, as well as simple low-pass, high-pass, and band-pass filters. In addition, any filter order is supported. Should a higher degree of accuracy be desired in the modeling of the laser driver, the mixed-level nature of iFROST easily allows for the simulation of the laser driver at the circuit level.

Laser diode

As previously mentioned, for laser waveform simulation, iFROST uses the traditional rate-equation approach. The current waveform that is output from the behavioral laser driver model is fed as an input to the lower-level, rate equation-based laser diode model. iFROST passes the input current signal to the rate equations; as a result, the optical output power waveform is generated.

Optical fiber

The next step in the waveform simulation of the optical link is simulation of the optical fiber. iFROST uses a frequency-domain description to simulate the fiber [8.69], [8.70]:

$$H(f) = \frac{1}{\sigma\sqrt{2\pi}}\exp[-(\sigma\pi f)^2 - j2\pi f t_d] \tag{8.111}$$

In this transfer function representation of the fiber, σ is the RMS impulse response width and t_d is the delay. The -3 dB optical bandwidth B_w is given to be $0.1874/\sigma$. The RMS impulse response can be further described by

$$\sigma^2 = \sigma_{mat}^2 + \sigma_{mod}^2 \tag{8.112}$$

where the pulse broadening due to material dispersion σ_{mat} and due to modal dispersion σ_{mod} are expressed as (8.113) and (8.114).

$$\sigma_{mat} = \sigma_{src} L D \tag{8.113}$$

$$\sigma_{mod} = \frac{1}{B_{mod}} \left(\frac{1}{L}\right)^{\gamma_{cutback}} \tag{8.114}$$

In these equations, D is the fiber dispersion, L is the fiber length, σ_{src} is the linewidth of the source, B_{mod} is the modal bandwidth of the fiber, and $\gamma_{cutback}$ is the cutback factor that takes into account mode coupling and mixing effects [8.76].

The delay time t_d has three parts:

$$t_d = (t_{prop} + t_{skew} + \Delta\lambda D) L \tag{8.115}$$

The first term in this equation is propagation delay through the fiber, the second term represents the variable delay due to fiber skew, and the third term models variable delay due to different lasing wavelengths when multiple channels (such as in optical buses) are being modeled [8.76]. For the simulations of this section, the propagation delay t_{prop} is assumed to be 5 ns/m. The second term t_{skew} is modeled as a random variable with a zero mean, a uniform distribution, and a maximum value of 2 ps/m. The last term $\Delta\lambda D$ represents the effect of chromatic dispersion on varying laser wavelengths; the chromatic dispersion D is assumed to be 120 ps/km/nm. The term $\Delta\lambda$ is the difference in lasing wavelengths between the channel under consideration and a reference channel.

In [8.69] and [8.70], the fiber loss is modeled as

$$L(\lambda) = \alpha\lambda^{-4} + b + c(\lambda) \tag{8.116}$$

where L is the loss in units of dB/km, α is the Rayleigh loss coefficient, b is the microbending loss factor, and c is the OH⁻ absorption factor. Obviously, in order to take advantage of the frequency-domain description of the fiber (8.111), a Fourier transform of the optical waveform generated by the rate equation model must be performed.

In addition to the optical fiber model, a model for fiber connectors will soon be implemented based on (8.117) and (8.118), where p is the probability density function and the subscripts g and e denote Gaussian and exponential distributions.

$$P_{e,lf,i} = \frac{1}{\sigma_{gi}\sqrt{2\pi}} \exp\left[\frac{-x^2}{2\sigma_{gi}^2}\right] \qquad \sigma_{gi}^2 = \gamma_l^2\left[\frac{1-\eta}{N\eta}\right] \tag{8.117}$$

$$P_{e,hf,i} = \frac{1}{\sigma_{ei}\sqrt{2\pi}} \exp\left[\frac{-|x|}{\sqrt{2}\sigma_{ei}}\right] \qquad \sigma_{ei}^2 = \frac{\gamma_h^2 k^2 (1-\eta)}{N\eta}\left[\frac{1-\sum a_k^2}{\sum a_k^2}\right] \tag{8.118}$$

This formulation uses the modal noise model described in [8.73] which separates the modal noise into high and low frequency components generated at each discontinuity in the link. In these equations, $\gamma_{l,h}$ are the low- and high-frequency fiber contrast, η is the magnitude of mode selective loss, N is the number of fiber modes, k is the mode partition coefficient, and $\sum a_k^2$ is the variance of relative mode power.

Photoreceiver

Finally, the last iFROST model in the link for the purposes of waveform simulation is the photoreceiver. Again, a high-level, frequency-domain model is used to describe the receiver, which is assumed to be a FET-based transimpedance (TZ) configuration [8.69], [8.70]:

$$H_{rec}(f) = \frac{\dfrac{R_F}{1 + j2\pi f C_T R_F}}{1 + \dfrac{1}{A(f)}} \cdot H_{post}(f) \tag{8.119}$$

$$A(f) = \frac{(g_m + j2\pi f C_{gd}) R_L}{1 + j2\pi f (C_{gd} + C_{gdl} + C_{gdb}) R_L} \tag{8.120}$$

In these equations, R_F is the feedback resistance of the TZ amplifier; C_T is the total input capacitance; $A(f)$ is the open loop gain; g_m is the FET transconductance; R_L is the FET load resistance; and C_{gd}, C_{gdl}, and C_{gdb} are the gate-to-drain capacitances of the input FET, the drive FET of the source follower, and the load FET of the source follower, respectively. The postamplifier transfer function $H_{post}(f)$ is modeled as a fourth-order Butterworth filter with gain. The photoreceiver model encompasses both shot and thermal noise. While a formulation for the noise is presented in Section 8.3.3, its full derivation is beyond the scope of this work; interested readers are referred to [8.74] and [8.75].

While iFROST waveform simulation uses several behavioral models for components (laser driver, optical fiber, photoreceiver), it is quite possible that in some situations, these models cannot provide adequate detail; indeed, that was the case for the laser diode. If a model with the desired detail is not included in iFROST, calls can always made to lower-level simulators. For example, the photoreceiver could always be simulated by passing the optical power waveform to an iSMILE MSM-HEMT photoreceiver simulation, such as that depicted in Chapter 4.

8.3.2 Link waveform simulation

An overview of the waveform simulation is presented in Figure 8.25. In this figure, there are seven main models: the deBruijn pseudorandom sequence

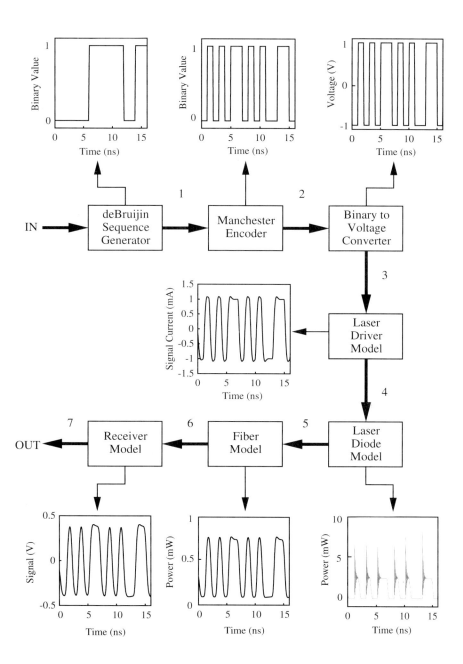

Figure 8.25: Waveform simulation of the entire optical link.

generator, the Manchester encoder, the binary-to-voltage converter, the laser driver, the laser diode, the optical fiber, and the photoreceiver.

Because of the way in which the iFROST models are structured, it is possible to statistically vary several of the component attributes simultaneously in order to assess their impact on overall system performance. In [8.70], Whitlock varies several parameters of both the laser diode and the fiber using a uniform distribution, although it should be noted that iFROST also allows the user to choose a Gaussian distribution. By superimposing multiple bits and statistically varying the parameters as described, an eye diagram of a single-channel optical link is depicted in Figure 8.26. In Figure 8.27, a simulated eye diagram is depicted in which the system outputs of a 32-channel fiber-optic bus (31 data channels and 1 clock channel) are superimposed. For the clock channel, the Manchester encoder and the binary sequence generator are replaced by a clock generator which is used to trigger the decision circuit of the data channels [8.76]. To produce these simulations, four attributes of the laser (carrier density at transparency, wavelength, linewidth, bias current) and five attributes of the fiber (attenuation per unit length, intermodal bandwidth, connector loss, coupling efficiency, delay/skew) were varied uniformly and the link outputs were superimposed. As seen in Figure 8.27, the eye is quite open. The capability to perform this type of eye-pattern analysis through simulation is critical to the efficient design of optical systems. Without simulation, an eye diagram can only be produced after the entire system has been integrated. By quantifying and characterizing the potential statistical variations of the components before the system is even built, a significant amount of time and engineering effort can be saved. In fact, not only can such simulations aid in system design, but they can also be used to help predict system reliability, provided that the parameter variations over time can be estimated.

8.3.3 Bit-error rate (BER) simulation

Several methods exist for simulation of the bit-error rate, such as importance sampling, the extreme value theory technique, and the tail extrapolation technique [8.71]. A commonly used method for BER estimation is Monte Carlo simulation. As discussed in Chapter 2, however, Monte Carlo simulation is typically very time consuming; indeed, in [8.76] it is stated that as a rule of thumb, a Monte Carlo approach to the simulation of noise will require 10/BER ~ 100/BER simulations for a target BER. Thus, for a target BER of 10^{-15}, 10^{16}~10^{17} simulation runs would be necessary.

For BER simulation, iFROST uses the quasianalytical technique for the modeling of noise [8.71], [8.76]. This method takes a hybrid approach by determining the (noiseless) signal waveform through simulation and the system noise through analytical techniques. In this context, the quasianalytical technique takes advantage of iFROST's ability to simulate signal waveforms throughout the

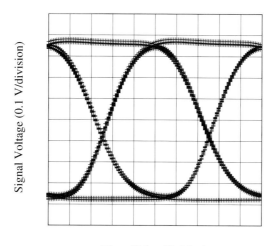

Time (0.2 ns/division)

Figure 8.26: System eye diagram — single optical link.

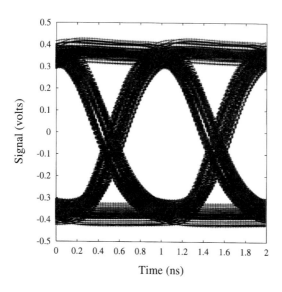

Time (ns)

Figure 8.27: Simulated eye diagram for 31 channels of a fiber-optic bus.

optical link. The noise is determined separately through analytical methods. As detailed in [8.71], as long as the noise is assumed to be additive with a known probability distribution function, it is possible to determine the BER quasianalytically. This combination of simulation and analysis is powerful because it affords both a high degree of accuracy and a computationally efficient solution.

In iFROST, it was determined that the dominant noise mechanisms are found at the photoreceiver front end. These noise sources are modeled through the well-known formulation

$$\sigma^2 = \frac{4kTI_2B}{R_L} + 2q(I_g + I_d + \Re P_{avg})I_2B + \frac{4kT\Gamma}{g_m}(2\pi C_T)^2(f_c I_f B^2 + I_3 B^3) \quad (8.121)$$

In this representation, I_g is the leakage current through the photoreceiver input transistor (a FET is assumed), I_d is the dark current of the photodetector, $\Re$ is the detector responsivity, P_{avg} is the average power of the received optical signal, R_L is the load resistance of the photoreceiver preamplifier, Γ is the FET channel noise factor, C_T is the total input capacitance, g_m is the transconductance of the input transistor, B is the bit rate, f_c is the FET $1/f$ noise corner frequency, and I_2, I_3, and I_f are the Personick noise bandwidth integrals [8.75], [8.77]; all other parameters take on their traditional meanings. While photoreceiver noise is considered the dominant mechanism here, modal noise could easily be added to (8.121).

Using (8.121) in parallel with iFROST's signal waveform simulation capability, the BER is estimated in [8.76] by

$$\text{BER} = \frac{1}{2N}\sum_{i=1}^{N}\left[\text{erfc}\left(\frac{V_i - V_{th}}{\sqrt{2}\sigma_1}\right)u(V_i - V_{th}) + \text{erfc}\left(\frac{V_{th} - V_i}{\sqrt{2}\sigma_0}\right)u(V_{th} - V_i)\right] \quad (8.122)$$

$$u(x) = \begin{cases} 1 & x \geq 0 \\ 0 & x < 0 \end{cases} \quad (8.123)$$

In these equations, N is the number of bits in the sequence, V_i is the signal value, V_{th} is the decision threshold, and σ_1 and σ_0 are the noise standard deviations in the 1 and the 0 state, respectively (in the case of (8.121), since the noise has no signal dependence, $\sigma_1 = \sigma_0$). In (8.123), it is clear that $u(x)$ is 1 if the signal value is above the predetermined threshold value and 0 otherwise. In [8.71], the hybrid nature of the quasianalytical technique is emphasized by pointing out that while (8.122) is an analytical function, the limits to the integral (the arguments of the erfc functions) are determined by simulation.

When analyzing optical buses, it is necessary to consider the total BER, or more appropriately the word-error rate, of the entire system. A single bit error on a single channel of the bus results in a word error for the entire system. In [8.76], this is taken into account by equating the word-error rate to be the worst single channel BER at each bit period. The overall average word-error rate is computed as the average of the individual computed word-error rates at all the bit periods. Whitlock points out in [8.76] that this overall average word-error rate is sometimes called the overall average BER of the bus. Thus, (8.122) can be extended to an M-channel bus:

$$
\text{BER} = \frac{1}{2N} \sum_{i=1}^{N} \left\{ max \left[\text{erfc}\left(\frac{V_{i,j} - V_{th}}{\sqrt{2}\sigma_1} \right) u(V_{i,j} - V_{th}) \right. \right.
$$

$$
\left. \left. + \text{erfc}\left(\frac{V_{th} - V_{i,j}}{\sqrt{2}\sigma_0} \right) u(V_{th} - V_{i,j}) \right] \Bigg|_{j=1}^{M} \right\}
\tag{8.124}
$$

where max symbolizes the act of taking the maximum value. As an example, for the values shown in Table 8.2, iFROST calculated a BER of 10^{-15}.

Parameter	Definition	Value
T	Temperature	300 K
R_L	Preamplifier Load Resistance	4 kΩ
I_g	Input FET Gate Current	10 nA
I_d	Detector Dark Current	170 nA
P_{avg}	Total Average Optical Power	321 μW
$\mathfrak{R}$	Detector Responsivity	0.35 A/W
Γ	FET Channel Noise Factor	1.74
g_m	FET Transconductance	3.5 mA/V
C_T	Total Input Capacitance	314 fF
f_c	1/f Noise Corner Frequency	100 MHz
B	Bit Rate	1 Gb/s
I_f	Noise Bandwidth Integral	9.61×10^6
I_2	Noise Bandwidth Integral	9.97×10^6
I_3	Noise Bandwidth Integral	1.83×10^7
N	Number of Data Bits	16
M	Number of Channels in Bus	31

Table 8.2: Parameters for optical link BER calculation.

8.4 Summary

The iFROST simulator was described in this chapter as a method for simulating entire optical interconnect and fiber-optic *systems*. It was continually emphasized that while conventional optoelectronic design methodologies utilize simulation at the component level, very few take advantage of system simulation. By using system simulation, the design of optical links can be performed both from the top down and from the bottom up. In the bottom-up approach, link design is driven by the available devices; in this case, system simulation is useful for predicting the *integrated* performance of these devices. In the top-down approach, system specifications are used to drive device design. In addition, top-down simulation can be used to pinpoint those aspects or characteristics of device performance that must be improved in order to improve the overall system performance. The capability of iFROST to model the rise time, bandwidth, and bit rate of an overall optical system was demonstrated; these system-level parameters were determined as functions of the characteristics of the individual components comprising the link. In addition, the use of iFROST for data-driven waveform simulation was also depicted. The particular example that was illustrated included new models for pseudorandom bit pattern generators, Manchester encoders, and signal level converters, as well as models for several optoelectronic and optical components. The statistical simulation capability of iFROST was also demonstrated by varying nine component parameters of a 32-channel fiber-optic bus. The capability of iFROST to generate eye diagrams and to calculate the BER of both individual channels and of entire optical buses was also illustrated.

In this chapter, a mixed-level circuit-system simulation environment for the system-level modeling and simulation of optical links was also presented. The general approach was to use a circuit simulator to compute difficult to evaluate terms, such as the electron density, the photon density, the gain, and the partial derivatives of the gain with respect to the electron and photon densities. Once these quantities were computed through circuit simulation, a mixed-level simulation interface was used which accepted circuit-simulation output parameters as its input. The interface extracted relevant quantities from the circuit simulation output and computed the frequency and decay rate of relaxation oscillations, the frequency- and time-domain RIN, the spectral width and center wavelength, the transient and maximum chirp, the threshold current, and the differential quantum efficiency using well-established theory. Finally, the interface passed these "system-level" quantities to iFROST for use as input parameters in its optoelectronic link simulation. In this manner, a mixed-level circuit-system CAD environment which combines the inherent accuracy of circuit simulation with the flexibility of system simulation was realized. The system-level laser input parameters for the laser of Table 8.1 are summarized in Table 8.3. In cases where the parameter is dependent on bias conditions, a power of 1 mW was chosen as a representative steady-state level. While most of the mixed-level simulations involved the laser diode (due mainly to

Parameter	Definition	Value
I_{th}	Threshold Current	10.3 mA
η	Differential Quantum Efficiency	0.40 mW/mA
E_{ph}	Photon Energy	1.56 eV
λ_{o}	Center Wavelength	0.80 μm
f_{o}	Center Frequency	106 THz
$\Delta\lambda$	Wavelength Spectral Width	0.16 pm
Δf	Frequency Spectral Width	78.9 MHz
$\delta\lambda_{max}$	Maximum Wavelength Chirp (Square Wave)	1.76 pm
δf_{max}	Maximum Frequency Chirp (Square Wave)	829.3 MHz
RIN(t)	Time-Domain Relative Intensity Noise	-11.6 dB
Γ_R	Relaxation Oscillation Decay Rate	$1.84 \times 10^9 \text{ s}^{-1}$
f_R	Relaxation Oscillation Frequency	4.42 GHz

Table 8.3: System-level quantities passed on as input parameters to iFROST.

the complexity of the device), it should be emphasized that this methodology is general. Indeed, it was emphasized that, for example, the system photoreceiver simulations could be performed at the circuit level using the iSMILE MSM and HEMT models, as shown in Chapter 4.

8.5 References

[8.1]   B. K. Whitlock, "Computer modeling and simulation of digital lightwave links using iFROST, illinois fiber-optic and optoelectronic systems toolkit," M. S. thesis, University of Illinois at Urbana-Champaign, 1990.

[8.2]   B. K. Whitlock, J. J. Morikuni, E. Conforti, and S. M. Kang, "Simulating optical interconnects," *IEEE Circuits and Devices Magazine*, vol. 11, no. 3, pp. 12-18, 1995.

[8.3]   S. M. Kang, B. K. Whitlock, J. J. Morikuni, and E. Conforti, "Simulation of optical interconnects in high-performance computing and communications systems," *SPIE OE/LASE Conference on Optical Interconnects II*, 1994.

[8.4]   G. D. Brown, "Bandwidth and rise time calculations for digital multimode fiber-optic data links," *IEEE Journal of Lightwave Technology*, vol. 10, pp. 672-678, 1992.

[8.5] H. Stark and J. W. Woods, *Probability, Random Processes, and Estimation Theory for Engineers*. Englewood Cliffs, NJ: Prentice Hall, 1986.

[8.6] G. P. Agrawal, *Fiber-Optic Communication Systems*. New York: John Wiley & Sons, 1992.

[8.7] L. B. Jeunhomme, *Single-Mode Fiber Optics: Principles and Applications*. New York: Marcel Dekker, Inc., 1983.

[8.8] E. Conforti, Private communication, University of Illinois at Urbana-Champaign, 1992.

[8.9] K. Ogawa, "Analysis of mode partition noise in laser transmission systems," *IEEE Journal of Quantum Electronics*, vol. QE-18, no. 5, pp. 849-855, 1982.

[8.10] K. Ogawa and R. S. Vodhanel, "Measurements of mode partition noise of laser diodes," *IEEE Journal of Quantum Electronics*, vol. QE-18, no. 7, pp. 1090-1093, 1982.

[8.11] W. R. Throssell, "Partition noise statistics for multimode lasers," *IEEE Journal of Lightwave Technology*, vol. LT-4, no. 7, pp. 948-950, 1986.

[8.12] J. C. Campbell, "Calculation of the dispersion penalty for the route design of single-mode systems," *IEEE Journal of Lightwave Technology*, vol. 6, no. 4, pp. 564-573, 1988.

[8.13] W. E. Stephens and T. R. Joseph, "System characteristics of direct modulated and externally modulated RF fiber-optic links," *IEEE Journal of Lightwave Technology*, vol. LT-5, no. 3, pp. 380-387, 1987.

[8.14] M. Shikada, S. Fujita, N. Henmi, I. Takano, I. Mito, K. Taguchi, and K. Minemura, "Long-distance gigabit-range optical fiber transmission experiments employing DFB-LD's and InGaAs-APD's," *IEEE Journal of Lightwave Technology*, vol. LT-5, no. 10, pp. 1488-1497, 1987.

[8.15] R. A. Linke, "Modulation induced transient chirping in single frequency lasers," *IEEE Journal of Quantum Electronics*, vol. QE-21, no. 6, pp. 593-597, 1983.

[8.16] C. Lin, T. P. Lee, and C. A. Burrus, "Picosecond frequency chirping and dynamic line broadening in InGaAsP injection lasers under fast excitation," *Applied Physics Letters*, vol. 42, no. 2, pp. 141-143, 1983.

[8.17] J. Gowar, *Optical Communication Systems*. Englewood Cliffs, NJ: Prentice Hall, 1984.

[8.18] S. Yamamoto, M. Kuwazuru, H. Wakabayashi, and Y. Iwamoto, "Analysis of chirp power penalty in 1.55 μm DFB-LD high-speed optical fiber transmission systems," *IEEE Journal of Lightwave Technology*, vol. LT-5, no. 10, pp. 1518-1524, 1987.

[8.19]   S. Yamamoto, H. Sakaguchi, M. Nunokawa, and Y. Iwamoto, "1.55 μm fiber-optic transmission experiments for long-span submarine cable system design," *IEEE Journal of Lightwave Technology*, vol. 6, no. 3, pp. 380-391, 1988.

[8.20]   G. J. Murakami, R. H. Campbell, and M. Faiman, "Pulsar: Non-blocking packet switching with shift-register rings," *Computer Commun. Rev.*, vol. 20.4, pp. 145-155, 1990.

[8.21]   J. W. Lockwood, H. Duan, J. J. Morikuni, S. M. Kang, S. Akkineni, and R. H. Campbell, "Scalable optoelectronic ATM networks: The iPOINT fully functional testbed," *IEEE Journal of Lightwave Technology*, vol. 13, no. 6, pp. 1093-1103, 1995.

[8.22]   D. E. McCumber, "Intensity fluctuations in the output of cw laser oscillators. I," *Physical Review*, vol. 141, no. 1, pp. 306-322, 1966.

[8.23]   H. Haug, "Quantum-mechanical rate equations for semiconductor lasers," *Physical Review*, vol. 184, no. 2, pp. 338-348, 1969.

[8.24]   D. J. Morgan and M. J. Adams, "Quantum noise in semiconductor lasers," *Physica Status Solidi*, vol. 11, pp. 243-253, 1972.

[8.25]   M. Lax, "Quantum noise IV: Quantum theory of noise sources," *Physical Review*, vol. 145, no. 1, pp. 110-129, 1966.

[8.26]   M. Lax, "Quantum noise VII: The rate equations and amplitude noise in lasers," *IEEE Journal of Quantum Electronics*, vol. QE-3, no. 2, pp. 37-46, 1967.

[8.27]   M. Lax and W. H. Louisell, "Quantum noise IX: Quantum Fokker-Planck solution for laser noise," *IEEE Journal of Quantum Electronics*, vol. QE-3, no. 2, pp. 47-58, 1967.

[8.28]   M. Lax, "Fluctuations from the nonequilibrium steady state," *Reviews of Modern Physics*, vol. 32, no. 1, pp. 25-64, 1960.

[8.29]   H. Haken, "A nonlinear theory of laser noise and coherence, I," *Zeitschrift für Physik*, vol. 181, pp. 96-124, 1965.

[8.30]   H. Haken, "A nonlinear theory of laser noise and coherence, II," *Zeitschrift für Physik*, vol. 182, pp. 346-359, 1965.

[8.31]   H. Haug, "Noise in semiconductor lasers," *IEEE Journal of Quantum Electronics*, vol. QE-4, no. 4, p. 168, 1968.

[8.32]   G. A. Korn, *Random-Process Simulation and Measurements*. New York: McGraw-Hill, 1966.

[8.33] A. van der Ziel, *Noise in Solid State Devices and Circuits*. New York, NY: John Wiley & Sons, Inc., 1986.

[8.34] K. Petermann, *Laser Diode Modulation and Noise*. AH Dordrecht, The Netherlands: Kluwer Academic Publishers, 1991.

[8.35] A. Yariv, *Optical Electronics*. Philadelphia, PA: Saunders College Publishing, 1991.

[8.36] W. R. Throssell, "Partition noise statistics for multimode lasers," *IEEE Journal of Lightwave Technology*," vol. LT-4, no. 7, pp. 948-950, 1986.

[8.37] R. H. Wentworth, "Noise of strongly-multimode laser diodes used in interferometric systems," *IEEE Journal of Quantum Electronics*, vol. 26, no. 3, pp. 426-442, 1990.

[8.38] K. Iwashita and K. Nakagawa, "Mode partition noise characteristics in high-speed modulated laser diodes," *IEEE Journal of Quantum Electronics*, vol. QE-18, no. 12, pp. 2000-2005, 1982.

[8.39] K. Ogawa and R. S. Vodhanel, "Measurements of mode partition noise of laser diodes," *IEEE Journal of Quantum Electronics*, vol. QE-18, no. 7, pp. 1090-1093, 1982.

[8.40] K. Ogawa, "Analysis of mode partition noise in laser transmission systems," *IEEE Journal of Quantum Electronics*, vol. QE-18, no. 5, pp. 849-855, 1982.

[8.41] M. Sargent III, M. O. Scully, and W. E. Lamb, *Laser Physics*. Reading, MA: Addison-Wesley Publishing Company, 1974.

[8.42] G. P. Agrawal and N. K. Dutta, *Long-Wavelength Semiconductor Lasers*. New York: Van Nostrand Reinhold, 1986.

[8.43] P. Langevin, "On the theory of Brownian motion," *Compt. Rend*, vol. 146, pp. 530-533, 1908.

[8.44] D. Marcuse, "Computer simulation of laser photon fluctuations: Theory of single-cavity laser," *IEEE Journal of Quantum Electronics*, vol. QE-20, no. 10, pp. 1139-1148, 1984.

[8.45] K. Vahala and A. Yariv, "Semiclassical theory of noise in semiconductor lasers – Part I," *IEEE Journal of Quantum Electronics*, vol. QE-19, no. 6, pp. 1096-1101, 1983.

[8.46] K. Vahala and A. Yariv, "Semiclassical theory of noise in semiconductor lasers – Part II," *IEEE Journal of Quantum Electronics*, vol. QE-19, no. 6, pp. 1102-1109, 1983.

[8.47] C. H. Henry, "Theory of the linewidth of semiconductor lasers," *IEEE Journal of Quantum Electronics*, vol. QE-18, no. 2, pp. 259-264, 1982.

[8.48] C. H. Henry, "Phase noise in semiconductor lasers," *IEEE Journal of Lightwave Technology*, vol. LT-4, no. 3, pp. 298-311, 1986.

[8.49] C. H. Henry, "Theory of the phase noise and power spectrum of a single mode injection laser," *IEEE Journal of Quantum Electronics*, vol. QE-19, no. 9, pp. 1391-1397, 1983.

[8.50] C. H. Henry, "Line broadening of semiconductor lasers," *Coherence, Amplification, and Quantum Effects in Semiconductor Lasers*. New York: John Wiley & Sons, Inc., 1991.

[8.51] L. Li, "Small-signal analysis for the time-averaged AM and FM noise power spectra of a directly modulated semiconductor laser," *IEEE Journal of Quantum Electronics*, vol. 29, no. 10, pp. 2625-2630, 1993.

[8.52] J. W. Nilsson, *Electric Circuits*. Reading, MA: Addison-Wesley Publishing Co., 1986.

[8.53] D. S. Gao, S. M. Kang, R. P. Bryan, and J. J. Coleman, "Modeling of quantum-well lasers for computer-aided analysis of optoelectronic integrated circuits," *IEEE Journal of Quantum Electronics*, vol. 26, no. 7, pp. 1206-1216, 1990.

[8.54] M. Osinski and J. Buus, "Linewidth broadening factor in semiconductor lasers - an overview," *IEEE Journal of Quantum Electronics*, vol. QE-23, no. 1, pp. 9-29, 1987.

[8.55] A. Mooradian, "Laser linewidth," *Physics Today*, vol. 38, no. 5, pp. 43-48, May 1985.

[8.56] A. Yariv, *Quantum Electronics*. New York: John Wiley & Sons, Inc., 1989.

[8.57] K. Kikuchi and T. Okoshi, "Estimation of linewidth enhancement factor of AlGaAs lasers by correlation measurement between FM and AM noises," *IEEE Journal of Quantum Electronics*, vol. QE-21, no. 6, pp. 669-673, 1985.

[8.58] M. G. Burt, "Linewidth enhancement factor for quantum-well lasers," *Electronics Letters*, vol. 20, no. 1, pp. 27-29, 1984.

[8.59] L. D. Westbrook and M. J. Adams, "Explicit approximations for the linewidth-enhancement factor in quantum-well lasers," *IEE Proceedings*, vol. 135, pt. J, no. 3, pp. 223-225, 1988.

[8.60] T. Yamanaka, Y. Yoshikuni, K. Yokoyama, W. Lui, and S. Seki, "Theoretical study on enhanced differential gain and extremely reduced linewidth enhancement factor in quantum-well lasers," *IEEE Journal of Quantum Electronics*, vol. 29, no. 6, pp. 1609-1616, 1993.

[8.61]   F. Kano, T. Yamanaka, N. Yamamoto, Y. Yoshikuni, H. Mawatari, Y. Tohmori, M. Yamamoto, and K. Yokoyama, "Reduction of linewidth enhancement factor in InGaAsP-InP modulation-doped strained multiple-quantum-well lasers," *IEEE Journal of Quantum Electronics*, vol. 29, no. 6, pp. 1553-1559, 1993.

[8.62]   G. P. Agrawal and G. M. Bowden, "Concept of linewidth enhancement factor in semiconductor lasers: Its usefulness and limitations," *IEEE Photonics Technology Letters*, vol. 5, no. 6, pp. 640-642, 1993.

[8.63]   G. P. Agrawal, "Intensity dependence of the linewidth enhancement factor and its implications for semiconductor lasers," *IEEE Photonics Technology Letters*, vol. 1, no. 8, pp. 212-214, 1989.

[8.64]   G. Morthier, P. Vankwikelberge, F. Buytaert, and R. Baets, "Influence of gain nonlinearities on the linewidth enhancement factor in semiconductor lasers," *IEE Proceedings*, vol. 137, pt. J, no. 1, pp. 30-32, 1990.

[8.65]   T. M. Shen and G. P. Agrawal, "Pulse-shape effects on frequency chirping in single-frequency semiconductor lasers under current modulation," *IEEE Journal of Lightwave Technology*, vol. LT-4, no. 5, pp. 497-503, 1986.

[8.66]   W. H. Press, B. P. Flannery, S. A. Teukolsky, and W. T. Vetterling, *Numerical Recipes in C*. Cambridge, MA: Cambridge University Press, 1989.

[8.67]   T. Ikegami and Y. Suematsu, "Large-signal characteristics of directly modulated semiconductor injection lasers," *Electronics and Communications in Japan*, vol. 53-B, no. 9, pp. 69-75, 1970.

[8.68]   H. Tsushima and Y. Suematsu, "Large-signal analysis of dynamic wavelength shift and carrier-density variation in directly modulated dynamic-single-mode lasers," *Transactions of the IEICE of Japan*, vol. E 67, no. 9, pp. 480-487, 1984.

[8.69]   P. K. Pepeljugoski, B. K. Whitlock, D. M. Kuchta, J. D. Crow, and S. M. Kang, "Modeling and simulation of the OETC optical bus," *IBM Research Report*, RC20084 (88831), 1995.

[8.70]   P. K. Pepeljugoski, B. K. Whitlock, D. M. Kuchta, J. D. Crow, and S. M. Kang, "Modeling and simulation of the OETC optical bus," *Proceedings of the Annual Meeting of the IEEE Lasers and Electro-Optics Society*, vol. 1, pp. 185-186, 1995.

[8.71]   M. C. Jeruchim, P. Balaban, and K. S. Shanmugan, *Simulation of Communication Systems*. New York: Plenum Press, 1992.

[8.72]   A. S. Tanenbaum, *Computer Networks*. Englewood Cliffs, NJ: Prentice Hall, 1988.

[8.73] R. J. S. Bates, D. M. Kuchta, and K. P. Jackson, "Improved multimode fibre link BER calculations due to modal noise and non-self-pulsating laser diodes," *Optical and Quantum Electronics*, vol. 27, pp. 203-224, 1995.

[8.74] J. J. Morikuni, A. Dharchoudhury, Y. Leblebici, and S. M. Kang, "Improvements to the standard theory for photoreceiver noise," *IEEE Journal of Lightwave Technology*, vol. 12, no. 7, pp. 1174-1184, 1994.

[8.75] T. V. Muoi, "Receiver design of optical fiber systems," in *Optical Fiber Transmission*. Indianapolis, IN: Howard W. Sams & Co., 1987.

[8.76] B. K. Whitlock, P. K. Pepeljugoski, D. M. Kuchta, J. D. Crow, and S. M. Kang, "Computer modeling and simulation of the OptoElectronic Technology Consortium (OETC) optical bus," *Submitted to IEEE Journal on Selected Areas in Communication,* May 1996.

[8.77] R. G. Smith and S. D. Personick, "Receiver design for optical communication systems," in *Semiconductor Devices for Optical Communication*. New York: Springer Verlag, 1980.

INDEX

A

Absorption
    detector 62, 65, 67, 69, 70, 159, 164,
        192, 212, 215, 217, 228
    laser 54, 256
AC: See Simulation analyses
Air bridge 136
AlGaAs 75, 76, 83, 128, 197, 201, 217,
    225, 226
AlSbAs 271
ATM 286
Autocorrelation 296, 297

B

Bandwidth
    circuit 12, 51, 97, 98, 99, 102, 103,
        283, 284, 287
    detector 119, 159, 176, 212, 219,
        221, 224, 226
    fiber 281, 282, 283, 324, 328
    Gaussian 278, 279
    laser 277
    LED 279, 280
    system 8, 276, 279, 284, 285, 287,
        320, 323
Bandwidth-efficiency product 217, 221,
    232
Behavioral simulation 8, 33–39, 324
Bernoulli equation 205
Bianchi 67–74, 106, 158, 170, 172
Bias
    circuit 19, 141, 225
    condition 318, 320, 332
    current 123, 136, 137, 142, 146,
        147, 328
    detector 62, 66–68, 72, 125, 161,
        164, 170, 172, 193, 207,
        208, 210, 219, 225, 226,
        228
    laser 136, 146
    point 19, 20, 21, 27, 57, 86, 102,
        103, 105, 109, 110, 111,
        112, 116, 119, 139, 147,
        208, 239, 241, 244, 258,
        260, 277, 293, 305
    transistor 80, 86, 124, 136, 137, 142,
        146, 147
Bit-error rate (BER) 275, 285, 328–331

ABOUT THE AUTHORS

James J. Morikuni received the Ph.D. degree in electrical engineering from the University of Illinois at Urbana-Champaign in 1994. He has held positions at IBM in Rochester, Minnesota; AT&T Bell Laboratories in Naperville, Illinois; and NEC in Tsukuba, Japan. His current research interests include the modeling, simulation, design, and analysis of optoelectronic integrated devices, circuits, and systems. Dr. Morikuni is a member of the IEEE and is currently with the Optical Interconnect Laboratory and Applied Simulation and Modeling Research (ASMR) at Motorola's Corporate Research and Development facilities in Schaumburg, Illinois.

Sung-Mo (Steve) Kang received the Ph.D. degree in electrical engineering from the University of California at Berkeley in 1975. Until 1985, he was with AT&T Bell Laboratories at Holmdel and Murray Hill and also served as a faculty member of Rutgers University. In 1985, he joined the University of Illinois at Urbana-Champaign where he is currently Department Head and Professor of Electrical and Computer Engineering; Professor of Computer Science; Research Professor of the Beckman Institute for Advanced Science and Technology and of the Coordinated Science Laboratory; and Associate Director of the NSF Engineering Research Center for Compound Semiconductor Microelectronics. His research interests include VLSI system design methodologies; optimization for performance, reliability, and manufacturability; modeling and simulation of semiconductor and optoelectronic devices, circuits, and systems; high-speed optoelectronic circuits; and fully optical network systems. Dr. Kang is an IEEE Fellow and Founding Editor-in-Chief of the *IEEE Transactions on Very Large Scale Integration (VLSI) Systems*. He has also co-authored five books: *Design Automation for Timing-Driven Layout Synthesis, Hot-Carrier Reliability of MOS VLSI Circuits, Modeling of Electrical Overstress in Integrated Circuits*, and *Physical Design for Multichip Modules*, from Kluwer Academic Publishers; and *CMOS Digital Integrated Circuits: Analysis and Design*, from McGraw-Hill.